ENGINE EMISSION CONTROL TECHNOLOGIES

Design Modifications and Pollution Mitigation Techniques

ENGINE EMISSION CONTROL TECHNOLOGIES

Design Modifications and Pollution Mitigation Techniques

G. Amba Prasad Rao, PhD
T. Karthikeya Sharma, PhD

Apple Academic Press Inc.
4164 Lakeshore Road
Burlington ON L7L 1A4
Canada

Apple Academic Press, Inc.
1265 Goldenrod Circle NE
Palm Bay, Florida 32905
USA

© 2020 by Apple Academic Press, Inc.

First issued in paperback 2021

Exclusive worldwide distribution by CRC Press, a member of Taylor & Francis Group

No claim to original U.S. Government works

ISBN 13: 978-1-77463-486-8 (pbk)
ISBN 13: 978-1-77188-835-6 (hbk)

Library and Archives Canada Cataloguing in Publication

Title: Engine emission control technologies : design modifications and pollution mitigation techniques / G. Amba Prasad Rao, PhD, T. Karthikeya Sharma, PhD.
Names: Rao, G. Amba Prasad, 1966- author. | Sharma, T. Karthikeya, 1985- author.
Description: Includes bibliographical references and index.
Identifiers: Canadiana (print) 20200200550 | Canadiana (ebook) 20200200712 | ISBN 9781771888356 (hardcover) | ISBN 9780429322228 (electronic bk.)
Subjects: LCSH: Internal combustion engines—Exhaust systems—Design and construction. | LCSH: Internal combustion engines—Exhaust gas—Measurement. | LCSH: Spark ignition engines—Exhaust gas—Measurement. | LCSH: Automobiles—Motors—Exhaust gas—Measurement.
Classification: LCC TD886.5 .R36 2020 | DDC 621.43/7—dc23

CIP data on file with US Library of Congress

Apple Academic Press also publishes its books in a variety of electronic formats. Some content that appears in print may not be available in electronic format. For information about Apple Academic Press products, visit our website at **www.appleacademicpress.com** and the CRC Press website at **www.crcpress.com**

About the Authors

G. Amba Prasad Rao, PhD
*Professor, Mechanical Engineering,
National Institute of Technology, Warangal,
India*

G. Amba Prasad Rao, PhD, is a Professor of Mechanical Engineering at the National Institute of Technology, Warangal, India. He completed his MTech (Design and Production of IC Engines and Gas Turbines) and PhD (Mechanical Engineering) from NIT Warangal (formerly REC Warangal) in the years 1992 and 2003, respectively. He has about 30 years of teaching and research experience.

He joined NIT Warangal as a Lecturer in the Department of Mechanical Engineering in 1996. His areas of research include internal combustion engines, combustion, energy, alternative fuels, and emissions control. He was promoted to an Assistant Professor and later as a Professor in the years in 2004 and 2011, respectively. Since he joined as a Lecturer, he was actively involved in the teaching and research activity.

He was the Principal Developer for the pedagogical course on internal combustion engines and automobile engineering. The project was funded by MHRD under NMEICT scheme sanctioned to IIT Kharagpur. He was co-investigator in three projects: (i) Technical Back-up Support Unit under the National Programme of Improved Chulha, (ii) Use of Jatropha Oil As and Alternative Fuel (vegetable-based) for Automobiles, and (iii) Development of Adiabatic Diesel Engines with Biodiesel as a Fuel.

He was the founder coordinator for the MTech in Automobile Engineering course at NIT Warangal in 2008-09. He has supervised eight PhD theses and currently three are in progress. He has guided over 50 MTech dissertations and over 20 BTech projects. He has published about 75 technical papers in reputed international journals and about 20 papers in reputed conferences. He has given a number of presentations in the area of internal combustion engines, alternative fuels, and emission control. He has organized several short-term programs.

He is a life member of professional bodies such as ISTE, IEI, Combustion Institute (Indian Section), Solar Energy Society of India (SESI), and an annual member of SAE (India) and ASME. He has organized the prestigious 1st International and 18th National Conference on Mechanical Engineering for Sustainable Development under the aegis of the Indian Society of Mechanical Engineers (ISME) during February 23–25, 2017 at NIT Warangal. Around 50 papers of the conference presentations have been brought out in the form of the book edited by him, by Apple Academic Press. He has also authored 3 chapters in the book.

He has held many administrative positions at NIT Warangal, including Warden for Hostels and Associate Dean-Academics. He has also worked a Professor in-charge Training and Placement Section at NIT Warangal and also as Associate Dean, Academics and Faculty Advisor for MTech in Thermal Engineering program and faculty advisor of Nature Club, Mechanical Engineering Association. He was a member of the Board of Studies and Faculty Selections. He has visited the USA, Malaysia, and the UAE in connection with the presentation of technical papers. He chaired many technical sessions. Presently, he is on deputation to NIT Andhra Pradesh at Tadepalligudem of WG District of State of Andhra Pradesh.

T. Karthikeya Sharma, PhD

Assistant Professor and Head, Department of Mechanical Engineering, National Institute of Technology Andhra Pradesh, India

T. Karthikeya Sharma, PhD, is an Assistant Professor and Head of the Department of Mechanical Engineering at the National Institute of Technology, Andhra Pradesh, India. He is specialized in internal combustion engines, refrigeration, and air conditioning and has been involved in computational fluid dynamic analysis of combustion in IC engines. He is a member of various prestigious professional bodies and a consultant engineer for several research laboratories. He has published research articles widely on pre-mixed HCCI combustion analysis. He, along with his student, exhibited a manual transmission system for hybrid vehicles at a festival of innovation held at Rastrapati Bhawan, New Delhi, India.

He received his PhD in Mechanical Engineering from the National Institute of Technology Warangal, India. Dr. Karthikeya has received awards and citations for excellence as a researcher and as an engineering educator. He is a past recipient of a Best Research Scholar Award of the Institution of Engineers (India), a Young Scientist Award of Andhra Pradesh Akademi of Sciences, and a Young Engineer Award of IIOS.

Contents

Energy Quotes

"Because we are now running out of gas and oil, we must prepare quickly for a third change, to strict conservation and to the use of . . . renewable energy sources, like solar power."

—Jimmy Carter, 1977

"I'd put my money on the sun and solar energy, what a source of power. I hope we don't have to wait until oil and coal run out before we tackle that."

—Thomas Edison, 1931

"The thermal agency by which mechanical effect may be obtained is the transference of heat from one body to another at a lower temperature."

—Sadi Carnot

"The total energy of the universe is constant; the total entropy is continually increasing."

—Rudolf Clausius

"I can never satisfy myself until I can make a mechanical model of a thing. If I can make a mechanical model, I can understand it."

—Lord Kelvin

"Bah! Genius is not inspired. Inspiration is perspiration. 2% is genius. 98% is hard work."

—Thomas Edison, 1898

"The cumulative amount of man-made global pollution that's in the atmosphere now traps as much extra heat energy every day as would be released by 400,000 Hiroshima-class atomic bombs exploding every day. It's a big planet, but that's an awful lot of energy."

—Al Gore

Abbreviations

AECC	Association for Emissions Control by Catalysts
AF	air-fuel ratio
AFs	alternative fuels
AFT	adiabatic flame temperature
Al_2O_3	aluminum oxide
ASTM	American Society for Testing Materials
BD	biodiesel-diesel
BDC	bottom dead center
BSFC	brake specific fuel consumption
BSNOx	brake specific nitrogen oxides
CCV	closed crankcase ventilation
CDA	cylinder deactivation
CFR	cooperative fuel research
CH_4	methane
CI Engines	compression-ignition engines
CI	compression-ignition
CLD	chemiluminescence detector
CNG	compressed natural gas
CO	carbon monoxide
CO_2	carbon dioxide
CPC	condensation particle counter
CR	compression ratio
CRBME	crude rice bran oil methyl ester
CRDI	common rail direct injection
CRT	continuously regenerating trap
CVS	constant volume sampling
DEE	diethyl ether
DI	direct injection
DISC	direct injection stratified charge
DME	dimethyl ether
DOCs	diesel oxidation catalysts
DPFs	diesel particulate filters
DPT	diesel particulate traps
EC	elemental carbon

ECE	Economic Commission for Europe
ece	external combustion engine
ECM	engine control module
ECU	electronic control unit
ecu	engine control unit
EFI	electronic fuel injection
EGR	exhaust gas recirculation
EICU	electronic ignition control unit
EIVC	early inlet valve closing
eivc	early intake valve closure
EMS	engine management system
EPA	Environmental Protection Agency
ET	evaporation tube
ETBE	ethyl tertiary butyl ether
ETC	European transient cycle
FA	fuel-air ratio
FFVs	flexible fuel vehicles
FID	flame ionization detector
FIP	fuel injection pressure
FTIR	Fourier-transform infrared
GC	gas chromatography
GDI	gasoline direct injection
GHG	greenhouse gas
GIE	gross indicated efficiency
GPS	global positioning system
H_2	hydrogen
H_2O	water
H_2S	hydrogen sulfide
HCCI	homogeneous charge compression-ignition
HCN	hydrocyanic acid
HCNG	hydrogen-enriched compressed natural gas
HCs	hydrocarbons
HC-SCR	hydrocarbon selective catalytic reduction
HEPA	high-efficiency particulate air
HLB	hydrophilic (water liking) balance
HPLC	high-performance liquid chromatography
HSDI	high-speed direct injection
HUCR	highest useful compression ratio
IC	internal combustion
ICEs	internal combustion engines

IDI	indirect injection
IF	inorganic fraction
ILUC	indirect land-use change
IR	infrared
ISFC	indicated specific fuel consumption
J	joule
LDV	light-duty vehicle
LHR	low heat rejection
LIVC	late inlet valve closing
livc	late intake valve closure
LNCs	lean NOx catalysts
LNG	liquefied natural gas
LNTs	lean NOx traps
LPG	liquefied petroleum gas
LTC	low-temperature combustion
MAHs	monocyclic aromatic hydrocarbons
MAP	manifold air pressure
MBT timing	maximum brake torque timing
MgO	magnesium oxide
MPGI	multipoint gas injection
MS	mass spectrometry
MSW	municipal solid waste
N	nitrogen
N_2O	nitrous oxide
N_2O_2	dinitrogen dioxide
N_2O_3	dinitrogen trioxide
N_2O_4	dinitrogen tetroxide
N_2O_5	dinitrogen pentoxide
NA	naturally aspirated
NDIR	non-dispersive infrared
NEDC	new European driving cycle
NGVs	natural gas vehicles
NH_3	ammonia
$NiFe_2O_4$	nickel-iron oxide
NiO	nickel oxide
NMHC	non-methane hydrocarbon
NO	nitric oxide
NO_2	nitrogen dioxide
NOx	nitrogen oxides
NPAHs	nitrated polycyclic aromatic hydrocarbons

OBD   on-board diagnostic
OPEC   Organization of Petroleum Exporting Countries
OEMs   Original Equipment Manufactures
PAHs   polycyclic aromatic hydrocarbons
PCCI   premixed charge compression-ignition
PCV   positive crankcase ventilation
PEM   proton-exchange-membrane
PEMS   portable emission measurement systems
PIV   particulate image velocimetry
PM   particulate matter
PMP   particle measurement program
PMT   photomultiplier tube
PN   particle number
PNC   particle number counter
PND1   first particle number diluter
PND2   second particle number diluter
ppbv   parts per billion per volume
PPC   partially premixed combustion
ppmv   parts per million by volume
PSZ   partially stabilized zirconia
PTD   port timing diagram
RCCI   reactivity-controlled compression-ignition
RI   ringing intensity
RPM   revolutions per minute
SADI   spark assisted stratified charge
SCR   selective catalytic reduction
SI   spark-ignition
SiC   silicon carbide
SiO_2 silicon oxide
SO_2 sulfur dioxide
SOF   soluble organic fraction
SOx   sulfur compounds
SR   steam reforming
T   temperature
TBCs   thermal barrier coatings
TDC   top-dead center
TEL   tetraethyl lead
THC   total hydrocarbons
TiO_2 titanium oxide

TKE	turbulent kinetic energy
TWC	three-way (catalytic) converter
UFPs	ultrafine particles
UHC	unburnt hydrocarbons
ULEV	ultra-low emission vehicle standards
ULSD	ultra-low sulfur diesel standards
UV	ultraviolet
VDTCE	variable displacement turbocharged engine
VE	volumetric efficiency
VGT	variable geometry turbo-charging
VM	velocity Magnitude
VOCs	volatile organic compounds
VPR	volatile particle remover
VTD	valve timing diagram
VVA	variable valve actuation
VVT	variable valve timing
WHTC	world harmonize test cycle
WiDE	water-in-diesel emulsion
ZnO	zinc oxide

Foreword

Dr. G. Amba Prasad Rao is my colleague at NIT Warangal, and I have known him closely for the past about 30 years. Right from his joining the Institute, he was very much interested in the subjects related to thermal engineering. He was acclaimed as a good teacher, and many of the students used to appreciate his way of teaching. He was quite patient with students when they approached for any clarifications in the subjects, such as internal combustion engines and alternative fuels.

I personally suggested that he to come out with a book on IC engines. Finally, he could make the time and contribute to AAP with this book, *Engine Emissions Control Technologies*, along with my junior colleague, Dr. T. Karthikeya Sharma.

Dr. Sharma did his PhD work under the supervision of Prof. Amba Prasad Rao on the new combustion concept, HCCI. These two published a good number of papers in reputed journals. Though I belong to the area of manufacturing engineering, I have glanced through quite a good number of books on IC engines, and I appreciate his choosing of topics related engine emissions control. The world cannot be thought of without an automobile, and such an important machine is subjected to scrutiny due to atmospheric pollution caused by automobiles. Every day in one part or other parts of the world, development of new model of automobile is taking place. Human population is growing, and so also the automotive population and new and stringent norms are being legislated around the world, and in this direction India leapfrogged from EURO IV to EURO VI and are poised to be implemented from April 2020 onwards.

The book covers the topics related to emissions, its formation in both versions, and its various pollution mitigation techniques. The book is covers the use of alternative fuels and new combustion concepts that would pave the way for complying with new emission concepts. Undoubtedly, the book is going to an asset to students to understand right from the basics of engines to methods of control of engine-out emissions.

— Prof. C. S. P. Rao
Director, National Institute of Technology Andhra Pradesh,
Tadepalligudem, Andhra Pradesh, India

Preface

Thanks to the inventors of internal combustion engines as they have made the mobility of goods, humans, etc. simpler when compared to external combustion engines. The IC engines have the advantage of being compact and have a high power-to-weight ratio and use liquid fuels. The real benefit of ICE-driven vehicles were realized after the discovery of petroleum fuels and pneumatic tires. The automobiles that extensively use these ICEs-prime movers have simplified the lifestyle of human beings. The ICEs easily fit into any application—whether it is public transport, power generation, etc. Indeed, transportation is much exploited by these prime movers. The automobile sector has played a key role in developing other sectors as well, such as manufacturing, etc. Civilization has seen stupendous development with the expansion of the automobile sector. The automobile has become an integral part of life of a human, whether rich or poor.

ICEs can be broadly classified as petrol/gasoline-fuelled or diesel-fuelled engines, i.e., spark-ignition engines and compression ignition engines. The spark-ignition engines are more compact than compression ignition engines. The former type of engine accommodates fuels that are volatile or gaseous fuels as opposed to later type viscous fuels. In earlier days, the SI engines were easily managed in the passenger sector whereas the CI engines were used for heavy-duty applications. However, with the advancements, CI engines have evolved from being bulky, smoky, and noisy to compact, cleaner, and quieter. The fuel injection system in diesel engines made these improvements possible. Also, fuel injection systems have made their way into SI engines too. All the credit of fuel efficiency of these engines goes to refined fuel available these days.

With the exponential growth in automobiles, the demand for petroleum products also increased. The nations that are have limited or no resources of liquid fuel are depending heavily on imports and thereby losing their hard-earned revenue. No doubt, the expansion of the auto sector has brought huge revenue, but at the same, it has also led environmental degradation issues and fuel/energy crisis. Nations such OPEC nations that have these resources are virtually ruling the world.

The smog problem, which was detected in the 1960s, has given rise to concern with environmental issues. With these, countries, especially in the

EU, have set the norm for emissions from automobiles. Now the focus has shifted from being more efficient to environmentally benign engine development. Both researchers and the manufacturing industry are perplexed with the kind of emissions that are contributing toward rise in CO_2 levels in the atmosphere. Automobiles have become partly to blame for the rise in pollutant emissions in the populated regions of several nations. Recently New Delhi witnessed heavy pollution levels that is not confined to only such cities.

The oil embargo of the 1970s has stimulated research on alcohol-fuelled engines, and of late biodiesels prepared mainly from vegetable oils are seeing its share in the vehicles. There are host of alternative fuels, such as CNG, LNG, biogas, alcohols, hydrogen, etc., that have been explored as future fuels. However, among these, each has its own advantages and limitations from being used in large scale.

The stringent emissions regulations that started in the 2000s in India in terms of Bharat Stage norms, which are in line with EURO norms, have to some extent brought environmental consciousness among the general public. However, skipping EURO V, India is poised to move to EURO VI-compliant vehicles from 2020 onward. There is a serious concern and work going on in meeting stringent norms by both engine manufacturers and the fuel refining industry. Ultra-low sulfur diesel is being developed for this purpose.

Researchers over a period of time developed new combustion concepts such as HCCI, PCCI, RCCI, stratified charge engine, LHR engines, variable compression, and dual-fuel engines. These have become important to addressing pollutant emission-related issues.

The subject of IC engines is taught at both undergraduate and postgraduate levels. There are very good books available in the market, such as those authored by John B. Heywood, W.W. Pulkrabek, R.S Benson, and W.D Whitehouse and C.F Taylor. However, the authors felt it essential to bring out a book exclusively focused on engine emission technologies. The authors have experienced that not many books are available that can precisely address the issues related to engine emissions and its mitigation techniques. This volume is a result of the authors' experience in teaching of subject on internal combustion engines for over 25 years, along with the knowledge they gained from research, colleagues, peers, experts, and scholars.

With this motivation, a keen attempt has been made to address the issues and is brought out in the form of this volume, *Engine Emission Control Technologies: Design Modifications and Pollution Mitigation Techniques.*

The book starts with an introduction of the issues related to emissions. Chapter 1 is a basic review of IC engine fundamentals. Chapter 2 describes emission formation mechanisms from both SI and CI engines. Emission measurement methods form the content of Chapter 3. Chapters 4, 5, 6, and 7 have been written describing the in-cylinder approaches and after-treatment techniques adopted to control the emissions from SI and CI engines. The role of alternative fuels and alternative combustion concepts have been covered in Chapters 8 and 9, respectively.

The book is written in lucid language and will be useful for both students and research scholars.

The authors are highly thankful to the publisher, Apple Academic Press, especially Mr. Ashish Kumar, Mr. Rakesh Kumar, and Ms. Sandra Sickels, for their immediate acceptance of our request of bringing out a book. Their timely cooperation is highly appreciated.

The authors are also thankful to their families, without whose support this small attempt couldn't have come to the present form.

Any suggestions toward its further improvement will be thankfully acknowledged.

— **G. Amba Prasad Rao**
T. Karthikeya Sharma

Inventors, Prominent Researchers, Manufacturers, and Authors in the Area of Internal Combustion Engines

Nicolas Léonard Sadi Carnot

He was the proponent of theoretical or ideal heat engine regarded as Carnot Engine based on reversible processes. He expressed maximum possible theoretical efficiency in terms of absolute temperature. He also defined the concept of entropy.

Nicholaus A. Otto

Inventor of a 4-stroke cycle engine (gasoline fuelled engine).

Rudolph Diesel

Inventor of the practical compression ignition engine.

	**Robert Bosch** Inventor of fuel injection pumps for diesel fuel used in Compression Ignition Engines.
	**Karl Benz** Creator of the Benz Patent Motorcar, considered the first practical automobile.
	**John Boyd Dunlop** Invented air inflated tire.
	**Henry Ford** Envisaged homegrown fuels for future engines.

	**Charles Franklin Kettering** Head of research at General Motors from 1920 to 1947.
	**Arie Haagen-Smit** In his lab, May 1961, he demonstrated smog.
	**Daimler-Chrysler** It is one of the oldest, reliable and largest manufacturer of automobiles in the world.
	**Atikinson** Inventor of over-expanded cycle.

	**Felix Wankel** Inventor of a rotary combustion engine.
	**Ralph H. Miller** Proposed a thermodynamic cycle with compression ratio less than expansion ratio.
	**Eugene Houdry** Inventor of catalytic converter.
	**John B. Heywood** Researcher and author of a book on Internal Combustion Engine Fundamentals.

	**Willard W. Pulkrabek** Author of engineering fundamentals of an internal combustion engine.
	**Edward F. Obert** Author of internal combustion engines and air pollution.
	The founder and chairman of The Climate Reality Project, a non-profit devoted to solving the climate crisis. In 2007, Mr. Gore was awarded the Nobel Peace Prize, along with the Intergovernmental Panel on Climate Change (IPCC), for "informing the world of the dangers posed by climate change."

Introduction

I.1 SCENARIO OF OIL CONSUMPTION

The use of *energy* in any form has become synonymous with the life of humans, and energy consumption of any country has become a yardstick to rate whether that country is a developed or developing one. The basic objective of a prime mover is to convert any form of energy to mechanical or electrical energy. The prime movers that are used in automobiles are internal combustion engines; the internal combustion engines have become an integral part of human life as they become indispensable. The real benefit gained over external combustion engines (popularly called steam-powered engines) is due to their high power-to-weight ratio and compact design. Human beings undoubtedly should be thankful to the inventors of internal combustion engines. The development of the internal combustion engine derives from knowledge of physics, chemistry, mathematics, thermodynamics, fluid mechanics, gas dynamics, heat transfer, heat transfer, and computational fluid dynamics. The machine appears to be so simple but uses a large number of components in its working. The combustion engines, especially piston engines, convert chemical energy into mechanical energy, available in fuel upon combustion.

The fuel energy, which is the main source for the majority of ICEs, is petroleum-derived fuel, and these fuels (fossil fuels) are serving this type of engine for over 150 years. No doubt, the invention of ICEs has generated a lot of employment and given rise to the development of other fields. The ICEs are broadly classified as spark-ignition engines and compression ignition engines fuelled by gasoline or petrol and diesel fuel respectively on a large scale. The former engines are dominating in passenger vehicle applications while the latter type is mostly used for goods transport and heavy-duty applications. The diesel engine fuel is little more viscous than gasoline, and their fuel ignition qualities are rated in terms of octane number and cetane number.

Therefore, the entire society is moving around three terms: energy, efficiency, and emissions (3Es). The emissions could be either evaporative emissions and combustion generated or tailpipe emissions. As far as automobiles are concerned, the combustion-generated tailpipe emissions are of interest. Any fuel (mostly hydrocarbons) upon its combustion produces emissions, but the concern is with regards to harmful emissions or pollutant emissions. Nowadays, it can be said here that there is a paradigm shift in 3Es; the concern now is more on the climate change issue. Thus, the Terms of importance are energy, environment, and climate. Environment-benign energy resources should be searched to mitigate the issues concerned with climate change.

I.2   ISSUES OF GLOBAL WARMING AND CLIMATE CHANGE

All around the world, the automobiles are powered by IC engines that are run by fossil-derived fuels because of their excellent power-to-weight ratio and smaller size. Major emissions produced by IC engines are oxides of nitrogen (NOx), carbon dioxide (CO_2), carbon monoxide (CO), hydrocarbons (HC), oxides of sulfur, and particulate matter (PM). All around the globe, the automobile is indicative of an improved quality of life. The mobility provided by automobiles expands the possibilities for employment, education and health care, as well as social and leisure activities. For automakers, one critical priority is to sustain the benefits that autos provide, while keeping them affordable, preserving the diversity of automobiles, and reducing their environmental impact.

Human activity has led to climate change, which is now concentrated in the atmosphere with the result of a reduction in thickness of ozone layer. The Earth's atmosphere is divided (called atmospheric stratification) into five main layers. Excluding the exosphere, the atmosphere has four primary layers: the troposphere, stratosphere, mesosphere, thermosphere, and exosphere. The highest region, which extends beyond about 500 km, is the exosphere, and from ground level, to an altitude of about 50 km, the ozone layer effect prevails, which is useful in preventing harmful rays from reaching the surface of atmosphere.

Climate scientists have observed that carbon dioxide (CO_2) concentrations in the atmosphere have been increasing significantly over the past century, compared to the pre-industrial era level of about 280 parts per million (ppm). In 2016, the average concentration of CO_2 (403 ppm) was about 40% higher than in the mid-1800s, with an average growth of 2 ppm/year in the last ten years. Significant increases have also occurred in the levels of methane (CH_4) and nitrous oxide (N_2O). If the same trend continues, the levels go beyond to about 450 ppm by 2020. Due to the emission of these manmade gases (anthropogenic causes), rise in levels of these gases, especially after the industrial era, are mainly responsible for the heating of the earth. This phenomenon is called the 'greenhouse effect.' It is due to the presence of these gases that trap the heat radiated from the earth's surface, the heating of an atmospheric, and thus the Earth experiences a rise in temperature because certain gases (water vapor, carbon dioxide, nitrous oxide, and methane) in the atmosphere allow incoming sunlight to pass through but are then trapped.

Anthropogenic activity made emissions of CO_2 represent less than 5% of the total, including CO_2 emissions from natural sources, but even this relatively small increase can shift the Earth's natural balance. That is why we believe it is important to continue to reduce CO_2 emissions from all sources, including those resulting from burning fuel in automobiles.

Bazylevych et al. (2014) in their work quoted that climate change is considered to be the greatest threat to nature and humanity in the twenty-first century. The warning made by former US Secretary of State John Kerry described climate change perhaps "the world's most fearsome weapon of mass destruction" and said that it was compelling us to act." The experts assume the average global temperatures to increase by 1.4–6.4°C by the end of the 21st century.

The consequences result in a sharp rise of the sea level, alteration of the rainfall distribution pattern, occurrence of floods and drought, leading to a decline in agricultural production, reduction in crop yields and livestock productivity, and food shortages and finally threats of extinction and disappearance of more than 25% of rare animal and plant species, and which also can may change migration processes. The ill effects of climate change are illustrated in Figure I.1.

A particularly critical situation may develop in Asia, Africa, and Latin America. According to UN estimates, in 2008, 20 million people were migrants due to climate change, and by 2050, these can be almost ten times more. Annual losses because of warming could are expected to reach almost about 5% of GDP.

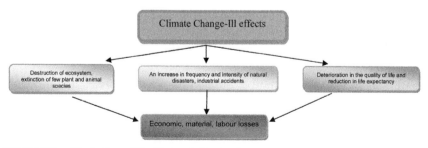

FIGURE I.1   Illustration of ill effects of climate change.

The International Energy Agency (IEA) is an apex body, and in its 2017, report it is illustrated that (Figure I.1) with fuel combustion there are two major sectors contributing to two-thirds of global CO_2 emissions, with 42% of CO_2 emissions contributed by electricity and heat and 24% by the transport sector.

Energy is important to get anything done, and in the process, the effectiveness of energy conversion is measured in terms of efficiency. However, to adhere to the second law of thermodynamics, losses will play a vital role, and in the process of conversion from fuel, there exist emissions. Therefore, the three are linked. If one wishes to reduce fuel consumption, the efficiency of system will improve and so also low emissions and vice versa.

The fossil fuel reserves (mainly oil) that have been concentrated in only certain pockets of earth's crust and in some countries, especially OPEC nations, those possessing large resources are virtually ruling the world. Thus, the nations that have minimal or no resources are losing their hard-earned revenue in importing such sources.

In the name of urbanization, civilization, lifestyle, and to keep the wheel moving, large sources are being spent indiscriminately. In one way lifestyles have improved but the exponential growth in automotive population has given rise to fuel crises, rise in crude prices, fluctuating fuel prices and a more serious one, degradation of the environment. The IPCC has established that human activity is mainly responsible for the rise of global temperatures, and rises in CO_2 levels in the atmosphere, and it has become concern for all nations. Unless the emissions are cut to significant levels, it is difficult to keep the global temperature under check; the Earth's planet is facing catastrophic effects.

I.3   ROLE OF AUTOMOBILES IN GLOBAL WARMING

Automobiles are the major source of harmful emissions, such as carbon monoxide, oxides of sulfur, oxides of nitrogen, and particulate matter. Though there are regulated and unregulated emissions, many countries are facing threats with regards to long-term availability of petroleum-derived fuels (Figure I.2).

An increase in the automotive population would increase the amounts of undesirable exhaust gases left in the atmosphere. Earlier, CO_2 was not treated as harmful as it is an ultimate product of complete combustion of any hydrocarbon fuel, but owing to the global warming phenomena, it started causing major concern for mitigation of levels of CO_2.

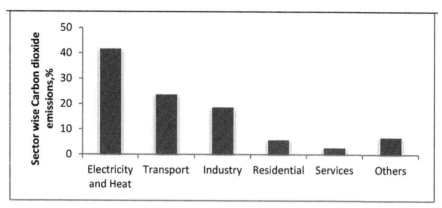

FIGURE I.2(A)   Sector-wise contribution of global greenhouse emissions.
Source: IPCC 2014.

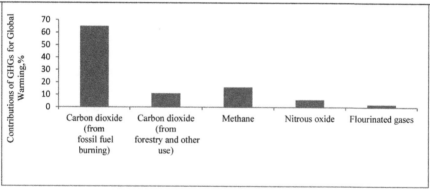

FIGURE I.2(B)   Contribution of CO_2 emission by transport sector.
Source: International Organization of Motor Vehicle Manufacturers-OICA.
Figure I.2(B) IEA estimated shares of global anthropogenic GHG, 2014.

Since the concentrations of these obnoxious emissions are increasing in the atmosphere at alarming levels, researchers are looking at the mitigation or control of their pollutant emissions from ICEs. It is becoming imminent to find pollutant emission mitigation methods for automobiles as these are causing a public nuisance and increasing ground-level ozone levels.

For the rise of atmospheric temperatures, the greenhouse gases (GHGs) are responsible. GHGs constitute carbon dioxide, oxides of nitrogen, methane, CFCs (especially fluorinated or F-gases). Among these, CO_2 contributes to a greater extent. There are many sectors of energy consumption through which the GHGs are produced. The carbon dioxide is contributed mainly from two broad sources: fossil fuel and industrial processes and forestry, and these account for about 76%, as confirmed by a report of IPCC 2014 (Figure I.2(a)). This is in line with the IEA 2014 report that established that anthropogenic activity is responsible for emission of these GHGs (Figure I.2(b)). It is obvious from the figure that the transport sector is the largest source for emission of carbon dioxide followed by other sectors. This is giving an impetus towards work to done for mitigation of vehicular engine pollutant emissions.

Statistics indicate that more than 60% of the cars are produced in Asia and Oceania, whereas Europe produces about 26%. This shows the nations that are major sources of GHGs in the world, as far as cars are concerned. In 2009 and 2014 respectively, the EU introduced legally binding CO_2 standards, for the first time setting a goal that, on average, new cars sold in Europe in 2015 should emit 130 g of CO_2 per kilometer, and 95g CO_2/km in 2021. The cap on the emission of CO_2 has become serious especially after the Kyoto Protocol and the Paris Protocol.

As far as the share of road transport in global CO_2 emissions, even though the estimates vary, but it can be said that up to about 16% of global man-made CO_2 emissions are contributed by motor vehicles (about 13% of total greenhouse gases). Therefore, automobiles are by no means the biggest CO_2 contributors, but they are a significant factor. As more and more people in emerging countries demand more and better mobility, the number of vehicles in the world is set to rise, offsetting progress already made in reducing fuel consumption of new vehicles. The International Energy Agency forecasts the increase in CO_2 emissions from traffic worldwide keeping pace with average increases from other sectors.

An American geophysicist, Mr. Hubbert, in 1956 proposed a Gaussian type correlation for resource utilization subjecting to three basic assumptions and the modal (Richard A Dunlap 2015). His model when applied to oil predicted an oil peak. However, there were many contradictions to his proposed theory. In the run for fast growth in different sectors, the nations China and India are competing neck to neck and, as per IEA report, while there is no peak oil demand in sight, the pace of growth will slow down to 1 mb/d by 2023 after expanding by 1.4 mb/d in 2018. There are signs of substitution of oil by other energy sources in various countries. Even though China imposed the world's most stringent fuel efficiency and emissions regulations, as the country recognizes the urgent need to tackle poor air quality in cities, efforts are intensifying. Sales of electric vehicles are rising, and there is strong growth in the deployment of natural gas vehicles, particularly of fleets of trucks and buses. The analysis shows that a rising number of electric buses and LNG-fuelled trucks in China will significantly slow gas oil demand growth.

The rise in global temperature and its harmful effects have been the subject of climate change as a result of human activity and is probably the greatest challenge facing society in the twenty-first century.

From the reports of IPCC, the Kyoto Protocol, and the Paris summit, it was asserted that developed nations typically emit high carbon dioxide emissions per capita, while some developing countries lead in the growth rate of carbon dioxide emissions. It is clear that uneven contributions to the climate problem are at the core of the challenges the world community faces in finding effective and equitable solutions.

SIAM is the apex industry body representing leading vehicle and vehicular engine manufacturers in India. SIAM is an important channel of communication for the automobile industry with government, national, and international organizations. According to SIAM reports, as illustrated in Figure I.3(a), there is a huge demand for 2-wheelers. The driving habits and driving cycles are different for 2-wheelers and more prone for emission of harmful pollutants. When gross total domestic sales trends of 4-types vehicles are together considered in India, as illustrated in Figure I.3(b) is obtained. There exists about 10–15% variation in production and sales.

I.4 POLLUTANT EMISSIONS FROM AUTOMOBILE VEHICLES

The major pollutants from gasoline-fuelled engines are CO, HC, and NOx emissions, whereas particulate matter (PM) (soot-carbonaceous matter or smoke) and NOx are from diesel engines. Among these pollutant emissions, PM is a visible pollutant. With large scale use of diesel vehicles, the smoke levels increased and combined with fog give rise to smog formation. The smog problem interfers with traffic on highways and has led to many accidents especially in cold climatic counties. In 1952, Dr. Arie J. Haagen-Smit's research findings asserted that automotive exhaust emissions are the major cause of degradation (i.e., atmospheric pollution) in Los Angeles in 1947. The majority of automobiles are fuelled with petroleum-derived fuels which are predominantly hydrocarbons. Under the conditions of complete combustion, each kg of hydrocarbon fuel when completely burnt produces mainly 3.1 kg of CO_2 and 1.3 kg of

H$_2$O. Since the conditions of vehicular engines would not permit either stoichiometric operation or complete combustion, then the exhaust species or products of hydrocarbon combustion always contain undesirable exhaust emissions which are produced even though in minute quantities (parts per million) such as: carbon monoxide (CO), unburnt hydrocarbons (HC), oxides of nitrogen (generally termed NOx), carbon dioxide (CO$_2$), lead salts, polyaromatics, soots, aldehydes, ketones, and nitro-olefins. Of these, only the first three are of major significance in the quantities produced.

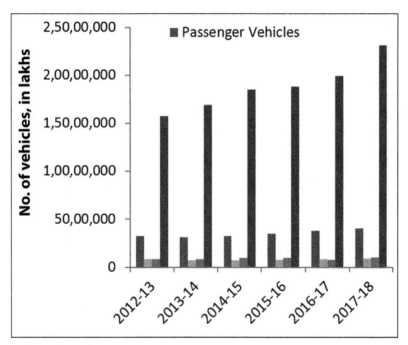

FIGURE I.3(A)   Year-wise production statistics of various types of vehicles in India.

For many years, manufacturers of catalytic converters pressed for unleaded petrol because lead deposits rapidly rendered their converters ineffective. Carbon monoxide is toxic because it is absorbed by the red corpuscles of the blood, inhibiting the absorption of the oxygen necessary for sustaining life.

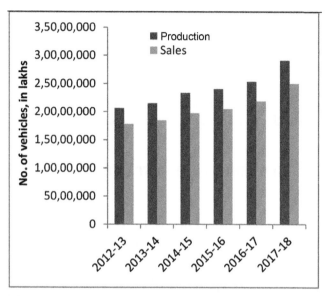

FIGURE I.3(B) Domestic sales trends of a gross total of 4-types of vehicles.
Source: Society of Indian Automobile Manufacturers (SIAM) report.

The toxicity of hydrocarbons and oxides of nitrogen, on the other hand, arises indirectly as a result of photochemical reactions between the two in sunlight, leading to the production of other chemicals. There are two main oxides of nitrogen: nitric acid and nitrogen dioxide, NO and NO_2, of which the latter is of greatest significance as regards toxic photochemical effects. Under the influence of solar radiation, the NO_2 breaks down into NO+O, the highly reactive oxygen atom, then combining with O_2 to make O_3, which of course is ozone. Normally, this would then rapidly recombine with the NO to form NO_2 again, but the presence of hydrocarbons inhibits this reaction and causes the concentration of ozone to rise. The ozone then goes on, in a complex manner, to combine with the other substances present to form chemicals which, in combination with moisture in an atmospheric haze, produce what has been described as the obnoxious smoky fog, now known as smog.

Unburnt hydrocarbons can come from evaporation from the carburetor float chamber and fuel tank vent as well as from inefficient combustion due in different instances to faulty ignition, inadequate turbulence, poor carburetion, an over-rich mixture, or long flame paths from the point of ignition. Other factors are over-cooling, large quench areas in the combustion chamber, the unavoidable presence of a quench layer of gas a few hundredths of a millimeter thick clinging to the walls of the combustion chamber, and quenching in crevices such as the clearance between the top land of the piston and the cylinder bore. As the conditions would not permit for complete oxidation of majority of intermediate species formed during combustion, and thus they rapidly exit into atmosphere. Some means of further oxidation is required to allow complete oxidation of species before letting into open air. The first technique was thought to be the use of thermal converters. Such techniques are often called exhaust gas aftertreatment methods.

I.5 AUTOMOBILE VEHICLE EXHAUST EMISSIONS CONTROL STRATEGIES

After the introduction of EURO norms, many other nations also followed, promulgating emissions, and have also periodically made even more stringent norms. The job of reduction of engine/vehicle emissions has been an integral part of researchers, Original Equipment Manufactures (OEMs), and leading manufacturers. In the process of evolving strategies to contain harmful emissions, both in-cylinder, geometrical, operational and exhaust gas aftertreatment methods have been adopted.

It is understood that the harmful emissions from gasoline-fuelled engines (spark ignition engines) are CO, HC, and NOx, while those from diesel-fuelled engines (compression ignition engines) are HC, PM and NOx emissions, even though the formation mechanisms for CO, HC, and NOx emissions from both types are one and the same. The PM emissions are specific to diesel-fuelled engines. Nowadays, many innovations are happening to see that all these emissions are maintained well below permissible limits. This has become a serious concern and urgent need to comply with EURO-VI norms from 2020 onwards. This is enticing both IC engine combustion research communities as well as auto manufacturers.

The Government of India is poised to introduce electric mobility from 2030 onwards, but at the same a big push is being given to enhance the use of alternative fuels, such as ethanol, biodiesel, CNG, etc. Also, it is raising the doubts of how long the ICE, driven vehicles will continue to serve mankind, and efforts from all corners are increasing. Another threat is with the increase in automation, enhanced use of sensors, and autonomous vehicles that are penetrating the market.

The equivalence ratio plays a vital role in controlling SI engine emissions, mainly CO, NOx, and HC. Even though the diesel engine works with lean mixtures, the PM-NOx trade-off works very differently. In general, the CO, HC, and PM emissions are predominantly of the equivalence ratio dependant while combustion temperatures and lean mixtures are dictating NOx emissions. The NOx emissions are fuel independent whereas CO, HC, and PM emissions are fuel-intensive.

Among these emissions, NOx emissions deteriorate the atmospheric ozone, and its control is much more important, and any effort made to reduce the in-cylinder peak temperatures and avoidance of lean mixtures helps towards the reduction in NOx emissions. Even at the cost of power down-sizing, the NOx emissions are being managed. For this either retarding fuel injection (or retarding spark timing) and/or circulation of burnt gases is most preferred. The method includes external and internal circulation techniques. With external circulation, exhaust gas recirculation (EGR) returns a small fraction of exhaust, i.e., products of combustion, to dilute the intake charge and thereby reduce the peak temperature and thus lower the amount of NOx formed. On the other hand, as soot/PM interferes with visibility issues, the use of diesel particulate filters or diesel particulate traps (DPF/DPT) are becoming mandatory with modern diesel that complies with EURO-VI emission norms.

With internal circulation, valve timing is controlled such that during intake there is an overlap period in which the exhaust valve is still opened along with the intake valve allowing internal exhaust gas recirculation. Three-way catalyst systems are also used to effectively reduce spark-ignition exhaust emissions by oxidizing CO and HC as well as providing for NOx reduction. However, the discharge control method requires a stoichiometric or oxygen-deficient exhaust gas composition to work. A computer-controlled closed loop air-fuel ratio feedback control system upstream of the three-way catalysts is required to monitor the oxygen content, i.e., stoichiometric conditions. The highly volatile nature of gasoline fuel predominates evaporative emissions, and non-engine HC emissions include the fuel tank, crankcase, and carburetor evaporative losses.

There are doubts, rumors, and debates on the future of IC engines in view of increased pollution levels. However, the debates suggest that fossil fuel-driven ICEs will exist for few more decades with the new combustion concept being developed, along with air intake technology, low carbon fuels, and fuel injection technology and improvised aftertreatment devices to meet EURO-VI norms.

Climate Action Tracker (CAT) is an independent scientific analysis that is involved in tracking climate action across 32 countries, covering about 80% global emissions. The analysis of CAT is appreciative of the policies initiated and implemented by India to limit the long-term rise average of world temperature to 2°C as a part of *Paris Agreement on Climate Change*.

The efforts are equally being put in by automotive industry and are all poised to meet EURO-VI targets by 2020, with the kind of innovations that are going on with electronic controls, electric mobility, use of alcohols, and hybrid propulsion that upgrading the present ICEs. The future is becoming uncertain whether there will be total shift from oil resources to lithium sources or fossil fuels will continue to exist for few more decades (Auto Tech Review, June 2018.).

In India in the north part of the country, especially in NCR region of New Delhi, the air pollution levels have been rising very high and are more serious in the harvesting season and cold weather conditions.

"That air pollution is a bigger contributor to the disease burden than tobacco should not come as a surprise," says Kalpana Balakrishnan, an author of the study and director at *WHO Collaborating Centre for Occupational and Environmental Health,* in Chennai, India. "Air pollution has been found to be associated with [the] same disease endpoints affected by smoking," she tells *SciDev.Net.* (https://www.scidev.net/asia-pacific/pollution/news/india-s-air-pollution-deadlier-than-tobacco-study.html)

I.5.1   NOISE POLLUTION

The increased number of automobiles apart from tail-pipe emissions are major sources of noise pollution. In the event of adopting any noise control technique, the identification of the source and its characterization plays a key role. In an automobile, researchers throughout the world have found by measurement and simulations that the primary noise sources are the power train, tire/road interaction and wind flow. The reasons for high levels of noise from automobiles could be a few manufacturing defects, improper handling of automobiles, longevity, poor maintenance, according to the Central Pollution Control Board (CPCB) of the Government of India. The pass-by noise levels measurements are made as per the BIS: 3028–1998 standard. This agency has fixed about 80 dBA as noise levels (the levels vary based on power rating) within a certain distance of vehicle movement. The noise levels, if left unattended to may lead to vibrations and harshness, and eventually the vehicles, would be broken down. Over a period of time many computer-aided engineering techniques have evolved to determine the vibration response of automobile components and also the noise radiated by noise emitting components (Mohanty and Fatima, 2014). The subject of vehicle refinement in terms of noise, vibration, and harshness is gaining a lot of attention. The subject is leading to the use of less weight components and reduction in components for preventing/mitigating noise levels.

This book deals with both SI and CI type engines. Emphasis is given on the combustion-generated pollutant emissions/tail-pipe emissions only. It starts with a basic review of internal combustion engines and goes on to dealing with emission formation mechanisms in both types, measurement of emissions, in-cylinder, and aftertreatment techniques adopted. Of late, many alternative fuels are emerging and other combustion concepts are evolving in the event of mitigation of harmful exhaust gases SI engines. The structure followed in the book is depicted in Figure I.4.

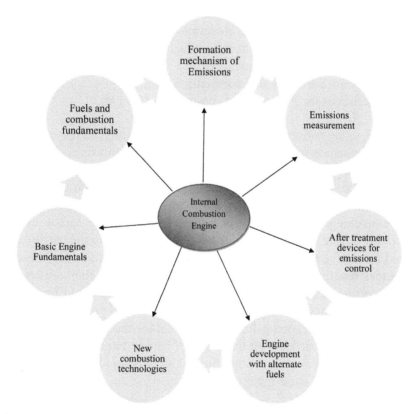

FIGURE I.4   Structure of various engine emission control techniques.

I.6   CLOSING REMARKS

The nations that do not have strict emission legislature may be encouraged to follow in line with the developed countries. Eventually, cooperation from customers is also very important.

It is observed that the people who have the knowledge of good vehicle maintenance are energy conservation bent of mind and environment are only following the government laid rules. The percentage of such people is fractional, and unless some serious steps are taken to save the environment, ultimately global temperatures, loss of revenue, and loss of life would only be awaiting in the days to come.

The following are the immediate measures expected from different sectors:

- o Enhanced research and development of new combustion engines with ultra-low emissions. Research on new and alternative fuels;
- o Policies to encourage alternative fuels;
- o Proper road infrastructure;
- o Fuels infrastructure development;
- o Incentives for fleet renewal eco-driving practices.

Of late, many nations are encouraging to use biofuels as they are attractive in many aspects. However, when well-wheel analysis is applied, the majority of these fuels are inferior to fossil fuels. When such biofuels are to encouraged, the treatment methods are being done with conventional sources. If bio-originated sources are treated with renewable sources, to make convenient fuels for existing automobiles, then the carbon cycle is effectively closed. Moreover, in the process of improving the performance of fuel or catalytic converters, nanomaterials are being added. These are sometimes reactive or unreactive, and eventually when they are emitted, they are prone to emit $PM_{2.5}$ level particulates, which could be harmful to life on the planet. The effect of nanoparticle added emissions, upon its release from tail pipe, on health is yet to be established on the lines similar to wide studies carried out on the effect of lead in tetraethyl lead, used to be added to improve octane rating. Similar studies should be carried out to ascertain the use of nano-materials addition.

REFERENCES

Auto Tech Review Magazine, Springer Vol.7, Issue 6,June 2018

Mohanty A. R. and Fatima S. *An overview of automobile noise and vibration control-* Noise & Vibration *Worldwide* Article *in* Noise Notes · March 2014 DOI: 10.1260/1475-4738.13.1.43

Richard A.Dunlap *Sustainable Energy* Cengage; 2015

CHAPTER 1

Basic Review of Internal Combustion Engines

1.1 INTRODUCTION: REVIEW OF IC ENGINES: COMBUSTION ENGINES DEVELOPMENT

Energy has become an essential component of human beings for meeting the requirements in day-to-day life. The lifestyle of modern days cannot be thought of without energy. Among different energy sources, fuels have been explored majorly to develop power, motive power, heat, etc. Motive power is essentially met by automobiles, trains, and aircraft. Automobiles are predominantly powered by petroleum-derived liquid fuels for around 150 years. These liquid fuels have been extracted from the crude obtained from the earth's crust. As the demand for mobility has increased, so also the automobile population has seen tremendous growth in its population. India is one among the fastest growing economies in the world, meets its fuel requirement for transportation sector majorly from liquid fossil sources and marginally through gaseous fuels. The situation exists among other nations too. The exponential growth in the automotive population undoubtedly brought a large market for the industries which would be supplying to the auto or transportation sector. Energy is synonymous with the consumption of fossil fuels. So for meeting energy demand in the transportation sector; the automobile is synonymous with fossil fuel. So also, the population has become synonymous with pollution.

Energy is a basic need to survival for any living being on the earth. However, the required energy is either be directly used or is transformed from one form to another for better and efficient usage. Energy consumption has become a yardstick to rate country a developed or an under-developing one. In general, the transformation of primary into secondary energy through a technical process in an energy conversion plant. Here, the energy conversion plant is the engine. For making an engine, firstly, the energy is derived from fuel (available as stored energy in the form of chemical energy) and is first burnt or combusted to release thermal energy and finally, thermal energy is converted to mechanical energy.

Thus, the energy conversion process takes place from primary energy to secondary energy as depicted in Figure 1.1.

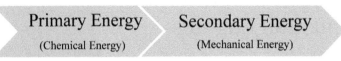

FIGURE 1.1   Primary energy to secondary energy.

The primary energy is obtained from sources such as liquid fuel and its derivatives, natural gas, etc. Similarly, hydro energy or solar energy, wind energy, etc. are used to obtain mechanical energy. But our main concern is from fuel energy to mechanical energy (Figure 1.2).

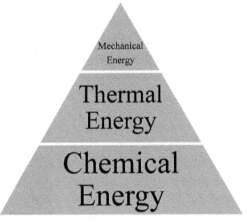

FIGURE 1.2   Pyramid depicting representative conversion efficiencies.

In general, the process of thermal energy conversion is done adhering to the laws of thermodynamics and can be described formally. The internal combustion engine (ICE) and the gas turbine are specialized energy conversion units, in which the chemical energy bound in the fuel is at first transformed into thermal energy in the combustion space or chamber, this being then transformed into mechanical energy by the electric motor. Preferably, for electrical energy production, the stationary gas turbine plants are normally used in addition to or in combination with steam turbines. Here, it can be said that owing to conversion losses, the quantities could be represented with the help of pyramid (Figure 1.2). The sizes are representative of conversion efficiencies. The conversion efficiencies are from thermal to mechanical, it can be regarded as thermal efficiency or fuel conversion efficiencies. Moreover, these are based on fundamentals of the first law of thermodynamics.

Basically, an engine is an energy conversion device that converts the chemical energy inherently available in fuel upon combustion with an oxidizer (mostly air) into mechanical energy. The engine used for powering automobiles called prime movers. Prior to the invention of ICE, external combustion engines (ECE) were in use such as steam engines. ICEs are quite different from ECEs, such as steam or stirling engines, in which the energy is delivered to a working fluid not consisting of, mixed with, or contaminated by combustion products. ICEs are usually powered by energy-dense fuels such as gasoline or diesel, liquids derived from fossil fuels. While there are many stationary applications, most ICEs are used in the dominant power supply for cars, aircraft, and boats.

The lifestyle of human beings has substantially changed with the invention of automobile. The development of automobile has undergone a multitude of developments in design since its discovery. This development is credited to the inventors and researchers who made available the present automobile. Basically, the energy required for motive power by the automobile is obtained from the heat engine or simply an engine. The engine performance is constrained by the second law of thermodynamics. The engine develops mechanical energy at the expense of thermal energy produced by combusting fuel with an oxidizer. The fuel could be either liquid or gas. Thus, engine in an automobile is simply a thermal energy conversion device.

In the early stages, human civilizations, people were largely depending on mobility of either human or goods on either human power or animal power. Subsequently, with the invention of steam engines, the scenario has shifted to machine-driven mobility. The steam engine, which is an ECE, is also a prime mover but it solely runs with steam as working fluid, for generating steam and consequent expansion of high pressure, high temperatures had to done through steam generators and steam engines respectively.

However, owing to the low power to weight ratio of ECEs and bulky, after the invention of ICEs started emerging and were preferred over the years. The mechanical energy thus produced is usually made available on a rotating output shaft. The name internal combustion refers also to gas turbines except that the name is usually applied to reciprocating internal combustion (IC) engines like the ones found in everyday automobiles. The chemical energy of fuel gets converted into thermal with consequent exothermic reactions.

The real benefit of ICE was realized after the discovery of petroleum crude. The resources of liquid fuels have been concentrated in limited pockets of earth, majorly dominated by OPEC nations. Due to this, the countries which have limited or no sources are heavily dependent on OPEC nations and thus, in turn, are losing their hard revenue in the import of petroleum-derived products. Therefore, there exists a mismatch between demand and supply. Moreover, there were instances of oil embargo (crisis) of the early 1970s and industry recession of 2008, fluctuating barrel prices in the international market. The studies by Hubbert sent shocks among many that supply of oil has reached a peak, thereby the existence of liquid fuels has been under severe threat as the demand is far more than supply.

Basically, the prime mover used in different sectors, does the job of energy conversion, considering a prime mover, energy requirement is met for the engine is obtained from combustion of fuel and air. Fuel has a large amount of chemical energy within it. This chemical energy is transformed into thermal energy and then mechanical energy. Therefore, engine is primarily an energy conversion device. In the process of developing mechanical energy, the engine loses some of its input energy to coolant, exhaust gas losses, etc. (Obert, 1973). Engine is a prime mover for automobiles and works under the constraints of second law of thermodynamics due to which, by producing prime source of energy i.e., propulsive power, it loses some of the energy in the form of exhaust.

The exhaust of the engine consists mainly of components which are obnoxious as the combustion may not take place under stoichiometric conditions. The more details of formation of these pollutant emissions and its control are primary objective of the present book which will be dealt in the following chapters. Like thermodynamics encompasses about 3Es (viz.; Energy, Equilibrium, and Entropy) (Van Wylen and Sonntag, 2008), the subject of ICEs broadly deals with 4Es from energy conversion of fuel energy to emissions.

The present book deals with engines used in automobile sector. The engine is primarily a heat engine, i.e., an ICE wherein power is produced by burning a fuel in the presence of an oxidizer. The power stored in the fuel is released when it is burned. The word internal means that the fuel is burned inside the engine cylinder itself. The most common fuels are gasoline/petrol and diesel. If gasoline is to burn inside the engine, there must be oxygen present to support the combustion. Therefore, the charge needs to be a mixture of gasoline and air. When ignited with an external source (electric spark), mixture of gasoline and air burns rapidly; it almost explodes, thereby producing mechanical energy. The engine is designed harness this energy.

From a small power to large value (0.01 kW to 20 MW) as per its swept volume (displacement volume); the ICE competes with gas turbines and electric motors (Ferguson and Kirkpatrick, 2001). This shows the versatility of ICEs in market share.

Historically, the IC engine was not fuelled by liquid fuels. It was the compaction of gun powder in a cylinder at atmospheric pressure led to explosion and rotation of wheels. This type of engine represented as cannon. The first 4-stroke engine concept was credited to Nikolaus A. Otto even though some similar effort was put in by Alphonse Beau de Rochas. Otto proposed systematically four thermodynamic processes representing 4-strokes of the piston with two revolutions of the crank shaft. In the process of making further simplification, 2-stroke idea was given by Sir Dougald Clerk; a 2-stroke engine was popularly called as Clerk engine. The engines were largely marketed and appreciated after the discovery of liquid fuels (petroleum-derived fuels) in Pennsylvania.

However, due to constraints on the ability of the gasoline (petrol) to high load operation, the problem of detonation was noticed. The detonation was more pronounced at higher compression ratios (CR). There was a limit on the CR for a gasoline-run engine and to overcome the difficulty, an anti-knocking agent was invented. The first of its kind was tetraethyl lead (TEL). The gasoline engines were further improved with the use of carburetor; a device that would help in preparing a homogeneous mixture of air and gasoline (petrol). With the large scale exploration of liquid fuels and to use denser fractions, a novel idea was given by Sir Rudolph Diesel; led to development of diesel cycle engine. The substance that enters during suction/intake stroke is called charge. The charge for the petrol run engine is air-fuel mixture whereas air alone is the charge for the diesel fuelled engines. In order to control the load/speed, the amount of fuel to be injected has to be only varied in case of diesel engines whereas it is the mixture that need to regulated in petrol engines. Therefore, there is a better control over fuel in diesel engines. The difficulty of high load operation and restriction on CR was not faced by diesel engines. However, the problem of smog (smoke +fog) was observed with increased use of diesel engines. With the scale of marketing of petroleum fuel derived fuels, Henry Ford envisioned the threat for large scale availability of fuels for future generations. Hence, attention was diverted to the home/backyard grown fuels such as alcohols, to overcome difficulties with petroleum-derived fuels and such fuels are renewable in nature.

With the exponential growth in the use of automotive vehicles, the problem of automotive pollution was recognized and stringent norms had been imposed to restrict the release of tail emissions such as CO, HC, NOx, and soot or particulate matter. In order to reduce the problems with noise pollution and tailpipe emissions, the emphasis is given on use of lightweight engine component materials and alternate fuels (such as alcohols, vegetable oils, hydrogen, CNG, etc.). With the development of the IC engine, there were also developments in allied fields that support the manufacturing of ICEs. Many OEMs have also grown. The milestones in the development of ICEs and its historical perspective in a chronological order are tabulated in Table 1.1.

TABLE 1.1   Milestones in the Historical Development of ICEs

1794	An engine with a mixture of vaporized turpentine and air was tried out by an Englishman, Robert Street, patented, but not built.
1857	A major development of Free Piston Engine was by two Italians-Eugenio Barsanti and Felice Matteucci. But observed to be large, noisy, and slow speed engine.
1859	Discovery of Liquid Petroleum crude in Pennsylvania, gave a birth to petroleum industry. Liquid fuels started emerging
1860	J.J. Etienne Lenoir-Belgian, developed a gas engine (without compression-atmospheric engine), double acting type with gas-air mixture, yielded thermal efficiency around 5%.
1851–1888	Sadi Carnot and Alphonse de Rochas-French people, define thermodynamic processes that laid the way for better understanding of ICE operation. They explained the importance of compression before ignition.
1867	Germans-Nicholaus Augustiene Otto and Eugen Langen developed 4 kW tall engines on the principles of Barsanti and Felice Matteucci produced more efficiency ~11%. Otto was the first inventor to name four thermodynamic processes-Suction, Compression, Expansion (power), and Exhaust.
1876	Otto developed a modern engine following 4-stroke principles and called it-silent engine, Compression ratio – 2.5:1, thermal efficiency rose to 14%.

Efforts were made to reduce the number of rotations for developing power and thus gave a birth to 2-stroke cycle concept.

1878–1881	Duggald Clerk- first to use an air standard cycle analysis for engines and adopted separate scavenging.
1877	Robson suggested scavenge compression and patented the principles.
1880	William Priestman and Hobert Acktroyed Stuart-Englishmen-developed a engine that used vaporizer for injecting kerosene and claimed thermal efficiency of ~15%

TABLE 1.1   *(Continued)*

1883	Gottlieb Daimler, German Engineer used a hot bulb concept and developed a high speed engine ~ 700 rpm.
1886	Karl Benz developed a tricycle
1888	Invention of pneumatic tire by John B. Dunlop further improved the working performance of automobile.
1892	Carburetor- Samuel Morey in 1826
1898	Robert Bosch invented a modern electric spark plug.

The principles of Carnot engine cycle made Rudolph Diesel to develop an engine with high distillates and without mixing fuel and air before combustion.

	Rudolph Diesel used a air-blast fuel injection system with a high pressure compressor (~7 Mpa)
1910	Mechanical Piston Pump with fine tolerances developed by McKechnie
	Acktroyl Stuart developed a pre-chamber (IDI) concept
	Anti-knocking Agent-TEL-
	Cycle innovations-Atkinson cycle-Miller cycle.
	Henry Ford envisaged the importance of backyard grown fuels

Majority of the inventions were over and inventors started to improve the thermal efficiency or devised methods to overcome difficulties with gasoline/diesel run engines.

Early 1920	Ricardo outlined the concept of Stratification and thus developed stratified charge engine.
1920	Midgley developed a knock suppressant-tetraethyl lead for SI engine to minimize combustion knock.
1930	Details of ignitions delay and diesel knock were related.
	A study on thorough thermodynamic cycle analysis for engines was done by Americans-Good enough and Baker
	Fuel injection pump-Robert Bosch
	Felix Wankel-Rotary Combustion Engine
	Opposed piston engines
	Automobile with IC Engine

Increased automotive population led to the degradation of air quality in Los Angeles. Researchers understood that CO, HC, NOx, and SOx emissions were partial oxidized compounds and started to mitigate the same. This has sprung up the new research for reduction of photochemical smog by NOx compounds.

1960s	Reduction in air quality is linked to automotives
1961	In California, positive crankcase ventilation was first introduced.
	CARB-California Air Resource Board was formed.
Mid-1970s	Japan, Canada, and Australia introduced emission controls

TABLE 1.1   *(Continued)*

1986	Stringent emission controls were introduced by the USA and European countries.
1973	Witnessed an oil crisis – Oil Embargo – emphasized the need for continuing high level of fundamental research into engine operation. Efforts were focused on simultaneous reduction of emissions and fuel consumption. Work started in the use of alternative fuels.
	Variable compression ratio engine
1990	Catalytic converter.
	Need for engine modeling studies was stressed.
	D. F. Caris and E. E. Nelson-optimized the compression ratio
	Fast burn combustion chambers
	Lean burn engines.
	Improved control and optimization of influencing operating and geometric parameters.
	Gasoline direct injection (GDI) and DISI concepts were started. Fuel Injection systems for SI engine-mechanical, electronic fuel injection and MPFI methods.
	Reduction of engine size-variable displacement engine
	Supercharging /Turbocharging and its other configurations.
	Low heat rejection/thermal barrier coatings-semi adiabatic Engine concept
	Sensors-electronic controls
1960s	CRDI-Common Rail Direct Injection by Robert Huber of Switzerland,
	New combustion concepts-HCCI/PCCI/RCCI
	Noise pollution
	New materials started emerging with a view to reduce overall weight of an engine.
1988	First production car diesel engines with direct injection (Fiat)
1989	First production car diesel engines with exhaust gas turbocharging and direct injection (Car Audi 100DI)
2000	First production car diesel engines with particulate filters (Peugeot)

1.2   ENGINE CLASSIFICATION

The versatility of ICE is that it can meet in any field power requirement and thus IC engine find applications in various fields such as land (automobiles, gen-sets, etc.), sea (marine) and air (aero planes). Though, the IC engines are classified in many aspects, prominently the engines are either Spark-ignition engines or compression engines and 2-stroke and 4-stroke cycle engines. IC engine do not undergo thermodynamic cycle, as the burnt gases again would not converted back to fresh air-fuel mixture but undergoes mechanical cycle (Heywood, 1988; Pulkrabek, 2006).

ICEs can be classified in a number of different ways:

1. Based on mode of ignition:
   i. Spark-ignition (SI);
   ii. Compression-ignition (CI).
      - SI Engines are further classified based on method induction of charge (mixture) or fuel:
        o Carbureted;
        o Multipoint port fuel injection;
        o Throttle body fuel injection;
        o Gasoline direct injection.
      - CI Engines are also further classified based on method induction of fuel either:
        o Direct injection; or
        o Indirect injection (IDI) type.
2. Based on working cycle:
   i. Four-stroke cycle;
   ii. Two-stroke cycle.
      - Developmental trials were also made into operation of engines on 3-stroke cycles and 6-stroke cycles.
3. Based on basic engine design:
   i. Reciprocating type;
   ii. Rotary type.
      - Reciprocating type also called piston engines.
4. Based on position and number of cylinders of reciprocating engines:
   i. Single cylinder;
   ii. In-line;
   iii. V- engine;
   iv. Opposed cylinder engine;
   v. W-engine;

      vi.   Opposed piston engine;

      vii. Radial engine.

5. Based on mode of charging: The substance that enter during intake stroke:
   - i.   Naturally aspirated;
   - ii. Supercharged/turbocharged;
   - iii. Crankcase compressed.

6. Based on type of fuel used:
   - i.   Gasoline or petrol (SI type);
   - ii. Diesel;
   - iii. Gas (natural gas/ methane);
   - iv. Alcohols;
   - v.   Dual fuel.

7. Based on application:
   - i.   The IC engines find its place in automobile;
   - ii. Locomotive;
   - iii. Stationary;
   - iv. Marine;
   - v.   Aircraft;
   - vi. Small;
   - vii. Portable power;
   - viii. Chain saw;
   - ix. Model airplane, etc.

8. Based on method of cooling:
   - i. Air cooled;
   - ii. Liquid cooled, (Oil/Water-cooled).

9. Based on method of governing:
   - i. Qualitative;
   - ii. Quantitative.

Invariably, one can find reciprocating piston engines as the common power source in transportation, i.e., in land, air, and sea and largely for land vehicles including automobiles, motorcycles, and locomotive ships. Moreover, if the power-to-weight ratios are required are very high, rotary engines, or combustion turbines can be used. Combustion turbines are mainly used in powering aircrafts. Airplanes can also use jet engines and helicopters can also employ turboshafts; both of these are types of turbines. As a stationary unit, ICEs are used to generate electric power. ICEs playing major a role in large scale electric power generation in the form of combustion turbines operated in combined cycle power plants with a power rage of 100 MW to 1 GW. In a smaller scale Diesel, generators are used for backup power and for providing electrical power to areas not connected to an electric grid.

1.3 TYPICAL GEOMETRY OF THE INTERNAL COMBUSTION ENGINE (ICE)

ICEs also sometimes called as piston engines as distinguished between rotary and reciprocating type engines as which component of the engine mainly involved in development/production of power.

Figure 1.3(a) shows a principle sketch of general type single cylinder reciprocating engines being built. The full description of engine details can be seen in Heywood (1988). The engine transforms the reciprocating or oscillating movement of the piston into the rotating movement of the crankshaft, see Figure 1.3(a).

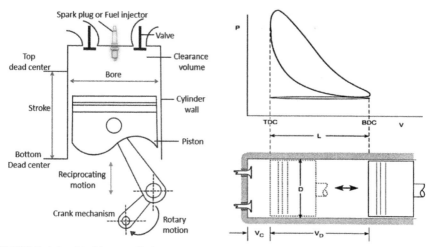

FIGURE 1.3 (a) Piston-cylinder assembly, (b) reciprocating type internal combustion engine with piston cylinder configuration.

1.4 ENGINE GEOMETRICAL PARAMETERS

For an engine with bore B (Figure 1.3(a)), crank radius a, stroke length L, turning at an engine speed of N:

$$L = 2a$$

Average piston speed is: $\overline{Sp} = 2LN$

where, N = Engine rotational speed, in RPM (revolutions per minute), S_p in m/sec, and a, and L in m.

Piston speed is an important parameter which helps in creating pressure differences to draw the charge and let the burnt gases out of the cylinder. Normally, the average piston speed for all engines will be in the range of 6 to 15 m/sec, with large diesel engines on the low end and high-performance automobile engines on the high end. Automobile engines usually operate in a speed range of 500 to 5000 RPM, with cruising at about 2000 RPM. Under certain conditions using special materials and design, high-performance experimental engines have been operated with average piston speeds up to 25 m/sec.

$$s = a \, \cos \theta + \sqrt{R^2 - a^2 \sin^2 \theta}$$

Normally, to allow higher flow rates with larger valves, high piston speeds are desired. Generally, the size of inlet valve is larger than exhaust valve. Cylinder bore sizes vary in the range of 0.5 cm (low) to 0.5 m (high) and bore to stroke ratios vary between 0.8 and 1.2. Larger ratios are used in SI engine and smaller ratios for CI engines (Figure 1.4).

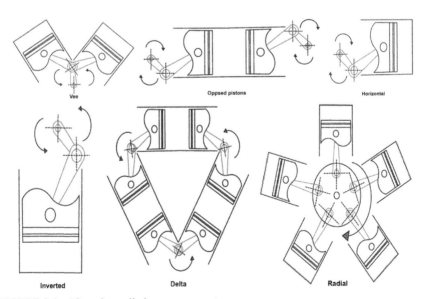

FIGURE 1.4   IC engine cylinder arrangement.

Based on the stroke to bore ratio, the engines are categorized as:

1. Square engines (bore = stroke);
2. Under square engines (bore < stroke);
3. Over square engines (bore > stroke).

It is customary to assume a square engine concept to initiate the design process. Under square engines are used in very large engines where in the stroke length is about 4 times the bore.

Referring to the typical engine geometry (Figure 1.5(b)), the following parameters are interrelated: a = Crank radius or crank offset; R = Connecting rod length; θ = Angle made by crank with vertical; B = Internal diameter of cylinder or piston head diameter; s = Instantaneous distance between piston pin to crank pin and it varies as the piston moves between two dead centers.

Following the geometry (similar to crank-slider mechanism):

$$s = a\ \cos\theta + \sqrt{R^2 - a^2\ \sin^2\theta}$$

Instantaneous piston speed can be obtained by differentiating s with time, i.e.:

$$\overline{Sp} = \frac{ds}{dt}$$

Engine is specified with cylinder volume or displacement volume or swept volume relating its bore and stroke as:

$$V_d = \left(\frac{\pi}{4}\right)B^2 L \quad \text{(for one cylinder)}$$

If n = number of cylinders, then its engine capacity $= n\left(\frac{\pi}{4}\right)B^2 L$.

As the modern engines are multi-cylinder engines, it is customary to express the engine displacement (cubic capacity) in terms of liters.

Specifications of the engines employed in versatile brands are given as annexure.

The cylinder volume V at any crank angle, with usual notation can be expressed as:

$$V = Vc + \left(\frac{\pi}{4}\right)B^2 L\ (R + a - s)$$

where, Vc = Clearance volume.

1.4.1 COMPRESSION RATIO (CR)

The compression ratio (CR) is an important parameter, which is normally expressed in terms of cylinder volumes at BDC and TDC, is used to relate the theoretical performance of engines in terms of thermodynamic efficiency.

$$r_c = \frac{V_{BDC}}{V_{TDC}} = \frac{V_d + V_c}{V_c}$$

$$V_{BDC} = \text{Volume at } BDC = V_d$$

$$V_{TDC} = \text{Volume at } TDC = V_c$$

Since the above definition of CR is given in terms of cylinder volumes, it is generally called as geometrical or static CR. However, in actual practice, engine's effective CR is taken into account and is given as at which the intake valve closes and continues either spark starts or fuel injection begins. As an engine runs, the static CR becomes secondary to the effective CR because compression cannot begin to build in the cylinder until the intake valve closes. Therefore, closing the intake valve later lowers the effective CR. It is obvious that the effective CR is considerably lower than the static CR.

The normal ranges of CRs for spark-ignition (SI) engines are 6.5 to 12, while CI engines have CRs in the range 12 to 24. The minimum value of CI engine is the maximum limit for SI engines. However, supercharged or turbocharged engines with work with lower CRs than respective naturally aspirated engines. The reasons may be attributed to use of limitations in engine materials, technology, and fuel quality. To derive maximum benefit from the fuel supplied to engines, attempts have also been made to develop engines with a variable CR.

> Ranges of Compression ratios: SI engine: 6.5 to 12 and CI engine: 12 to 24.

The piston reverses its movement at the top-dead-center (TDC) and at the bottom dead center (BDC). These dead centers are two extreme positions or point within which piston makes its motion. At both of these dead point positions, the speed of the piston is equal to zero, whilst the acceleration is at the maximum. Between the top-dead-center and the underside of the cylinder head, the compression volume V_c remains (also the so-called dead space or clearance volume in the case of reciprocating compressors).

Combustion is different from burning in a sense that when the burning of fuel-air mixture occurs inside the container (in which the mixture is contained). Burning of any material may also happen in an open atmosphere such as burning of firewood or dry leaves. The combustion may be initiated with the help of ignition source such as spark, etc. or by virtue of self-ignition upon attaining suitable conditions. Thus, in an ICE, the burning of fuel-air mixture happens inside the engine cylinder and thus it derived its name as internal combustion engine. This explanation brings a clarity in the terms-Combustion, Burning, and Ignition.

1.5 IMPORTANT TERMINOLOGY ASSOCIATED WITH ENGINE

The basic terms or terminology associated with engine components, operation, and geometrical features are detailed below and should be well versed in the better understanding of engine working (Sean, 2010):

1. **Internal Combustion:** Burning of fuel-air mixture or contents, which forms the working medium, takes place in a confined space in general or more specifically inside the engine cylinder in producing thermal for subsequent conversion to mechanical energy. The burning of cylinder contents in a confined space is called combustion. The burning could be initiated with an external source of ignition i.e., a spark plug or by virtue of self ignition of fuel upon compression.

2. **Spark-Ignition (S1):** When combustion process is initiated in each cycle by a spark plug, such an engine is called spark assisted engine or spark ignited or simply spark-ignition engine. Since the external source is used, such ignition is called forced ignition. Since the external source is used for initiating combustion, such ignition is also called forced ignition. Typically, SI engines use gasoline or petro as a fuel. Popularly SI engines are known as gasoline or petrol engines. Essentially, in SI engines, combustion is caused by an electric spark-ignition.

3. **Compression-Ignition (CI):** In each cycle, owing to high CR of engine, the fuel at times could auto-ignite or self-ignites, by mixing with air with the attainment of high pressures and temperatures. Such engine is termed as compression ignited or CI engine.

CI engines are often called diesel engine as mostly commonly diesel fuel is used. Therefore, in CI engines, combustion is initiated by self-inflammation due to compression heat.

Components	Description
Head	The component that exists at the top of the cylinder, usually containing part of the clearance volume of the combustion chamber. The head is usually cast iron or aluminum, and bolts to the engine block. In some less common engines, the head is one piece with the block. The engine head houses the spark plugs in SI engines and the fuel injectors in CI engines and few SI engines. Most modern engines have the valves in the head, and many have the camshaft(s) positioned there also (overhead valves and overhead cam).
Piston	The cylindrical-shaped mass that reciprocates back and forth in the cylinder, transmitting the pressure forces in the combustion chamber to the rotating crankshaft. The top of the piston is called the crown and the sides are called the skirt. The face on the crown makes up one wall of the combustion chamber and maybe a flat or highly contoured surface. Some pistons contain an indented bowl in the crown, which makes up a large percent of the clearance volume. Pistons are made of cast iron, steel, or aluminum. Iron and steel pistons can have sharper corners because of their higher strength. They also have lower thermal expansion, which allows for tighter tolerances and less crevice volume. Aluminum pistons are lighter and have less mass inertia. Sometimes synthetic or composite materials are used for the body of the piston, with only the crown made of metal. Some pistons have a ceramic coating on the face.
Piston Rings	Metal rings that fit into circumferential grooves around the piston and form a sliding surface against the cylinder walls. Near the top of the piston are usually two or more compression rings made of highly polished hard chrome steel.
	The purpose of these is to form a seal between the piston and cylinder walls and to restrict the high-pressure gases in the combustion chamber from leaking past the piston into the crankcase (blow by). Below the compression rings on the piston is at least one oil ring, which assists in lubricating the cylinder walls and scrapes away excess oil to reduce oil consumption. Their length as part of a pressurized lubrication system.
Push Rods	These are rod mounted between cam and rocker arm used to actuate the valves in 4-s cycle engines

Components	Description
Valves	Used to allow flow into and out of the cylinder at the proper time in the cycle. Most engines use poppet valves, which are spring-loaded closed and pushed open by camshaft action (Figures 1–12). Valves are mostly made of forged steel. Surfaces against which valves close are called valve seats and are made of hardened steel or ceramic. Rotary valves and sleeve valves are sometimes used, but are much less common. Many two-stroke cycle engines have ports (slots) in the side of the cylinder walls instead of mechanical valves.
Head Gasket	Gasket which serves as a sealant between the engine block and head where they bolt together. They are usually made in sandwich construction of metal and composite materials. Some engines use liquid head gaskets.
Bore	Diameter of the cylinder or diameter of the piston top face.
Cylinder	This is a major component in which piston is housed wherein contents or fuel-air mixture undergoes different variations in pressure, temperature, and composition. Cylinders The circular cylinders in the engine block in which the pistons reciprocate back and forth. The walls of the cylinder have highly polished hard surfaces. Cylinders may be machined directly in the engine block, or a hard metal (drawn steel) sleeve may be pressed into the softer metal block. Sleeves may be dry sleeves, which do not contact the liquid in the water jacket, or wet sleeves, which form part of the water jacket. In a few engines, the cylinder walls are given a knurled surface to help hold a lubricant film on the walls. In some very rare cases, the cross section of the cylinder is not round.
Cylinder Block	Body of engine containing the cylinders, made of cast iron or aluminum. In many older engines, the valves and valve ports were contained in the block. The block of water-cooled engines includes a water jacket cast around the cylinders. On air-cooled engines, the exterior surface of the block has cooling fins.
Crankcase	Part of the engine block surrounding the rotating crankshaft. In many engines, the oil pan makes up part of the crankcase housing.
Clearance Volume	The space between cylinder head inner side and TDC where the cylinder contents occupy minimum volume during upward motion of piston. At this volume, called the clearance volume the whole cylinder contents are momentarily compressed in such a small space, also called cushion volume, i.e., when the piston is at TDC. This is a minimum volume in the combustion chamber.

Components	Description
Top-dead center (TDC) or Top Center	In a vertical cylinder engine, the extreme position a piston reaches during its upward motion and near to the cylinder head beyond which no further movement to go up. This piston reaches a top most point and stops there.
Bottom Dead Center (BDC) or Bottom Center	In a vertical cylinder engine, the extreme position a piston reaches during its downward motion and near to the crank beyond which no further movement to go down. This piston reaches a bottom most point and stops there. The cylinder contents would occupy maximum volume when piston reaches BDC or BC
Stroke Length	The linear distance covered by piston between TDC and BDC or BDC to TDC.
Stroke Volume or Swept Volume or Displacement Volume	The volume occupied by cylinder contents between TDC and BDC. The name is derived as the piston sweeps between TDC and BDC. Generally, the engine's capacity is mentioned in terms of displacement volume or cubic capacity.
Compression Ratio	This is important parameter and is defined as the ratio of maximum cylinder volume to minimum cylinder volume. Also, it is defined as the ratio between volumes of cylinder contents at BDC to volume of cylinder contents at TDC.
Combustion Chamber	The end of the cylinder between the head and the piston face where combustion occurs. The size of the combustion chamber continuously changes from a minimum volume when the piston is at TDC to a maximum when the piston is at BDC. The term "cylinder" is sometimes synonymous with "combustion chamber" (e.g., "the engine was firing on all cylinders"). Some engines have open combustion chambers which consist of one chamber for each cylinder. Other engines have divided chambers which consist of dual chambers on each cylinder connected by an orifice passage.
Connecting Rod	Rod connecting the piston with the rotating crankshaft, usually made of steel or alloy forging in most engines but may be aluminum in some small engines.
Connecting Rod Bearing	Bearing where connecting rod fastens to crankshaft.
Cooling Fins	Metal fins on the outside surfaces of cylinders and head of an air-cooled engine. These extended surfaces cool the cylinders by conduction and convection.

Components	Description
Crankshaft	Rotating shaft through which engine work output is supplied to external systems. The crankshaft is connected to the engine block with the main bearings. It is rotated by the reciprocating pistons through connecting rods connected to the crankshaft, offset from the axis of rotation. This offset is sometimes called crank throw or crank radius. Most crankshafts are made of forged steel, while some are made of cast iron.
Direct Injection (DI)	Typically, CI engines are classified as a DI type engine in which fuel is directly injected onto the top of the piston or into the main chamber. Such engine is also called open chamber engines.
Camshaft	Rotating shaft used to push open valves at the proper time in the engine cycle, either directly or through mechanical or hydraulic linkage (push rods, rocker arms, tappets). Most modern automobile engines have one or more camshafts mounted in the engine head (overhead cam). Older engines had camshafts in the crankcase. Camshafts are generally made of forged steel or cast iron and are driven off the crankshaft by means of a belt or chain (timing chain). To reduce weight, some cams are made from a hollow shaft with the cam lobes press-fit on. In four-stroke cycle engines, the camshaft rotates at half engine speed.
Indirect Injection (IDI)	On the contrary to DI type, if the fuel injection commences into a secondary chamber of an engine with a divided combustion chamber.
Square Engine	An engine in which bore is equal to stroke.
Under Square Engine	In this type, stroke is larger than the bore of the engine.
Over Square Engine	In this engine bore is larger than the stroke of the engine.
Smart Engine	Nowadays, modern automobiles are being equipped with electronic controls assisted by sensors and actuators for precisely regulating air-fuel ratio for control emissions and other parameters. Engine management system (EMS) is designed to work with computer-controlled systems.
Wide-Open-Throttle	Operation of the engine with fully open position of throttle valve. Normally engines are operated under WOT for deriving maximum power or speed.
Air-Fuel Ratio (AF)	Ratio of mass of air to mass of fuel input into engine.
Fuel-Air Ratio (FA)	Ratio of mass of fuel to mass of air input into engine.
Maximum Brake Torque (MBT)	Maximum torque produced at a given speed at spark or injection timing.
Overhead Valve Engines	Valves are located on the head of the cylinder.

Components	Description
Intake Manifold	Piping system which delivers incoming air to the cylinders usually made of cast metal, plastic, or composite material. In most SI engines, fuel is added to the air in the intake manifold system either by fuel injectors or with a carburetor. Some intake manifolds are heated to enhance fuel evaporation. The individual pipe to a single cylinder is called a runner.
Exhaust Manifold	Piping system which carries exhaust gases away from the engine cylinders, usually made of cast iron.
Exhaust System	Flow system for removing exhaust gases from the cylinders, treating them, and exhausting them to the surroundings. It consists of an exhaust manifold which carries the exhaust gases away from the engine, a thermal or catalytic converter to reduce emissions, a muffler to reduce engine noise, and a tailpipe to carry the exhaust gases away from the passenger compartment.
Fan	Most engines have an engine-driven fan to increase airflow through the radiator and through the engine compartment, which increases waste heat removal from the engine. Fans can be driven mechanically or electrically, and can run continuously or be used only when needed.
Flywheel	Rotating mass with a large moment of inertia connected to the crankshaft of the engine. The purpose of the flywheel is to store energy and furnish a large angular momentum that keeps the engine rotating between power strokes and smooth out engine operation. On some aircraft engines, the propeller serves as the flywheel, as does the rotating blade on many lawnmowers.

1.6   ADDITIONAL ENGINE COMPONENTS

The following is a list of major components found in most reciprocating ICEs (see Figures 1–15):

1.   **Carburetor:** It consists of a Venturi flow device which meters the proper amount of fuel into the airflow by means of a pressure differential. For many decades, it was the basic fuel metering system on all automobile (and other) engines. It is still used on low-cost small engines like lawnmowers, but is uncommon on new automobiles.

2.   **Catalytic Converter:** Chamber mounted in exhaust flow containing catalytic material that promotes reduction of emissions by chemical reaction.

3. **Fuel Injector:** A pressurized nozzle that sprays fuel into the incoming air on SI engines or into the cylinder on CI engines. On SI engines, fuel injectors are located at the intake valve ports on multipoint port injector systems and upstream at the intake manifold inlet on throttle body injector systems. In a few SI engines, injectors spray directly into the combustion chamber.

4. **Fuel Pump:** Electrically or mechanically driven pump to supply fuel from the fuel tank (reservoir) to the engine. Many modern automobiles have an electric fuel pump mounted submerged in the fuel tank. Some small engines and early automobiles had no fuel pump, relying on gravity feed.

5. **Glow Plug:** Small electrical resistance heater mounted inside the combustion chamber of many CI engines, used to preheat the chamber enough so that combustion will occur when first starting a cold engine. The glow plug is turned off after the engine is started.

6. **Main Bearing:** The bearings connected to the engine block in which the crankshaft rotates. The maximum number of main bearings would be equal to the number of pistons plus one, or one between each set of pistons plus the two ends. On some less powerful engines, the number of main bearings is less than this maximum.

7. **Oil Pan:** Oil reservoir usually bolted to the bottom of the engine block, making up part of the crankcase. Acts as the oil sump for most engines.

8. **Oil Pump:** Pump used to distribute oil from the oil sump to required lubrication points. The oil pump can be electrically driven, but is most commonly mechanically driven by the engine. Some small engines do not have an oil pump and are lubricated by splash distribution.

9. **Oil Sump:** Reservoir for the oil system of the engine, commonly part of the crankcase. Some engines (aircraft) have a separate closed reservoir called a dry sump.

10. **Radiator:** Liquid-to-air heat exchanger of honeycomb construction used to remove heat from the engine coolant after the engine has been cooled. The radiator is usually mounted in front of the engine in the flow of air as the automobile moves forward. An engine-driven fan is often used to increase airflow through the radiator.

11. **Spark Plug:** Electrical device used to initiate combustion in an SI engine by creating a high-voltage discharge across an electrode gap. Spark plugs are usually made of metal surrounded with ceramic

insulation. Some modern spark plugs have built-in pressure sensors which supply one of the inputs into engine control.

12. **Speed Control-Cruise Control:** Automatic electric-mechanical control system that keeps the automobile operating at a constant speed by controlling engine speed.

13. **Starter:** Several methods are used to start IC engines. Most are started by use of an electric motor (starter) geared to the engine flywheel. Energy is supplied from an electric battery.

On some very large engines, such as those found in large tractors and construction equipment, electric starters have inadequate power and small IC engines are used as starters for the large IC engines. First, the small engine is started with the normal electric motor, and then the small engine engages gearing on the flywheel of the large engine, turning it until the large engine starts. Early aircraft engines were often started by hand spinning the propeller, which also served as the engine flywheel. Many small engines on lawn-mowers and similar equipment are hand started by pulling a rope wrapped around a pulley connected to the crankshaft. Compressed air is used to start some large engines. Cylinder release valves are opened, which keeps the pressure from increasing in the compression strokes. Compressed air is then introduced into the cylinders, which rotates the engine in a free-wheeling mode. When rotating inertia is established, the release valves are closed and the engine is fired.

14. **Supercharger:** Mechanical compressor powered off of the crank-shaft, used to compress incoming air of the engine. Throttle Butterfly valve mounted at the upstream end of the intake system, used to control the amount of airflow into an SI engine. Some small engines and stationary constant -speed engines have no throttle. Turbocharger Turbine-compressor used to compress incoming air into the engine. The turbine is powered by the exhaust flow of the engine and thus takes very little useful work from the engine. Water jacket System of liquid flow passages surrounding the cylinders, usually constructed as part of the engine block and head. Engine coolant flows through the water jacket and keeps the cylinder walls from overheating. The coolant is usually a water-ethylene glycol mixture.

15. **Water Pump:** Pump used to circulate engine coolant through the engine and radiator. It is usually mechanically run off of the engine.

16. Wrist Pin: Pin fastening the connecting rod to the piston (also called the piston pin).

1.7 SPARK-IGNITION AND COMPRESSION-IGNITION ENGINES (CI ENGINES): COMPARISON, GEOMETRY, AND VALVE TIMING DIAGRAMS

The most commonly used engine for person mobility is spark-ignition engine for either in 2-wheeler or 4-wheeled vehicles. The SI engine uses fuel- highly volatile and low viscous fuel. On the contrary, relatively high viscous fuel is employed in CI engines. Typically, the CRs adopted in SI engines are lower than that used in CI engines, an external source of ignition is used in SI engines-spark plug, and the fuel should have a high auto-ignition temperature that is the fuel would not take part in combustion effectively owing to compression pressure and temperatures alone. Therefore, the SI engines are also called as forced ignition engines. Mostly, gasoline or petrol is used for SI engines and also CNG, LPG, bio-gas, etc. are widely being used in SI engines.

On the other hand, fuel of relatively low self-ignition temperature is employed in CI engine. The pressure and temperatures prevailed at the end of compression would just be enough to undergo combustion. Normally, all such fuels that have low auto-ignition temperature are used in compression-ignition engines (CI engines). Fuels such as diesel, vegetable oils (Biodiesel) are commonly used in CI engine.

Typically, in a 4-stroke cycle engine, the piston makes reciprocating motion 2 times from TDC to BDC and 2 times from BDC to TDC and so is 4-stroke engine. During these 4-strokes, the crank makes two revolutions to develop power. Therefore, in 4-stroke cycle engine, the cycle of operations are completed in 4-strokes of the piston or 2-revolutions of the crankshaft. Alternatively, in a two-stroke cycle engine, the cycle of operations are completed in 2-strokes of the piston or in one revolution of the crank shaft.

Also, during these physical 4-strokes, piston completes the working processes such as suction, compression, expansion (power) and exhaust to develop power. The 4-processes/operations are diagrammatically represented in the form of valve timing diagram (VTD) and port timing diagram (PTD) in the case of 4-stroke and 2-stroke cycle engines respectively.

Theoretically speaking, a two-stroke engine develops double the power that of an equivalent 4-stroke engine. 2-stroke engines are mostly found in compact applications such as two wheelers, portable gen sets.

However, due to overlapping of processes, the combustion would not be complete and there will be more fuel consumption and there by 2-stroke engines are less fuel efficient are also cause of more pollutant emissions compared to 4-stroke engines. Nowadays, 2-wheelers are also being incorporated with 4-strokecycle engines. In this text book, only 4-stroke cycle engines are discussed (Table 1.2).

TABLE 1.2 Comparison between 4-Stroke Cycle and 2-Stroke Cycle Engines

1.	Four Stroke Cycle Engine	Two Stroke Cycle Engine
2.	It completes 4-strokes of piston or 2 revolutions of crankshaft and develops one power stroke.	It undergoes 2-strokes of piston or 1 revolutions of crankshaft and develops one power stroke.
3.	Since the power developed in non uniform, it employs a large size heavy flywheel.	Lighter flywheel is required and turning moment is more uniform to one power stroke for each revolution.
4.	Engine is heavy	Engine is light
5.	Mechanical valves are used to allow charge into engine and let out the burnt gases; an extra valve mechanism is used.	Since piston covers and uncovers the ports, no such mechanism is used and hence an engine operation is simple.
6.	It is relatively expensive.	It is economical.
	Engine mechanical efficiency is low due to more friction on many parts.	More mechanical efficiency due to less friction on a few parts.
7.	Since dedicated strokes are provided for engine operation, engine develops practically more output due to full fresh charge intake and full burnt gases exhaust.	Less output due to mixing of fresh charge with the hot burnt gases.
8.	Engine runs cooler.	Engine runs hotter.
9.	Mostly Engine is water cooled.	Engine is air cooled.
10.	The specific fuel consumption of engine is low due to near complete combustion of fuel.	More fuel consumption and fresh charge is mixed with exhaust gases.
11.	Engine is bulky.	Engine is compact.
12.	Complicated lubricating system.	Simple lubricating system.
13.	Less noise is created by engine.	More noise is created by engine.

TABLE 1.2   *(Continued)*

14.	Engine consists of inlet and exhaust valve.	Engine consists of inlet and exhaust ports.
15.	More thermal efficiency.	Less thermal efficiency.
16.	Relatively less consumption of lubricating oil.	There is continuous development of power, engine is always subjected to high temperature and hence it consumes more lubricating oil.
17.	Less wear and tear of moving parts.	More wear and tear of moving parts.
18.	Used in cars, buses, trucks, etc.	Used in mopeds, scooters, motorcyclist.

1.8   ENGINE PERFORMANCE PARAMETERS

It is better to get aquatinted with the terms of engine performance parameters as they commonly used while describing an engine performance and comparing different engines. The engine performance is specified in terms of parameters and is therefore called the metrics that are used to characterize or compare performance of one engine with another.

1.8.1   WORK

The primary objective of an engine is to obtain work and is the work out which is produced by the gases in the combustion chamber of the engine cylinder. It is the result of an object is moved against a resistance or opposing force. Work is measured in watts. In the metric system, work is measured in Newton-meters or joules. Work is spent or produced by the gases (working substance in an engine) and are represented four different strokes of a naturally aspirated (NA) engine:

1. **Intake Stroke:** Work is done on system of gases (mixture) to get it into the cylinder (NA type).
2. **Compression Stroke:** Work is done on the mixture when it is compressed.
3. **Power Stroke:** The expanding air-fuel mixture produces work on the piston and crankshaft.
4. **Exhaust Stroke:** Work is performed as the exhaust gas is expelled from the engine.

As the piston moves from one dead center to other, in an engine cycle, force due to gas pressures on the moving piston is expressed as:

$$\int F\,dx = \int PA_p\,dx$$

where, P = pressure in the combustion chamber; A_p = piston surface area, on which the gas pressure acts; x = distance the piston moves.

On the other hand, Energy is the ability to do work, or the ability to produce a motion against a resistance. Inertia is the tendency of a body to keep its state of rest or motion. The larger the mass, the more it is effected by inertia. Inertia and energy are stored in the engine's flywheel. When a body is in motion, it has momentum. Momentum is a product of a body's mass and speed.

1.8.2 POWER AND TORQUE

- **Power:** It is how fast work is done or how fast motion is produced against a resistance.
- **Torque:** It is the ability to make power. It is defined as the tendency of force to rotate a body on which it acts.

The definitions for torque and work are similar; both are a force being multiplied by a distance. However, torque and work are very different quantities. Work is force times the distance moved, and torque is force times leverage, which is the distance from a pivot point to the applied force. When distinguishing work from torque, the metric unit of measurement for torque is the Newton-meter (Nm) and the metric work unit is the joule (J).

Engine torque varies with rpm. The pulling ability of a car from a standing start depends on its engine's torque. This means that torque should be high at lower speeds.

Horsepower is a familiar term associated with engine/vehicle performance. is the measurement of an engine's ability to perform work. James Watt described 1horsepower as 33,000 foot-pounds of work per minute or the amount of power described by a horse pulling a weight of 330 pounds across a distance of 100 feet in 1 minute. One horsepower is the amount of work required to lift 550 pounds 1foot in 1 second. In the metric system, horsepower is measured as watts. One watt is the power to move 1 Nm per second. Because this is so small a measurement, kilowatts (kW) are used. One horsepower equals 0.746 kW.

Horsepower is a measure of work performed in a straight line in a specified time. Torque measures force in a rotating direction. Power produced at the crankshaft is called gross horsepower. Accessories that absorb power include the alternator (charging system), air conditioning, coolant pump, cooling fan, power steering, etc. These absorb about 25% of the power available at the crankshaft. The power that remains for use is called net horsepower. Power is also lost through friction in the driveline (transmission and differential) and due to wind resistance, vehicle weight, tires, and weather.

1.8.3   DYNAMOMETER

An engine's output can be measured using a dynamometer, commonly called a dyno or dyne. The engine must be loaded to measure the torque it can produce. Depending on the type of dynamometer, braking can be done by electricity, hydraulics, or friction. A simple dynamometer that uses friction is called a prony brake. An arm pushes on a scale to provide a reading in pounds. When the length of the arm is known, the measurement can be converted to foot-pounds or Newton-meters.

1.   **Engine Dynamometer:** An engine dynamometer measures horsepower coming out of the engine. The horsepower measured is called brake horsepower because the dynamometer acts as a brake on the engine's crankshaft.
2.   **Chassis Dynamometer:** These are used when a complete vehicle is needed for the measurement of fuel consumption, noise, or emissions. A chassis dynamometer measures horsepower available at the vehicle's drive wheels. This is called road horsepower. It is always less than brake horsepower because of friction losses through the driveline.

Chassis dynamometers can be driven by the wheels or connected directly to the hubs. Wheel-driven dynos have a single or double roller.

A dual roll dynamometer, with smaller rollers (typically 11" diameter), is sometimes called a cradle roll. One roller is attached to the power absorption unit (covered later) and the other is an idle roller.

1.8.4   *MEASURING TORQUE AND POWER*

To make an engine perform work during a dynamometer test, the engine is put under load using a power absorption device. Automotive dynamometer power absorption units are one of two types: electromagnetic (eddy current) or water brake (hydraulic), controlled by the amount of water that enters the device.

Water brake units are more popular in the engine performance field. An EC, or eddy current, dynamometer has a magnetic eddy current brake like those used on the brakes of many municipal busses. The EC power absorber is used in emissions and research because it is more efficient at low rpm and is easier to control accurately than a water brake. Eddy current engine dynos are water cooled and precise but expensive. Water brake dynamometers deal with higher loads effectively. With higher power and speed requirements, the water brake is a more economical alternative to the eddy current dynamometer.

A fluid power absorption unit is a fluid coupling consisting of two members: a turbine and a stator. The turbine tries to move the water, but the stator prevents it from moving. The load unit is like a torque wrench that measures the load applied. The load is varied by the amount of water that is put into the fluid coupling.

Torque can be measured at the flywheel or the rear wheels, but horsepower is a calculation made from the torque measurement. The amount of load put on the engine and the amount of torque it produces are used to calibrate horsepower. A typical dynamometer control panel that displays torque.

1.8.5   *ENGINE EFFICIENCY*

To rate an engine in terms of efficiency, both the output and the input must be expressed in a common value. There are three types of engine efficiency measurements: mechanical efficiency, volumetric efficiency (VE), and thermal efficiency.

An efficiency measurement is a value less than 100%. The difference between efficiency measurement and 100% is the amount of loss.

The term thermal efficiency gross indication of how much the supplied fuel energy is utilized to develop brake power. This is sometimes called first law efficiency or fuel conversion efficiency.

$$\text{Thermal efficiency} = \eta_t = \frac{\text{Power developed}}{\text{Heat supplied}}$$

If instead, indicated power is used in place of brake power, it is indicated thermal efficiency.

Mechanical efficiency describes all of the ways friction is lost in an engine. Horsepower is a value that can be used to compare the mechanical efficiency of two engines. Brake Power divided by the indicated power gives the mechanical efficiency of the engine.

The formula is:

$$\text{Mechanical efficiency} = \eta_m = \frac{\text{Brake power}}{\text{Indicated power}}$$

The difference between indicated power and brake power is friction power. The smaller the difference between indicated power and brake power, the lower is the friction power and the higher is the mechanical efficiency.

1. **Air-Fuel Ratio (AF):** The AF of the engine is measured by the ratio of mass of air to mass of fuel.

$$(A/F)_{actual} = \frac{\text{Mass of air}}{\text{Mass of fuel}}$$

The actual AF how different from theoretical or stoichiometric air-fuel is the equivalence ratio.

$$\text{Equivalence ratio} = \phi = \frac{\text{Actual fuel-air ratio}}{\text{Stoichiometric fuel-air ratio}}$$

2. **Volumetric Efficiency (VE):** It shows the relation between the amounts of charge drawn into the cylinder at intake conditions relative to engine displacement. Thus, the measurement comparing the mass of airflow actually entering the engine with the maximum that theoretically could enter it corresponding to its displacement is called VE. It is indication of engine's breathing capacity or represents the effectiveness of engine's induction process.

$$\text{Volumetric efficiency} = \eta_v = \frac{m_a}{\rho_{a,i} V_d N}$$

where, $\rho_{a,i}$ = inlet air dentisy; V_d = engine displacement volume; N = Engine speed.

VE determines the engine's maximum torque output.

The speed at which the engine does its best breathing usually determines its maximum torque. Engine is just like a big air pump.

Theoretically, the engine should draw charge or air equivalent to its displacement volume. This represents a horizontal line on the coordinates of VE versus speed. However, in practice, it doesn't follow this trend. VE is a strong function of engine speed. As the speed increases, the amount of air drawn into engine also increases. Due to higher speed, friction also increases and thereby due to rubbing friction, the amount of charge drawn into decreases.

Engine performance maps provide a means of representing complete engine characteristics over a wide range of speed, power, and torque. The general nature of these curves can be ascertained from relations developed.

ICE power is directly related to its fuel-air mixture and, for particular engine geometry and specific fuel-air ratio (FA), the ideal charge per cylinder per cycle should be independent of speed. The ideal and actual induction processes were related by the use of a VE η_v. VE is strongly influenced by heat transfer, fuel parameters, the thermodynamic state of both intake charge and residual gases, and engine parameters such as intake and exhaust manifolds and valve design. For a particular condition of engine load, valve timing, and fuel-air input, the actual VE versus engine speed yields the characteristic curve. At low engine (piston) speed, an incoming charge has little kinetic energy influence on the intake process, i.e., ramming. At higher speeds, inertial ramming will increase the charge input to a maximum valve but, as speed is, increased, frictional effects will tend to reduce the magnitude of ηv from its mid-speed optimum valve. Changing the engine operational regime may shift the VE curve.

Specific power or power-to-weight ratio is a measure of performance for an engine in a vehicle or in a power plant. It is defined as the power output by it divided by its mass, typically in units of W/kg. This value allows for a clear metric of the power that is independent of the vehicle or power plant's size.

1.8.6 *SPECIFIC EMISSIONS AND EMISSION INDEX*

1. **Specific Emission:** It is the amount of pollutant emitted per unit amount power developed by the engine. It is expressed in g/kW-h or kg/kW-h.

   Specific emission of Carbon monoxide = SP_{CO} = mass flow rate CO emitted /Brake Power

2. **Emission Index:** It defined as the amount of emissions released per unit of fuel consumed.

Emission Index of Carbon monoxide = Mass flow rate of Carbon monoxide emitted/Mass flow rate of fuel.

> *In earlier days, the quantification of pollutant emissions used to expressed in ppm or ppb or %v/v. This practice fail to condition of engine loading, etc. Hence, engine emissions are being given in terms of g/kW-h or kg/kW-h. Moreover, the use of Emission Index is even better as it relates directly the amount of fuel consumed.*

1.8.7 SPECIFIC POWER

It is an important normalized parameter is a measure of use of available piston areas to develop power independent of cylinder size.

$$\text{Specific Power} = P/A_p = P \,/\, A_p = \frac{\eta_i \eta_v \, S_p \, Q_{HV} \, \rho_{a,i} \left(\dfrac{f}{a} \right)}{4}$$

Besides relating power with piston area, it also reflects the important parameter that affect to obtain higher power. The parameters that directly proportional are density of charge, heating value, piston speed, rich mixture and thermal and VE.

$$\text{Specific Power} = \text{Engine Power/Piston face area} = \text{bp}/A_p$$

Turbines typically have the largest specific power for an engine, which makes them useful for airplanes in order to achieve sufficient lift. Sports car engines will also typically have a fairly high specific power, allowing them to have increased acceleration and handling.

The power-to-weight ratio (Specific Power) formula for an engine (power plant) is the power generated by the engine divided by the mass ("Weight" in this context is a colloquial term for "mass." To see this, note that what an engineer means by the "power to weight ratio" of an electric motor is not infinite in a zero-gravity environment).

A typical turbocharged V8 diesel engine might have an engine power of 250 kW and a mass of 380 kg, giving it a power-to-weight ratio of 0.65 kW/kg.

$$\frac{P}{A_p} = \frac{\eta_f \eta_v \overline{S}_p Q_{HV} \rho_{a,i} \left(F/A \right)}{4}$$

It can be seen that as the adoption of high calorific value fuel, rich mixtures, dense charge and efficient induction process and efficient energy conversion process all leads to higher power from a given size of piston area.

Examples of high power-to-weight ratios can often be found in turbines. This is because of their ability to operate at very high speeds. For example, the Space Shuttle's main engines used turbopumps (machines consisting of a pump driven by a turbine engine) to feed the propellants (liquid oxygen and liquid hydrogen) into the engine's combustion chamber. The original liquid hydrogen turbopump is similar in size to an automobile engine (weighing approximately 352 kilograms and produces 54 MW) for a power-to-weight ratio of 153 kW/kg (Figure 1.5).

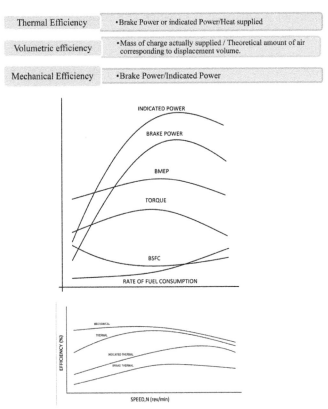

FIGURE 1.5   (a) Definitions of non-dimensional parameters, (b) Performance characteristics of a typical engine.

The performance parameters which are defined and described above can be grouped into three broad categories as detailed below:

The performance parameters are classified as:

- practical parameters;
- normalized parameters; and
- non-dimensional or dimensionless parameters.

1. **Practical Parameters:** These are generally or most commonly used one even by the customer.
   e.g., Brake power, Torque, piston displacement (cubic capacity), etc.
2. **Normalized Parameters:** These are the parameters which normalized with respect to brake power:
   e.g., Specific fuel consumption (brake or indicated), mean effective pressure, brake specific emissions, etc.
3. **Non-Dimensional or Dimension Less Parameters:** The parameters are non-dimensional in sense that majority of the parameters or ratios.

e.g., Brake thermal efficiency (fuel conversion efficiency), VE, mechanical efficiency, swirl ratio, equivalence ratio, emission index, etc.

While ordering a vehicle equipped with an IC Engine, within the operating range of engine, generally a customer of an engine aims at the following factors:

i.   The engine's performance within the allowed operating range;
ii.   The cost of fuel and the fuel consumption rate within this operating range;
iii. The air pollutant emissions and noise levels within this operating range;
iv.   The initial cost of vehicle with the engine and installation charges;
v.   The maintenance requirements, reliability, and durability of the engine.

Thermal efficiency is generally defined as the ratio of output (power) and heat supplied:

$$\text{bmep}\left(\text{brake mean effective pressure}\right) = \eta_f \eta_v Q_{HV} \rho_{a,i} \left(\text{fuel/air ratio}\right)$$

1.9   THERMODYNAMIC PERFORMANCE OF ENGINE-BASIC ENGINE CYCLES

1.9.1   THERMODYNAMIC CYCLES-CLOSED CYCLES

For producing mechanical work from the engine, a sequence of operations is required to follow. For example in a 4-stroke engine, the engine operations -suction, compression, expansion, and exhaust are to be related with basic thermodynamic processes.

The simplest models for the actual engine process are closed, internally reversible cycles with heat supply and removal, which are characterized by the following properties:

The chemical transformation of fuel as a result of combustion is replaced by a corresponding heat addition or heat supply. The charge changing process is replaced by a corresponding heat removal. Air, seen as an ideal gas, is chosen as a working medium.

An ideal engine is assumed to undergo closed cycle with instantaneous heat addition and rejection. All processes are allowed in a reversible manner. Thus, the ideal engine undergoes a thermodynamic cycle. However, not all real engines undergo reversible processes as dictated by laws of thermodynamics, practical engine undergo mechanical or open cycle. However, for arriving at the evaluation of a particular, the knowledge of thermodynamic is used to obtain the maximum efficiency of an engine. The operation of engine are related with thermodynamic processes and assumed to take place in a closed cycle. Most ICEs both spark-ignition and CI; operate on either a four-stroke cycle or a two-stroke cycle. These basic cycles are fairly standard for all engines, with only slight variations found in individual designs.

1.9.2   THE CARNOT CYCLE-THE PROCESSES OF THE PERFECT ENGINE

As described above, if engine operations are sequentially related with the help of thermodynamic processes, an engine cycle is obtained. The Carnot cycle, represented in Figure 1.6, is the cycle with the highest thermal efficiency and thus represents combination of the ideal processes arranged in a cyclic manner.

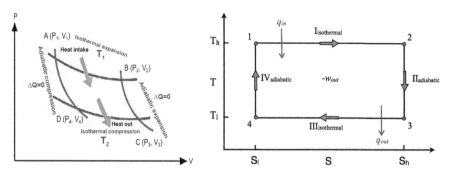

FIGURE 1.6 Carnot cycle on p-v and T-s plots.

The Carnot cycle represents following four processes:

T_1-T_2	Isothermal heat addition at T_h (highest temperature)
T_2-T_3	Isentropic expansion
T_3-T_4	Isothermal heat rejection at T_1 (lowest temperature)
T_4-T_1	Isentropic compression

On T-s coordinate system, the Carnot cycle represents a rectangle. Thermodynamically, Carnot Cycle sets a limit on the highest possible efficiency when it is defined in terms of source temperature and sink temperature alone.

$$\eta_{Carnot} = \frac{[T_{Highest} - T_{Lowest}]}{T_{Highest}}$$

The Carnot cycle cannot, however, be realized in ICEs, because. The isothermal heat addition at T_h and isentropic expansion, are not practically feasible.

The highest possible temperature could be adiabatic flame temperature (AFT) and the lowest possible are ambient. However, due to metallurgical considerations, thus, the temperature cannot be attained.

1.9.3 IDEAL AND ACTUAL THERMODYNAMIC CYCLES FOR SI ENGINE

Nicholaus A. Otto invented the 4-stroke engine and has been in practice in the form of gasoline/petrol-fuelled engine. The Otto cycle represents the following 4-processes:

p_1–p_2	Isentropic compression, pv^γ = constant
p_2–p_3	Constant volume heat addition
p_3–p_4	Isentropic expansion, pv^γ = constant
p_4–p_1	Constant volume heat rejection

Extending the similar way calculations for obtaining expression for Otto cycle efficiency, the ratio of work output to the heat supplied and making use of expressions using knowledge of thermodynamics (Van Wylen and Sonntag, 2008) and making use of CR, r_c (Figure 1.7).

$$\eta_{Otto} = 1 - \frac{1}{r_c^{\gamma-1}}$$

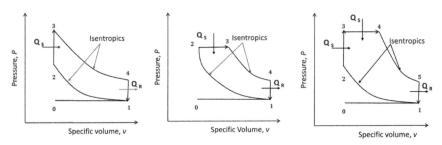

FIGURE 1.7 (a) The thermodynamic cycle for ideal or Otto cycle and actual cycle for SI engine, (b) the thermodynamic cycle for Ideal cycle and actual for CI engine, (c) The thermodynamic cycle for Ideal cycle and actual for CI engine.

1.10 WORKING OF A TYPICAL FOUR-STROKE SI ENGINE CYCLE

Four-stroke SI engine cycle (Figure 1.8):

1. **Suction Stroke or Intake Stroke or Induction Stroke:** Suction stroke is the first operation than an engine undergoes to induct charge into the engine cylinder. During this, intake valve kept open and exhaust valve closed and the piston traverses from TDC to BDC. As piston descends, low pressure (vacuum-in case of naturally aspirated engines) is created due to which charge (charge is air in case ideal engines) flows into the cylinder thus create an increasing volume in the combustion chamber. The resulting pressure differential through the intake system from atmospheric pressure on the outside to the vacuum on the inside causes charge to be pushed into the cylinder.

Practical engines induct mixture of air-fuel mixture prepared by carburetor.

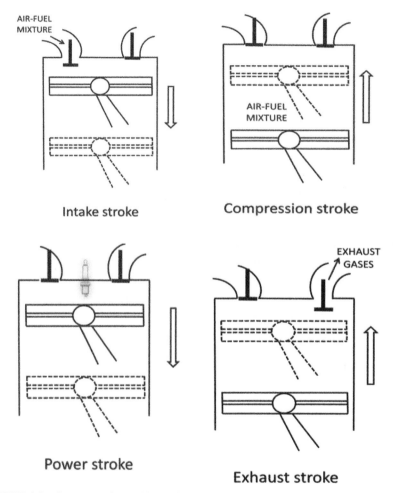

FIGURE 1.8 Sequence of actual SI engine operations in a cycle.

2. **Compression Stroke: Second Stroke:** The intake valve closes as the piston travels back to TDC with valves closed. During the upward movement of the piston, the inducted charge the air-fuel mixture gets compressed, raising both the pressure and temperature in the cylinder. Near the end of the compression stroke, the spark plug is fired and combustion is initiated.

- Combustion of the air-fuel mixture occurs in a very short but finite length of time with the piston near TDC (i.e., nearly constant-volume combustion). It starts near the end of the compression stroke slightly bTDC and lasts into the power stroke slightly aTDC. Combustion changes the composition of the gas mixture to that of exhaust products and increases the temperature in the cylinder to a very high peak value. This, in turn, raises the pressure in the cylinder to a very high peak value.

3. **Expansion Stroke or Power Stroke or Third Stroke:** With both (or all) valves closed, the high pressure created by the combustion process pushes the piston away from TDC. This is the stroke which produces the work output of the engine cycle. As the piston travels from TDC to BDC, cylinder volume is increased, causing pressure and temperature to drop.

   - **Exhaust or Blow Down Stroke:** Late in the power stroke, the exhaust valve is opened and exhaust blowdown occurs. Pressure and temperature in the cylinder are still high relative to the surroundings at this point, and a pressure differential is created through the exhaust system which is open to atmospheric pressure. This pressure differential causes much of the hot exhaust gas to be pushed out of the cylinder and through the exhaust system when the piston is near BDC. This exhaust gas carries away a high amount of enthalpy, which lowers the cycle thermal efficiency. Opening the exhaust valve before BDC reduces the work obtained during the power stroke but is required because of the finite time needed for exhaust blow-down.

4. **Fourth Stroke: Exhaust Stroke:** By the time the piston reaches BDC, exhaust blowdown is complete, but the cylinder is still full of exhaust gases at approximately atmospheric pressure. With the exhaust valve remaining open, the piston now travels from BDC to TDC in the exhaust stroke. This pushes most of the remaining exhaust gases out of the cylinder into the exhaust system at about atmospheric pressure, leaving only that trapped in the clearance volume when the piston reaches TDC. Near the end of the exhaust stroke bTDC, the intake valve starts to open, so that it is fully open by TDC when the new intake stroke starts the next cycle. Near TDC, the exhaust valve starts to close and finally is fully closed sometime aTDC. This period when both the intake valve and exhaust valve are open is called valve overlap (Figure 1.9).

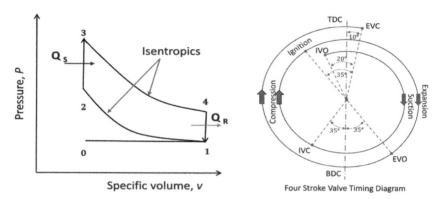

FIGURE 1.9 Theoretical p-θ diagram and typical valve timing of a 4-stroke cycle engine.

The following calculation shows the typical value of air-standard Otto cycle efficiency.

From the information of above sections:

$$\text{Air-standard efficiency} = 1 - \left(\frac{1}{r_c}\right)^{\gamma-1} = 1 - \left(\frac{1}{8}\right)^{1.4-1} \quad \therefore r_c = 8$$

$$\text{Relative efficiency} = \frac{\text{Thermal efficiency}}{\text{Air-standard efficiency}}$$

∴ Indicated thermal efficiency = 0.6 × 0.5647 = 0.3388

Also,

$$\text{Indicated thermal efficiency} = \frac{1}{\text{ISFC} \times \text{CV}}$$

∴ Calorific value of fuel (C.V)$= \dfrac{1}{ISFC \times \eta_{in}} = \dfrac{3600}{0.3 \times 0.3388} = 35.417 \; kJ \, / \, kg$

1.11 WORKING OF A TYPICAL FOUR-STROKE CI ENGINE CYCLE

1. **First Stroke: Intake Stroke:** The same as the intake stroke in an SI engine with one major difference: no fuel is added to the incoming air.

2. **Second Stroke: Compression Stroke:** The same as in an SI engine except that only air is compressed and compression is to higher pressures and temperature.

Late in the compression stroke, fuel is injected directly into the combustion chamber, where it mixes with the very hot air. This causes the fuel to evaporate and self-ignite, causing combustion to start.

- Combustion is fully developed by TDC and continues at about constant pressure until fuel injection is complete and the piston has started towards BDC.

3. **Third Stroke: Power Stroke:** The power stroke continues as combustion ends and the piston travels towards BDC.

- Exhaust blowdown same as with an SI engine.

4. **Fourth Stroke: Exhaust Stroke:** Same as with an SI engine (Figure 1.10 and Table 1.3).

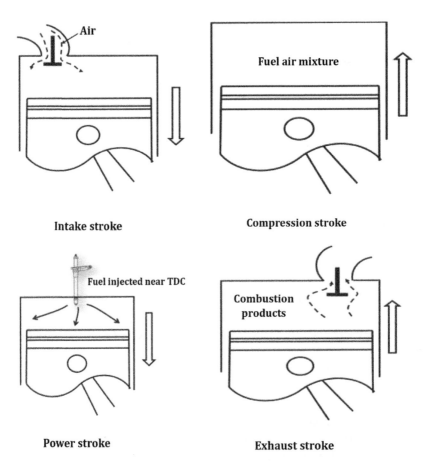

FIGURE 1.10   Sequence of CI engine operations in a cycle.

If $\phi < 1$, the mixture is lean mixture and if $\phi . 1$, it is rich mixture. $\phi = 1$ represents the stoichiometric mixture

TABLE 1.3 Typical Ranges

Parameter	SI Engine	CI Engine
Brake mean effective pressure	Naturally Aspirated = 850–1050 kPa	Naturally Aspirated = 700–900 kPa
	Turbocharged = 1250–1700 kPa	Turbocharged = 1000–1200 kPa
Brake specific fuel consumption	~270 g/kW-h	~200 g/kW-h
Compression ratio	6.5–12	12–24
Air-fuel ratio	12–18	18–70
Mechanical efficiency	Naturally aspirated ~ 85%	Naturally aspirated ~ 90%
Volumetric efficiency	~85%	~90%

1.11.1 COMPRESSION RATIO (CR) AND ENGINE POWER

During combustion, the potential energy of the air-fuel mixture is turned into thermal (heat) energy and kinetic energy. CR affects the amount of power an engine can produce by increasing the thermal efficiency of the engine. Squeezing the mixture into a smaller space results in higher combustion pressure and more expansion of the mixture throughout the power stroke. More of the heat energy of the fuel is converted to work, as less heat is allowed to escape from the engine. Increasing pressure in the cylinder by raising compression can make a relatively big difference in engine power.

Very early gasoline engines had compression ratios of about 2.5:1. Compression ratios of about 4:1 were common until the mid-1940s. Some engines had a 6:1 compression ratio after the mid-1920s because lead was added to the fuel to give it higher octane. During World War II, higher octane fuels were required for high-performance airplanes, and refining technology improved. By the 1950s, 10:1 compression ratios were available on high horsepower cars. In the late1960s, compression ratios on gasoline engines rose as high as 13.5:1. In the 1970s, two environmental reasons called for compression ratios to be lowered to about8:1. The first was higher emissions. Higher compression results in higher temperatures of combustion. This caused more oxides of nitrogen (NOx). Leaded fuels were phased out, too. Tetraethyl lead was an economical way to protect the valve seats and raise the octane of gasoline, but it was a pollution problem. Computerized fuel and ignition timing controls have allowed compression ratios to climb once again in newer cars.

1.11.2   *EFFECTIVE COMPRESSION RATIO (CR)*

The point at which the intake valve closes determines an engine's effective CR. As an engine runs, the static CR becomes secondary to the effective CR because compression cannot begin to build in the cylinder until the intake valve closes. Therefore, closing the intake valve later lowers the effective CR. This complicates the cam designer's job because leaving the intake valve open longer is a proven way to bring more air into the cylinder at high engine rpm. The effective CR is considerably lower than the static CR. When the throttle plate is partly opened, the engine needs more compression to run efficiently. A higher static CR, therefore, will help part-throttle operation, as will using VVT to close the intake valve sooner (Figure 1.4).

TABLE 1.4   Comparison Between 4-Stroke Cycle and 2-Stroke Cycle Engines

S. No.	Four Stroke Engine	Two Stroke Engine
1.	It has one power stroke for every two revolutions of the crankshaft.	It has one power stroke for each revolution of the crankshaft.
2.	Heavy flywheel is required and engine runs unbalanced because turning moment on the crankshaft is not even due to one power stroke for every two revolutions of the crankshaft.	Lighter flywheel is required and engine runs balanced because turning moment is more even due to one power stroke for each revolution of the crankshaft.
3.	Engine is heavy	Engine is light
4.	Engine design is complicated due to valve mechanism.	Engine design is simple due to absence of valve mechanism.
5.	More cost.	Less cost than 4 strokes.
6.	Less mechanical efficiency due to more friction on many parts.	More mechanical efficiency due to less friction on a few parts.
7.	More output due to full fresh charge intake and full burnt gases exhaust.	Less output due to mixing of fresh charge with the hot burnt gases.
8.	Engine runs cooler.	Engine runs hotter.
9.	Engine is water cooled.	Engine is air cooled.
10.	Less fuel consumption and complete burning of fuel.	More fuel consumption and fresh charge is mixed with exhaust gases.
11.	Engine requires more space.	Engine requires less space.
12.	Complicated lubricating system.	Simple lubricating system.
13.	Less noise is created by engine.	More noise is created by engine.
14.	Engine consists of inlet and exhaust valve.	Engine consists of inlet and exhaust ports.

TABLE 1.4 *(Continued)*

S. No.	Four Stroke Engine	Two Stroke Engine
15.	More thermal efficiency.	Less thermal efficiency.
16.	It consumes less lubricating oil.	It consumes more lubricating oil.
17.	Less wear and tear of moving parts.	More wear and tear of moving parts.
18.	Used in cars, buses, trucks, etc.	Used in mopeds, scooters, motorcyclist.

The typical theoretical valve timing diagrams of a 4-stroke CI engine CI engine are depicted in Figure 1.11.

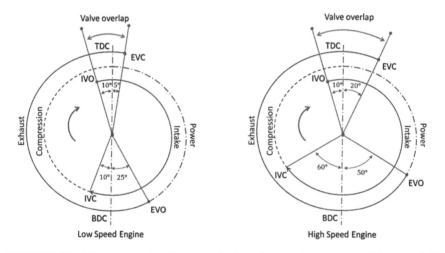

FIGURE 1.11 The valve timing diagrams of a 4-stroke cycle low-speed and high-speed CI engine.

1.12 DIFFERENCES BETWEEN AIR-STANDARD AND FUEL-AIR CYCLES

1.12.1 AIR-STANDARD CYCLE

1. The gas mixture in the cylinder is treated as air for the entire cycle, and property values of air are used in the analysis.

2. The real open cycle is changed into a closed cycle by assuming that the gases being exhausted are fed back into the intake system.
3. The combustion process is replaced with a heat addition term Qin of equal energy value.
4. The open exhaust process, which carries a large amount of enthalpy is denoted by Qout of the equal energy value.
5. Actual engine processes are approximated with ideal processes:
    i.   The intake and exhaust strokes are assumed to be constant pressure.
    ii.   Compression strokes and expansion strokes are approximated by is entropic processes.
    iii. The combustion process is idealized by a constant-volume process (SI cycle), a constant-pressure process (CI cycle), or a combination of both (CI dual cycle).
    iv.   Exhaust blow down is approximated by a constant-volume process.
    v.   All processes are considered reversible.

1.12.2   FUEL-AIR CYCLE

1. The fuel-air cycle take into consideration the following: The actual composition of the cylinder contents.
2. The variation in the specific heat of the gases in the cylinder.
3. The dissociation effect.
4. The variation in the number of moles present in the cylinder as the pressure and temperature change.
5. Compression and expansion processes are frictionless.
6. No chemical changes in either fuel or air prior to combustion.
7. Combustion takes place instantaneously at top-dead-center.
8. All processes are adiabatic.
9. The fuel is mixed well with air.
10. Subsequent to combustion, the change is always in chemical equilibrium.

1.12.3   THERMODYNAMIC CYCLES

Closed Cycles: The simplest models for the actual engine process are closed, internally reversible cycles with heat supply and removal, which are characterized by the following properties:

The chemical transformation of fuel as a result of combustion is replaced by a corresponding heat supply. The charge changing process is replaced by a corresponding heat removal. Air, seen as an ideal gas, is chosen as a working medium.

1.13 THERMODYNAMIC ANALYSIS OF ENGINE PERFORMANCE

As per thermodynamics, a cycle encompasses a series of processes in which the working substance undergoes changes and finally returns to the original state. A precise analysis of such thermodynamic cycle is difficult due to complex chemical reactions occurring during combustion and heat transfer. It is therefore essential to analyze the cycle making certain assumptions. Though such analysis helps one to get an overall performance and thus an ideal one which is far from realistic performance. However, the thermodynamic analysis is useful in deciding the upper limit of engine's performance and helps to compare the performance of two engines. In actual practice, the working fluid of an engine is a mixture of fuel, air, and some residual gases. Moreover, depending on the type of engine, load/speed variations the proportions of air and fuel vary a lot. Also, under cyclic variations, the properties of working substance differ. It is very difficult to account exactly such variations. Therefore, the analysis is initially done by assuming air as a working substance.

As enunciated earlier units, the engine does not undergo thermodynamic cycle. As the working substance does not return to its original state, it gets exhausted after a completion of operation (suction, compression, expansion, and exhaust) and therefore undergoes a mechanical cycle.

The following assumptions are made in analysis of engine performance with air as working substance:

1. The working medium (substance) is air alone, treated as perfect gas (ideal gas), and maintains uniform composition throughout.
2. The ideal gas obeys the equation of state, $Pv = mRT$.
3. The working substance has a fixed mass flow rate.
4. The air assumed to be diatomic and ratio of specific heats γ (to be equal to 1.4).
5. The working gas has constant specific heats. The heat addition and heat rejection processes reversible and are instantaneous.
6. The compression and expansion processes are isentropic.

7. The working gas does not undergo any chemical changes throughout the cycle.
8. Frictional losses of the working fluid are neglected.
9. The changes kinetic energy and the potential energy of working gas are neglected.

The following are the reference cycles for the engines of concern:

1. **Otto Cycle:** It is a reference cycle for analyzing Spark-ignition type engine comprising of two constant volume processes and two isentropic processes. The compression and expansion are isentropic and heat addition and heat rejection processes occur at constant volume.
2. **Diesel Cycle:** It is a reference cycle for analyzing CI type engine comprising of one constant pressure and one constant volume process. The compression and expansion are isentropic and heat addition and heat rejection processes occur at constant pressure and constant volume respectively.
3. **Mixed Cycle or Dual Cycle:** It is also a reference cycle for analyzing CI type engine comprising of two constant volume processes and one constant pressure process. The compression and expansion are isentropic processes and heat addition occurs at both at constant pressure and constant volume and heat rejection process occur at constant volume.

The air standard theory gives an estimate of engine performance which is much greater than the actual performance.

For example, the actual indicated thermal efficiency of a petrol engine of, say 7:1 CR is, of the order of 30% whereas the air standard efficiency is of the order of 54%. This large divergence is partly due to non-instantaneous burning of and valve operation, incomplete combustion, etc. But the main reason of divergence is the over simplification in using the values of the properties of the working fluid for cycle analysis. In the air, standard cycle approximation it was assumed that the working fluid is nothing but air and the air was a perfect gas and had constant specific heats. In actual engine, the working fluid is not air but a mixture of air, fuel, and residual gases. Furthermore, the specific heats of the working fluid are not constant but increase as temperature rises, and finally, the products of combustion are subjected to dissociation at high temperature.

If the actual physical properties of the cylinder gases before and after burning are taken into account, a much closer approach to actual performance is obtained.

The analysis of cycle based on the actual properties of the cylinder gases (working substance), is called the fuel-air cycle analysis.

The fuel-air cycle calculations are based on the following considerations:

1. The contents of the cylinder are fuel, air, water vapor in air and residual gases.
2. The FA is varied as per actual load/speed variations.
3. The value of γ is assumed to vary with temperature.
4. Fuel-air mixture does not combine completely at high temperatures.
5. The changes in pressure and temperature change the number of molecules present in the cylinder.

However, the analysis obtained through air standard and fuel-air cycle concept are on the higher side compared to actual engine performance. Therefore, it is customary to understand the factors affecting or making deviations in analyses. The actual engine performance is better known by performing tests on the engine. The engine air standard cycle efficiency is sometimes called indicated efficiency as it is found from the area of the indicator diagram (P-v diagram). The indicated thermal efficiency/ cycle efficiency is also called as fuel conversion efficiency is about 80% of the engine efficiency predicted by air standard cycle efficiency (Figure 1.12).

1.14 DIFFERENCES BETWEEN IDEAL CYCLES AND ACTUAL CYCLES

Though thermodynamics based air-standard cycle analysis gives rise maximum attainable efficiency but in is impractical due to the following reasons. There exists certain factor that makes deviation between actual and ideal cycle performance, to satisfy second law of thermodynamics.

1. The basic assumption heat addition and heat rejection are instantaneous and external makes quite difference. The working substance being mixture of fuel, air, and residual gases and are allowed undergo combustion process which is of finite duration.
2. These gases are no longer diatomic and hence the ratio of specific heats is not constant but would strongly depend on pressure and temperature.
3. The burnt gases are at higher temperature and engine cylinder components need to properly cooled but heat transfer plays a major role in restricting engine efficiency.

4. The burnt gases at higher in-cylinder temperature prone for dissociation certain species and allows endothermic reactions.
5. The narrow regions called crevices, existing trap certain gases and hence lead to loss of power.
6. Exhaust blow down lead to loss of energy being utilized.

It can be approximated that the efficiency of actual efficiency is 80% of indicated efficiency obtained through thermodynamic cycle analysis.

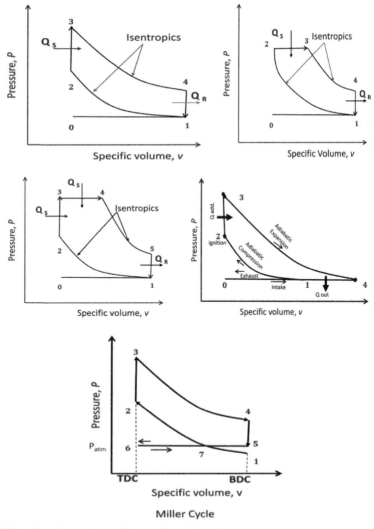

FIGURE 1.12   Comparison of thermodynamic cycles.

1.15 THERMO-CHEMISTRY OF COMBUSTION-COMBUSTION CHEMISTRY

The combustion process is a result of rapid chemical reactions (oxidation reactions) that happens with combination of fuel with an oxidizer. These reactions are chemically called exothermic reactions with the release of large amount of heat. The maximum temperature that obtains under ideal condition in such combustion reactions is called AFT. Fuel is an important component of reactants participating in combustion reactions. Fuels could be solid, liquid, and gaseous in phase. Over the years, liquid fuels have dominated the IC engines. The charge for the SI engine is fuel and air mixture. The load or speed is controlled by varying the quantity of air and fuel mixture. Carburetors were predominantly used in conventional SI engines, however, in modern SI engines; Fuel injection systems are being used. The spark-ignition engines employ forced ignition source spark plug for initiating combustion. Whereas fuel of relatively lower self-ignition temperature is used in CI engines. It is fuel ignition quality that plays vital role in avoiding abnormal combustion. Fuels of high heating value help in obtaining better fuel conversion efficiency. CRs of the order of 6–12 and 12–24 are adopted in SI and CI engines respectively. Generally, SI engines are quiet engines but due to abnormal process, high combustion noise is generated.

The charge for the CI engines is air alone and load/speed is regulated by varying the amount of fuel injected into the engine. It is the rigorous motion that is imparted to the air that would help in thorough mixing of fuel and air mixture in CI engines. The patterns of swirl, tumble, squish, and turbulence that help in obtaining thorough mixture of air and fuel.

The factors that promote detonation or knocking are different in SI and CI engines. It is usually termed as a good SI engine fuel is poor CI engine fuel vice versa. Design of combustion chambers signify in getting most out of the energy supplied to the engines. Also, good designs yield gradual pressure rise and low tendency for knocking or detonation. Cetane number and Octane number are the respective fuel ignition quality indicators of CI and SI engines. The numerical values of ignition quality indicate the intensity of knocking in respective engines.

Fuel forms the important ingredient of the requirements of engine that is required to produce energy. A fuel is substance that majorly consists of hydrocarbons and mostly extracted from petroleum crude. The fuel has energy in it and is expressed in terms of kCal or kJ per kg or cu.m. A fuel upon combustion with oxidizer releases thermal energy (the reactions or exothermic in nature).

SI and diesel engine fuels are each mixtures of several hundred different hydrocarbons of various groups ($CxHyO_z$) (Van Wylen and Sonntag, 2008). These components differ with reference to molecular size and structure and as a result have sometimes strongly varying properties. Moreover, their composition diverges and with it the properties of gasoline, diesel, and potentially alternative fuels (AFs), making a separate treatment necessary. The limit values of the most important material properties of engine fuels adhere to standards (DIN, EN, etc.) in order to guarantee consistent quality and composition as well as reliable engine operation. In the following, the structure and function of the most important hydrocarbon compounds in engine fuels will be described.

Besides the alkanes, which have no double bond and are the most common group in engine fuels, the aliphatic hydrocarbon group also contains alkenes (formerly: olefins) with (at least) one double bond as well as alkines (formerly: acetylenes) with a triple bond. The alkane group is further subdivided into the n-alkanes, which have a straight chain structure, and isoalkanes, which have a branched chain structure. Because of the intermolecular forces which increase with the length of the chain, boiling temperature, viscosity, etc. increase along with the chain length. Fuels with oxygen in its molecular structure are especially relevant for ignition and combustion processes. Thus, aliphatic hydrocarbons, besides alkanes, which have no double bond are the most common group in engine fuels. The aliphatic hydrocarbon group also contains alkenes (formerly: olefins) with (at least) one double bond as well as alkines (formerly: acetylenes) with a triple bond.

The characteristic structure of the aromatics is described by a delocalized charge cloud in the center of the molecule, which is also their distinguishing feature from the cyclo-alkanes. The benzene ring represents the basic building block for all aromatic compounds. Aromatics with more than one benzene ring are designated Auto-ignition and the combustion sequence (Figure 1.13).

In a reaction when fuel and air mixture is allowed to enter under stoichiometric conditions and datum conditions (25°C and 1 atm), applying first law of thermodynamics, with no heat release and no work transfer, the products will be leaving at datum conditions (25°C and 1 atm), in such the reactants will be undergoing reaction and giving rise to highest temperature. The maximum temperature that results under these ideal conditions is the AFT.

A flame is a thermal wave front that sustains combustion as long as it is mixing with oxidizer in definite proportion as per flammability limits.

Generally, a fuel emits light while burning in presence of air but there are exception. Thus, luminosity is not a characteristic of existence of a flame. Fuels can be solid, liquid, and gaseous in phase. Mostly automobiles are fuelled by liquid fuels. Fuel distribution structure is well developed for the handling of liquid fuels for automobiles.

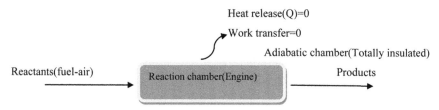

FIGURE 1.13   Illustration of stoichiometric combustion.

The following are the important characteristics of fuels:

1.   Heating or calorific value;
2.   Viscosity;
3.   Flash and fire points;
4.   Carbon residue;
5.   Cloud point, pour point;
6.   Reid vapor pressure, etc.

The heating value of solid fuels and liquid fuels are determined using bomb calorimeter, whereas Junker's gas calorimeter is used for finding the heating value of gaseous fuel. Fuels of high viscosity are not preferred as residue upon combustion interfere with working of intricate components such as fuel injection pump, fuel injector, etc. However, gaseous are preferred as they form homogeneous mixture with air. Flash and fire points are important as far as safety issues are concerned. Liquid fuels also exhibit lubrication properties.

When any hydrocarbon fuel undergoes complete combustion, it leaves behind only carbon dioxide and water vapor. However, under all working conditions of engine, fuel would not undergo complete combustion and thus leaving behind certain incomplete products of combustion. The exhaust gases such as carbon monoxide, oxides of sulfur, oxides of nitrogen and soot or particulate matter are becoming increasingly important of gases of concern in the wake of atmospheric pollution which are causing health hazards, loss of vegetation and animals, etc. Carbon dioxide gas is the main cause of global warming.

Since the threat posed by these harmful gases is attracting the researchers to find viable option in the long run. These days, research attention is being given to the large scale use of fuels of low carbon content (such as methane, etc.), alcohols, hydrogen, biogas, and vegetable oils (biodiesel), etc.

Fuel's ignitions quality plays a vital role in promoting knock or detonation. The SI engine's fuel ignition quality is given in terms of octane number. The higher the Octane number to better is the anti-knocking tendency of the SI engine. Cetane number is indicative of CI engine's fuel ignition quality. Generally, few additives are added to enhance the anti-knocking tendency of fuels.

1.16   COMBUSTION PROCESS, MIXTURE FORMATION, MODES OF OPERATION

1.16.1   SI COMBUSTION FUNDAMENTALS: COMBUSTION OF FUEL IN A TYPICAL SI ENGINE

The combustion process is a rapid oxidation of fuel with air. Combustion is different from burning in a sense that former happens in a confined space. Therefore, the combustion is burning of fuel and air mixture in a combustion chamber. The charge for SI engine is a homogeneous mixture of fuel and air. The charge is inducted in the required quantity and combustion is initiated by triggering a spark before the end of compression stroke. Therefore, a thermal wavefront is established called flame front. If the flame front traverses at subsonic speeds, it is called deflagration wavefront. In normal combustion process, the flame propagation takes place and length of flame travel or conversion unburned charge to burned charge (mass burned fraction) is indication of completion of charge. This is in-line with the heat release pattern and simultaneous conversion to mechanical energy. The flame propagation pattern and in-cylinder pressure versus crank angle diagram. Gradual rise in pressure rise is indication of normal combustion process. The initiation of spark timing also influences the heat release pattern. There were many theories proposed to understand the combustion process in a SI engine. The contribution of Lewis and Von Elbe greatly helped in understanding the process. The summary of theories helped in grouping the combustion process into two phases, viz.; preparation phase and transposition phase or the three stages, Flame initiation, flame propagation and after burning or late burning are well-known combustion stages of a fuel in SI engine.

Advancing of spark timing generally results high peak pressures. Whereas retarding spark timing delays the occurrence peak pressure and thereby reduced work. A spark timing that result in maximum brake torque (MBT) is regarded as MBT timing. In certain conditions, even before the spark initiates the combustion, the combustion occurs and this mostly with the end charge being attaining critical conditions of pressure and temperature. If end charge initiates the combustion, then the flame front traverses in an opposite direction. This results in sudden release of heat lease and pressures would be high and exceeds the safe limits of stiffness of intricate components of engine. These issues results in abnormal combustion process and called detonation. The detonating wavefront is highly unstable and can be seen in the occurrence of fluctuations in pressure crank angle diagram. The detonating wavefront traverses at supersonic speeds. Such abnormal combustion results in deterioration of engine components, high heat transfer, etc. In some condition, the detonation results in piston seizure. Also, SI engine works quietly whereas with abnormal combustion process results in audible sound called knock, ping, thud, etc. The pre-ignition and autoignition are precursors for the normal combustion process. The parameters such as pressure, temperature, density, time, and composition affect the combustion process. Any engine geometrical or operational parameters that influence the above factors greatly impair the SI engine combustion process. The combustion in SI is a strong function of CR and the major parameters can be related to the CR. The compression that be safely handled by SI engine without detonation is called highest useful compression ratio (HUCR).

Fuel ignition quality (Octane number) and design of combustion chamber influence the occurrence of combustion process. The determination of Octane number and sensitivity of fuel to be used in SI engines is done on a cooperative fuel research engine (CFR).

The important requirements for combustion in a SI engine to take place are:

1.   Combustible mixture;
2.   Source of ignition.

The combustible mixture forms from a mixture of fuel-air mixture prepared in a typical SI engine carburetor. Thus, the combustible mixture is pre-mixed charge or homogeneous charge. The mixture should be within flammable limits to regard as combustible mixture capable propagating within a flame zone. Generally, iso-octane is taken as a reference fuel and its flammability limits are 8:1 to 22:1, whereas the stoichiometric ratio being 14.7:1. The combustible mixture (charge) is drawn into the engine during suction stroke, is compressed during compression stroke, before the completion of compression stroke; high pressure and high temperature mixture is ignited with a external source of ignition (electric spark), however, mixture would not burn immediately, it takes some time to catch fire called ignition delay (mostly chemical delay) is over, then the charge quickly burns and a flame front initiates and it starts propagating across a combustion chamber thus heat gets released as combustion proceeds.

Typically, the SI engine fuel undergoes three phases during combustion process:

1.   Ignition lag;
2.   Flame propagation phase; and
3.   After burning.

And thus the broadly, the combustion phases are:

1.   Preparation phase;
2.   Transposition phase.

The occurrence of peak cylinder pressure takes place a little after TDC and this position can be altered by initiation of spark before TDC to obtain work from burned charge.

The ignition lag or preparation phase can be reduced by improving the quality of fuel in terms of ignition quality or octane rating of fuel.

This phase depends on the pressure, temperature, density, composition, and time factors.

The pressure, temperature, density can be related to each other and these factors make the mixture more active and effectively reduces the ignition lag. The time factors can be related to length travel of flame front and speed of operation of engine. Apart from these, fast burn combustion chambers are preferred for allowing quick combustion.

Flame speed, a characteristic of fuel-engine interactions, is a maximum near stoichiometric proportions that for gasoline engines can achieve values in the range of 20–40 m/sec. increasing engine speed will promote greater turbulence and higher flame speeds, but spark-ignition must be advanced to maintain combustion near the TDC position. Spark advance is also required at part-load operation, a regime characterized by lower power output, lower total energy release, and lower flame speed.

Derivatives of SI engine combustion methods which are presently employed in series or are in development by some manufacturers can be classified as follows:

1.16.1.1 *HOMOGENEOUS COMBUSTION*

The salient feature of SI engine combustion is homogeneous due to thorough mixture formation. For reasons of optimal exhaust gas treatment with the three-way catalytic converter, the air-fuel ratio (λ) is nearly maintained 1 in almost the entire characteristic map. Mixture formation takes place at an early time during the working cycle, either by injection of the fuel into the intake manifold or – inmost new designs – by direct injection (DI) into the combustion chamber during the intake stroke. The mixture thereby attained exists ideally in a homogeneous form with the same in the entire combustion chamber. Stratified combustion: Mixture formation takes place via DI into the combustion chamber during the late compression phase. The wall and air guided combustion methods still used a few years ago have in the meantime been replaced by spray-guided combustion methods. The spark plug must be positioned on the edge of the spray such that it reaches a locally narrow area with an ignitable mixture directly on the edge of the injected fuel spray. As soon as this area is ignited, the flame front can burn even under extremely lean zones throughout the combustion chamber in its broad propagation, allowing for global combustion air-fuel ratios nearly of 1. For this reason, an additional exhaust gas treatment for nitrogen oxide (NOx) with an excess of air is required in addition to the three-way catalytic converter.

Spark-ignition combustion processes can be classified as normal, i.e., a spark-initiated process; abnormal, i.e., pre-ignition via hot-spot processes, such as carbon deposits or hot spark plugs; and self-ignition, i.e., auto-catalytic detonation processes.

Under the normal combustion process, the pressure rise is gradual during combustion.

Typically, SI engine combustion requires a homogeneous combustion process, while current efforts are directed toward non-homogeneous or stratified charge engine chemistry.

Basic design objectives of classic spark-ignition combustion include:

- Developing a high level of turbulence within the charge.
- Promoting a rapid but smooth rise in pressure versus time (crank angle) during burn.
- Achieving peak cylinder pressure as close to TDC as possible.
- Establishing the maximum premixed flame speed.
- Burning the greatest portion of the charge early in the reaction process.
- Attaining complete fuel-air charge mixture combustion.
- Precluding the occurrence of detonation or knock.
- Minimizing combustion heat loss.

Today, most SI engines are designed to operate in high-speed conditions, within the 1,000–6,000-rpm range, with spark firing required from 8 to 50 times each second. Engine power results from spark-initiated chemical kinetic and related thermo-chemical events occurring with a given piston-cylinder head configuration.

Spark-ignition engine output, therefore, is strongly influenced by the combustion chamber geometry at TDC and the most obvious shape is hemi-spherical at TDC, that allow high surface-to-volume with quick heat loss and with a marginal reduced power output. However, combustion chamber configurations should not be chosen that can promote the development of detonation waves and combustion knock and with crevice regions, cold wall quench zones ultimately cause incomplete combustion.

Part-load performance, with its greater residual gas containing charge and hotter running cylinders, makes the engine more sensitive to spark knock. Flame speed for extremely lean burn can be so slow that a flame may still be progressing through the chamber when inlet valves open, leading to backfire into the inlet manifold. Spark advance is set at minimum ignition advance for best torque (MBT).

The thermodynamic cycle (an ideal cycle) for assessing typical SI engine is Otto cycle (Figure 1.9) wherein heat addition is at constant volume. Following such a cycle, if the p-v diagram is converted into an equivalent p-θ diagram, the constant volume heat addition coincides with the vertical axis (Figure 1.14(a)). In a practical SI engine, the reference fuel for a typical SI engine is iso-octane and air is an oxidizer. As per flammability limits, a combustible mixture as shown in Figure 1.14(b) is allowed during suction stroke in accordance with load and speed requirements.

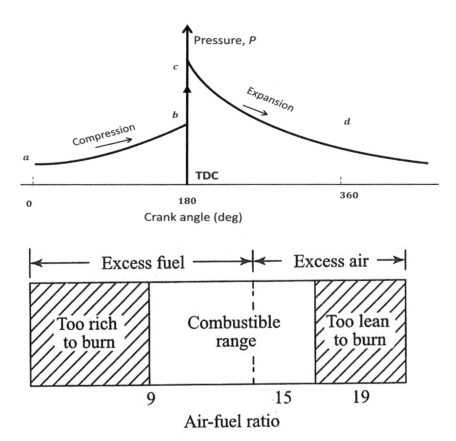

FIGURE 1.14 (a) The theoretical p-θ diagram of SI engine, (b) flammability Limits of iso-octane.

The inducted charge is compressed during compression stroke and before the end of compression stroke, the high pressure, high temperature, and dense charge is ignited with the help of a spark. The spark spreads through whole charge, the charge overcomes its ignition lag or ignition delay, mostly chemical delay and there would be gradual rise in pressure as shown in Figure 1.15 As the spark spreads or flame front develops and propagates all along the combustion chamber, the charge is converted from unburnt to burnt charge. The peak pressure occurs a little later TDC i.e., after TDC. The occurrence of peak pressure depends on where exactly the spark initiates. During normal combustion process, the flame propagation is in line with conversion of unburnt to burnt charge and the flame travels at sub-sonic speeds and there lies gradual conversion of thermal energy to mechanical energy. The wavefront or thermal wavefront traverses at sub-sonic speeds is called deflagration. The transfer of energy to piston is also smooth without any abnormal noise. Thus in normal combustion process, the pressure energy of gases smoothly transferred to piston via connecting to crank shaft. It is obvious from Figure 1.15 that there will initiations or preparation phase, flame propagation phase and late burning phase. There will be about 10% loss of flame travel with no effective conversion, and 80% travel realizes in transfer of energy and remaining 10% results in late burning phase which again doesn't give rise to work energy. Totally, there will be three definite phases of combustion in a SI engine.

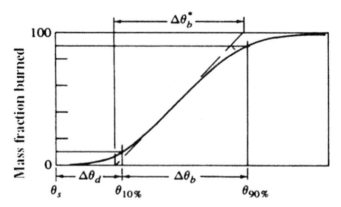

FIGURE 1.15   Variation of mass burned fraction with distance travel across combustion chamber.

If air alone is compressed and the rise in pressure will be low enough, the p-θ diagram is referred to as motoring curve or no combustion curve, as depicted in Figure 1.16.

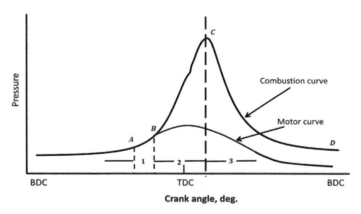

FIGURE 1.16   Comparison of motoring and fuel combustion curves- p-θ diagrams.

The charge would instantaneously start igniting as there would be delay called-ignition lag. There will be some lapse of crank travel, in the meantime, spark initiation, flame development flame propagation and flame termination will occur. The gradual rise in pressure will result in conversion of thermal energy into mechanical energy. The effect of spark timing on the p-θ diagram is more important. The active charge after the initiation of spark would overcome ignition lag. The timing at which optimum yield of pressure occurs with maximum torque is referred to as MBT timing.

However, sometimes, the spark is advanced and the charge is ignited and ignited charge is further compressed and then there would be rise pressures and temperatures that will shoot up so high that there will be jerks in the piston and connecting rod. The occurrence of peak pressure would be just after TDC. During the advancement of spark, the pressure and temperatures are very high and whole charge is oxidized fully with maximum conversion of hydrocarbons and absolutely there wouldn't be high and thus very low HC and CO emissions. However, the temperatures are so high that with the lean mixtures prevailing, the nitrogen in the air is dissociated into nitrogen atoms and oxygen is also gets dissociated. Thus, it forms complex compounds of oxides of nitrogen emissions (NOx).

On the other hand, if the initiation of spark is retarded and is done just before TDC and it so happens that the peak pressure occurs later in the expansion stroke with small rise in peak pressure. With the retarding of spark there would be loss of energy which obvious from the p-θ diagram Figure 1.17.

During retarding of spark, the temperatures are low enough then the probability of formation of NOx emissions but there would be partial oxidation and therefore HC and CO emissions will be higher.

Therefore, there exists trade-off in HC, CO, and NOx emissions.

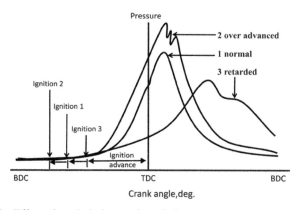

FIGURE 1.17 Effect of spark timing on the p-θ diagram.

It so happens under certain conditions that the engine components will be maintained hot during the extended hours of operation and improper coolant supply and all leading to hot combustion chamber. If the hot chamber, keeps hot spots and eventually combustion or spark like points will traverse in opposite direction. The combustion thus occurs at different points without the interference of spark, thus auto-ignition takes place. The travel of flame front thus opposite direction take place leading abnormal noise such as detonation. The noise is called ping or thud. The flame travels at supersonic speeds and occurrence of such conditions deteriorates engine intricate components and such combustion is not normal and detonation as against deflagration wavefront in case of normal combustion process. The charge attains critical conditions of pressure, temperature, and density and leads to auto-ignition temperature.

From Figure 1.18, the difference in pressure-crank (p-θ diagram) diagram is clearly illustrated. The abnormal combustion process late in the expansion stroke.

The traces of pressure are to obtained to analyze the normal and abnormal combustion process in a typical SI engine. The in-cylinder pressures are obtained with the help of a piezoelectric or piezo-resistive pressure transducer.

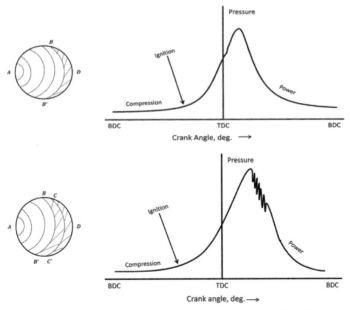

FIGURE 1.18   Comparison of p-θ diagrams of normal and abnormal combustion processes.

Many simulation studies have been conducted to observe the effect of spark initiation on the flame propagation. A typical mass burnt fraction versus flame travel resembles S-type curve and the probability obtaining such a curve can be studied with the help of number of cases. It is observed that a very active mixture will quickly take part in combustion and it represent an almost a vertical line and on the other hand at in-active mixture will slowly progress during combustion with little release of energy. In between, we get different curves. The probability of obtaining such combustion is represented by Wiebull function or Weibe function as illustrated in Figure 1.19 (a) and (b).

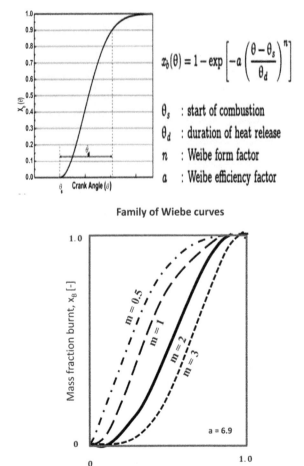

$$x_b(\theta) = 1 - \exp\left[-a\left(\frac{\theta - \theta_s}{\theta_d}\right)^n\right]$$

θ_s : start of combustion
θ_d : duration of heat release
n : Weibe form factor
a : Weibe efficiency factor

Family of Wiebe curves

FIGURE 1.19 (a) Wiebe function and its application to SI combustion flame propagation, (b) family of Wiebe curves.

1.16.2 *MIXTURE FORMATION IN SPARK-IGNITION ENGINES-CARBURETION AND FUEL INJECTION*

Charge for the SI engine is a mixture of air and fuel. Generally, fuel of high volatility, low viscosity, and high self-ignition temperature are used for SI engine. The reference fuel is isooctane. The fuel's ability to resist knock is ignition quality and is expressed in terms of Octane number. The stoichiometry demands the chemically correct amount of air that is required for complete combustion fuel. Under stoichiometric conditions, the complete products of combustion are CO_2 and H_2O and N_2.

The following the stoichiometric combustion equation for isooctane with air:

$$C_8H_{18} + 12.5\ O_2 + 47N_2 \rightarrow 8CO_2 + 9H_2O + 47\ N_2.$$

The AF works out to be 15:1 (approx.). It does represent that about 15 parts of air is required for complete combustion of one mole of air. The richness or leanness of fuel is known by Equivalence ratio. The ratio of actual fuel/air ratio to stoichiometric fuel/air ratio is called an Equivalence ratio. It is represented by ϕ (phi). If ϕ is greater than 1, then it is rich mixture (fuel rich) and if it is less than 1, then it is lean mixture (fuel lean). As far as flammability limits for isooctane, suggest that flame can propagate between 8:1 and 20:1.

Normally, the mixing device should supply air-fuel mixture within the above limits for sustaining the flame (Figure 1.20).

The performance of a SI engine strongly depends on AF or equivalence ratio. The governing system in a SI engine is quantity governing in which the well prepared is regulated and allowed in a SI engine by varying its quantity. When engine is allowed to run with different AFs. When its performance is monitored, the best power or maximum power is obtained at a slightly at a richer side and best economy or lowest fuel consumption is obtained at a slightly leaner mixture. However, it is not practical always to run at either of mixture strengths aiming for either lower fuel consumption or highest power. The trends are represented in Figure 1.20.

Therefore there should be some arrangement that regulates the charge (mixture of air and fuel) to allow the engine to run at part loads or at any specified loads.

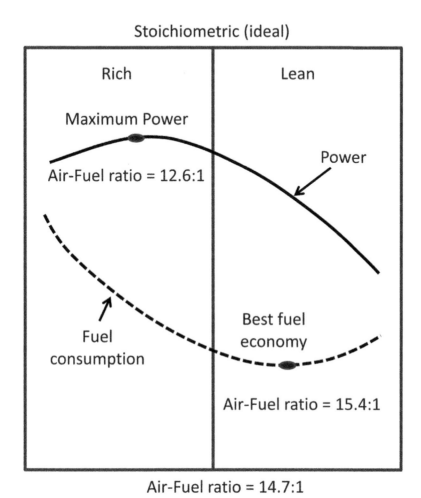

FIGURE 1.20 Variation of engine power and fuel economy with air-fuel ratio.

When a SI engine is fitted with a carburetor sort of device (Figure 1.21), which has a throttle valve that permits the required amount of charge to enter the engine cylinder. In such a case if, an engine is allowed to with different AFs at different throttle opening, a trend as shown in Figure 1.22 is obtained. In the case of single-cylinder engine fitted with a simple carburetor, the AF variation starts with a rich mixture and attains a minimum value and then rises to richer side covering from idling, cruising, and acceleration respectively. The values will be little different for a multi-cylinder engine, the pattern trends will be similar. Figure 1.22 resembles a basket-type curve.

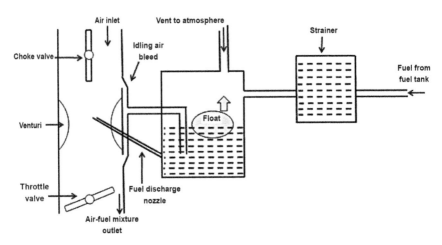

FIGURE 1.21   Simple or elementary carburetor of a SI engine.

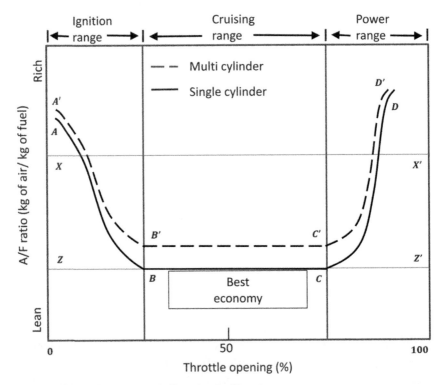

FIGURE 1.22   Mixture strength diagram of a SI engine.

The device that does the job of mixing and metering the amount of fuel-air mixture is called the carburetor. The process is called carburetion. There are three types of carburetors: Up draught type, down draught type and horizontal draught type. Down draught carburetors are the most commonly used ones. Zenith, Solex, Carter, and SU are the well known automotive type carburetor.

An elementary type carburetor in its simple form is simple carburetor, which offers limitations on the part of multi-cylinder engines. Even though such a carburetor is provided with compensating devices, the problems are not totally solved.

Therefore, these days modern petrol/gasoline multi-cylinder engines are incorporated with fuel injection systems for precise control of mixture along with electronic controls.

1.16.3 ROLE OF COMBUSTION CHAMBER IN SI ENGINES

A spherical combustion chamber having a radius (r_0) can be used to describe essential elements of SI premixed homogeneous charge combustion. Assume that such geometry contains reactants initially at uniform conditions T2 and P2. Central ignition of this fuel-air mixture produces a spherically shaped premixed flame front that burns outward toward the combustion chamber wall. A premixed flame speed depends on the initial thermodynamic state, FA, chemical kinetics of the fuel-air mixture, and aerodynamics of flow.

Initial rates of chemical reactions and flame propagation at the center of the constant-volume chamber will be relatively slow because of low turbulence, initial temperature, and pressure within the original charge. This is why efforts are made to initiate swirl and turbulence within a cylinder during charge induction and prior to charge ignition. Too much turbulence, however, could contribute to increased convective heat transfer loss, flame extinction, and/or reduction in indicated power output.

As the flame front progressively burns through the charge, burned gas product temperature T3, pressure P3, and specific volume v3 behind the flame front will be greater, causing thermal compression of the remaining unburned charge ahead of the flame. In addition, heat transfer and diffusion of some reactive radical elements across the traveling reaction zone from the burned product gases will raise local temperature and reactivity in the charge ahead of the flame. This means that flame speed will increase as it moves into these gases. Note that this differs from the one-dimensional steady-state and steady-flow subsonic reaction propagation process. A reaction front was burning in an open duct with constant upstream conditions; whereas, in an SI engine, flame propagation occurs within confined spaces having upstream conditions that are not constant.

The highest kinetic rates, gas temperatures, and hence flame speeds are achieved in the middle region of the sphere. Since flame speeds are greatest in this area, the largest part of the charge is burned during this stage of combustion. Thermal compression of the unburned charge ahead of the flame, however, continues during this portion of the reaction.

Flame extinction and termination of combustion occur near the wall. High heat transfer rates between the moving flame front and cold wall cause rapid reduction in gas temperature and flame speed approaching the cavity wall. Quenching of the flame occurs prior to its arrival at the wall, leaving a boundary layer region of incomplete reaction within the volume. Note that flame quench in narrow gaps and crevice geometries of an actual chamber can play a major role in emissions formation in an operating engine.

Higher temperatures and pressures are found in the unburned charge at the end of burn as a result of thermal compression and heating. The potential for knock, or detonation, therefore, exists in a spherical cavity near the end of burn, where conditions may allow gas phase autocatalytic reactions to propagate at a rate that causes oxidation chemistry to self-accelerate. This phenomenon of engine knock can be observed by the use of a cylinder pressure-crank-angle diagram. Central ignition and combustion of a homogeneous fuel-air charge using a constant-volume spherical chamber produces the following general traits:

Low initial burn rate with little flame expansion and turbulence. An intermediate burn rate having the greatest flame speed, percent of burn, and distance of travel. Terminal burn rate having the maximum flame front area and a thermally compressed end charge that can potentially detonate (Figure 1.23).

1.16.4   IGNITION QUALITY OF A SI ENGINE FUEL

The design and development of more efficient ICEs will require a fuller understanding of the general characteristics of fuel-engine compatibility.

It is essential to determine whether an engine will detonate for a given CR or fuel-air mixture, at certain conditions in both spark-ignition (SI) and CI engines. This explosive nature is termed auto-ignition, or knock. Knock characteristics are a function of the chemical nature of a fuel and the combustion processes in a particular engine. Since knock performance of fuels must be determined experimentally, a variety of standardized tests for both SI and CI fuels have been established by the American Society for Testing Materials (ASTM). These tests involve the use of specific engines and include research, motor, or vehicle tests.

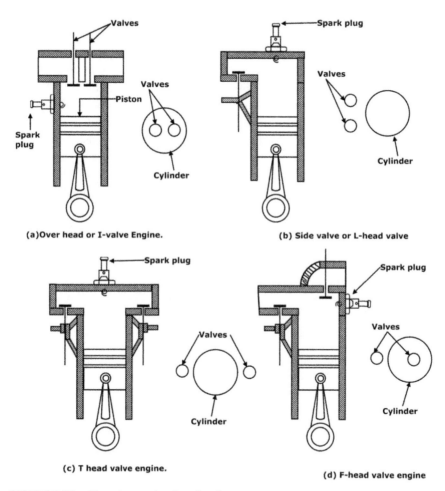

(a)Over head or I-valve Engine.

(b) Side valve or L-head valve

(c) T head valve engine.

(d) F-head valve engine

FIGURE 1.23 SI engine combustion chambers.

The research method for testing SI and CI fuels for knock uses a single-cylinder CFR engine. This unique 3.25 in × 4.5 in. four-stroke engine has several parameters that can be varied while the engine is in operation, including the CR (3:1 to 15:1), AF, and/or fuel type. Variable compression is accomplished by means of a hand crank and worm gear mechanism, which allows the cylinder and head assembly to be raised or lowered with respect to the crankshaft. A unique overhead valve design maintains constant-volume clearance with varying CR.

Three separate gravity-fed fuel bowls are connected to a horizontal draft, air-bled jet carburetor, allowing any of three different fuels to be used and/ or interchanged while the engine is running. The FA can also be changed by simply raising or lowering each of the three fuel bowls.

Engine speed is maintained at 900 rpm by means of an engine-belted, synchronous AC power generator. Also, particular SI versions of the CFR engine have components that allow spark timing to be changed, while CI configurations have the capability of varying fuel injection rate and timing.

Detonation, or knock, in an engine will result in a rapid rise in peak engine pressure. Cylinder pressure in the research test is measured using a pressure transducer with filtered voltage output displayed on a knock meter. The knock rating of a particular fuel is obtained by matching the knock intensity of a sample at fixed operating conditions to that produced by blends of reference fuels.

Four primary fuels are used as standards for making mixtures of reference blends Spark-ignition engine fuel ratings are measured in terms of an octane rating, while CI engine fuel ratings are specified by a cetane scale.

Octane numbers are based on isooctane having an octane value of 100 and n-heptane having an assigned value of 0 octane. Thus, a 90-octane reference blend would be 90% isooctane and 10% n-heptane mixture by volume or:

$$\text{Reference octane no.} = \% \text{ isooctane} + \% \text{ n-heptane.}$$

Cetane numbers are based on n-cetane (nexadecane) having a cetane value of 100 and n-methylnaphthalene ($C_{11}H_{10}$) has an assigned value of 0 cetane. The octane scale for a reference blend by volume is given as:

$$\text{Reference cetane no.} = \% \text{ n-cetane} + \% C_{11}H_{10}$$

In general, the higher the octane rating, the lower the tendency for the fuel-air mixture to auto ignites in an SI engine. Also, the higher the cetane rating of a fuel, the greater the tendency for autoignition in a CI engine. Thus, a high-octane fuel will be a low-cetane fuel, and a low-octane fuel will have a high cetane index. Knock rating of IC engine fuels based on the research test method is often different from values obtained using the motor method. Octane and/or cetane rating variations according to knock testing by research and motor methods are specified by fuel sensitivity. In general, the research method yields octane ratings higher than those obtained from the motor test. Current practice requires that automotive fuels be based on an average of the research and motor octane numbers.

Additives, such as TEL, were used in the past to raise the octane values of gasoline, but environmental impact forced the development of unleaded fuels and alternate octane boosters, such as the efforts with anhydrous ethanol-gasoline blends, or gasohol. In addition, some fuels may actually yield octane numbers in excess of 100.

1.17 FUEL SUPPLY IN COMPRESSION-IGNITION ENGINES (CI ENGINES)

The charge for the CI engines (mostly diesel-fuelled engines) is air alone and speed/load adjustments are done through amount of fuel injected per cycle per cylinder. In a way, the fuel and mix does not mix together as in SI engine and the mixture is mostly heterogeneous. The amount air inducted is a strong function of speed. That's why the CI engines are commonly called as unthrottled engines. Also, CI engine are called quality governed engines. This makes these engines to become more fuel efficient than their counterparts. The fuel normally used in such is relatively of high viscosity. The fuel is injected normally when air gets compressed nearly to the self/ autoignition temperature of fuel; near to the end of compression stroke. In a small period of time, fuel has got to be mixed thoroughly with dense and high temperature air. Therefore, there lies a great job on the part of fuel injection system to atomize the fuel and supply right amount of fuel as per load and speed requirement. Therefore, the fuel has to be pressurized to the level of air for thorough mixing.

The fuel is allowed to mix with air in two ways:

1. Air injection here is blast of compressed air is utilized to split the liquid fuel into small droplets; and
2. Solid injection fuel alone is pressurized in a component called fuel injection pump, where in due to high pressure, the fuel split into small droplets and kept ready for passing through an injector.

Partially, the job of carburetor is done through fuel injection pump or fuel pump and fuel injector. Normally, it is treated that fuel injection system as heart of CI engine. The important job of fuel injection system is to obtain a smaller and smaller size of droplets. The smaller is the droplet; The better is the mixing of fuel with air in an engine cylinder. Injector or simply a nozzle is used to convert high pressure energy into high kinetic energy so as to allow the fuel to penetrate into dense air.

Ideally, each cylinder should have fuel pump and an injector. The size of droplet obtained depends on the size of nozzle hole or orifice size. If the surface to volume ratio of each droplet is equivalent to the surface to volume ratio of whole spray, the diameter of such a droplet is called as Sauter Mean Diameter.

A modern fuel injection in a diesel engine—common rail direct injection system (CRDI) (Figure 1.24).

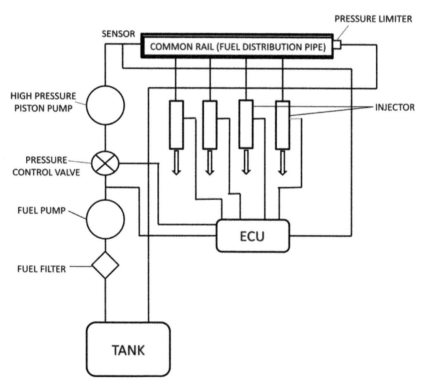

FIGURE 1.24   Common rail direct injection of a CI engine.

1.17.1   IN-CYLINDER MOTION

In CI engine, the fuel and air do not mix together before their entry into the engine. The higher the rigorous motion imparted to the air that help in thorough mixing of fuel and air. Swirl, squish, tumble, and turbulence are the patterns of air motion that are imparted to air. Induction induced swirl, compression induced swirl, combustion generated swirl are the flow patterns. Swirl intensity is better known compare to engine speed. The factor that signifies the swirl intensity is the swirl ratio. Tangential velocity imparted through swirl generation is also another term that differentiates from one pattern to another. Swirl and tumble are the most common forms of charge motion in the combustion engine.

Swirl flow is a swirling motion of in-flowing air around the vertical axis of the cylinder, which can be obtained by means of asymmetrically formed inlet ducts or by switching off one of the two inlet valves. However, a swirling flow often remains at least partially in its original form until the time of ignition instead of dissipating completely into turbulence.

Tumble flow is defined as a swirling motion of cylinder air around the transverse axis of the cylinder in the direction of the crankshaft axis. It is often created, for example, in supercharged SI engines by relatively flat inlet ducts. Since such a duct shape entails certain charging disadvantages as well, high-performance naturally aspirated engines usually omit such design. As long as the combustion method requires an intensified tumble flow, it can be created as needed by closing a tumble valve in the inlet duct, which closes the lower half of the cross-section. Combustion accelerated by turbulence is a basic prerequisite for realizing high engine speeds because only in this way can the flame front burn though from the spark plug to the combustion chamber walls completely in the short time available. Values between 0.3 and 1 m/s are quoted in the literature for purely laminar flame speeds of the reference fuel iso-octane depending on pressure and temperature. Specifications for gasoline vary depending on the combustion air ratio in a similar way, whereby the maximum speeds are indicated for the case of a rich mixture (about l ¼ 0.83).

1.18 COMBUSTION IN COMPRESSION-IGNITION ENGINES (CI ENGINES)

Combustion process of a fuel in CI engine is different from that of fuel of SI engine the charge for CI engine is air alone. The charge is inducted without throttling in the intake or suction stroke. Fuel in the required quantity is injected into hot and dense air in the form of fine spray at the end of compression process. And combustion is initiated by triggering a spark before the end of compression stroke. The way fuel and air allowed mix makes lot of difference inefficiency of combustion in CI engine. There is no definite flame front as in the case of SI engine. Therefore, combustion initiates at many places inside the combustion chamber. Generally, the combustion process of a fuel undergoes four stages of; Ignition delay, uncontrolled combustion (premixed phase of combustion), controlled combustion (mixing controlled combustion) and after burning.

The fuel undergoes a sort of delay in taking part in combustion immediately after its injection into hot and dense air at the end of combustion. The fuel before it undergoes combustion; It experiences atomization, penetration, dispersion, evaporation or vaporization, ignition, and combustion. The delay period or ignition delay depends on the factors that influence the above mentioned issues. The time period between start of injection and start of combustion is termed as ignition delay. During the ignition delay process, the injected fuel gets accumulated and once the suitable conditions are prevailed, sudden combustion occurs, results in high peak pressures. The more the delay period the higher is the magnitude of peak pressure. During the delay period, fluctuations in pressure-crank angle diagram can be observed. In a CI engine, the fluctuations in the initial phase of combustion whereas SI engine the fluctuations can be seen in the end of combustion.

Advancing of injection timing generally results in high peak pressures. Whereas retarding injection timing delays the occurrence peak pressure and thereby reduced work. Generally, CI engine works with high CRs and the intensity of knock could not be audible. The parameters such as pressure, temperature, density, time, and composition affect the combustion process. Any engine geometrical or operational parameters that influence the above factors greatly impair the CI engine combustion process. Fuel ignition quality (Cetane rating) and type of combustion chamber influence the occurrence of combustion process.

Combustion process in diesel engines is heterogeneous and is different from homogeneous combustion process of gasoline SI engines and thus is very complex.

The understanding was difficult despite the availability of modern tools such as high speed photography used in "transparent" engines, computational power of contemporary computers, and the many mathematical models designed to mimic combustion in diesel engines. The application of laser-sheet imaging to the conventional diesel combustion process in the 1990s was key to greatly increasing the understanding of this process.

In conventional SI engines, there is carburetor that helps in mixture formation of air-fuel required for combustion. However, the process is different in CI engines, the mixing of fuel and air, which is often referred to as *mixture preparation takes place within the cylinder.*

In diesel engines, fuel is often injected into the engine cylinder near the end of the compression stroke, just a few crank angle degrees before top-dead-center *(*Heywood, 1988*)*. The liquid fuel is usually injected at high velocity as one or more jets through small orifices or nozzles in the injector tip. It atomizes into small droplets and penetrates into the combustion chamber. The atomized fuel absorbs heat from the surrounding heated compressed air, vaporizes, and mixes with the surrounding high-temperature high-pressure air. As the piston continues to move closer to top-dead-center (TDC), the mixture (mostly air) temperature reaches the fuel's ignition temperature.

Rapid ignition of some premixed fuel and air occurs after the ignition delay period. This rapid ignition is considered the start of combustion (also the end of the ignition delay period) and is marked by a sharp cylinder pressure increase as combustion of the fuel-air mixture takes place. Increased pressure resulting from the premixed combustion compresses and heats the unburned portion of the charge and shortens the delay before its ignition. It also increases the evaporation rate of the remaining fuel. Atomization, vaporization, fuel vapor-air mixing, and combustion continue until all the injected fuel has combusted.

Diesel combustion is characterized by lean overall A/F ratio. The lowest average A/F ratio is often found at peak power conditions. To avoid excessive smoke formation, A/F ratio at peak torque is usually maintained above 25:1, well above the stoichiometric (chemically correct) equivalence ratio of about 14.4:1. In turbocharged diesel engines the A/F ratio at idle may exceed 160:1. Therefore, excess air present in the cylinder after the fuel has combusted continues to mix with burning and already burned gases throughout the combustion and expansion processes.

At the opening of the exhaust valve, excess air along with the combustion products are exhausted, which explains the oxidizing nature of diesel exhaust. Although combustion occurs after vaporized fuel mixes with air, forms a locally rich but combustible mixture, and the proper ignition temperature is reached, the overall A/F ratio is lean. In other words, the majority of the air inducted into the cylinder of a diesel engine is compressed and heated, but never engages in the combustion process. Oxygen in the excess air helps oxidize gaseous hydrocarbons and carbon monoxide, reducing them to extremely small concentrations in the exhaust gas.

The following factors play a primary role in the diesel combustion process:

- The *inducted charge air*, its temperature, and its kinetic energy in several dimensions.
- The *injected fuel's* atomization, spray penetration, temperature, and chemical characteristics.
- While these two factors are most important, there are other parameters that may dramatically influence them and therefore play a secondary, but still important role in the combustion process. For instance:
- *Intake port design*, which has a strong influence on charge air motion (especially as it enters the cylinder) and ultimately the mixing rate in the combustion chamber. The intake port design may also influence charge air temperature. This may be accomplished by heat transfer from the water jacket to the charge air through the intake port surface area.
- *Intake valve size*, which controls the total mass of air inducted into the cylinder in a finite amount of time.
- CR, which influences fuel vaporization and consequently mixing rate and combustion quality.
- *Injection pressure*, which controls the injection duration for a given nozzle hole size.
- *Nozzle hole geometry* (length/diameter), which controls the spray penetration as well as atomization.
- *Spray geometry*, which directly impacts combustion quality through air utilization. For instance, a larger spray cone angle may place the fuel on top of the piston, and outside the combustion bowl in open chamber DI diesel engines. This condition would lead to excessive smoke (incomplete combustion) because of depriving the fuel of access to the air available in the combustion bowl (chamber). Wide cone angles may also cause the fuel to be sprayed on the cylinder walls, rather than inside the combustion bowl where it is required. Fuel sprayed on the cylinder wall will eventually be scraped downward to the oil sump where it will shorten the lube oil life. As the spray angle is one of the variables that impact the rate of mixing of air into the fuel jet near the outlet of the injector, it can have a significant impact on the overall combustion process.
- *Valve configuration*, which controls the injector position. Two-valve systems force an inclined injector position, which implies uneven spray arrangement that leads to compromised fuel/air mixing. On the other hand, four-valve designs allow for vertical injector installation, symmetric fuel spray arrangement and equal access to the available air by each of the fuel sprays.

- *Top piston ring position*, which controls the dead space between the piston top land (area between top piston ring groove and the top of the piston crown), and the cylinder liner. This dead space/volume traps air that is compressed during the compression stroke and expands without ever engaging in the combustion process.

It is therefore important to realize that the combustion system of the diesel engine is not limited to the combustion bowl, injector sprays, and their immediate surroundings. Rather, it includes any part, component, or system that may affect the final outcome of the combustion process.

1.18.1   CI ENGINE COMBUSTION CHAMBERS

Design requirements for diesel combustors are similar to many of the design objectives set for SI configurations.

- Development of turbulence within the charge;
- Promoting a rapid but smooth rise in pressure versus time (crank angle) during burn;
- Achieving peak cylinder pressure as close to TDC as possible;
- Reducing the occurrence of detonation or knock;
- Promoting complete combustion;
- Minimizing heat loss.

Several illustrative chamber geometries that have been developed in an effort to achieve proper CI combustion, as shown in Figures 1.25 and 1.26.

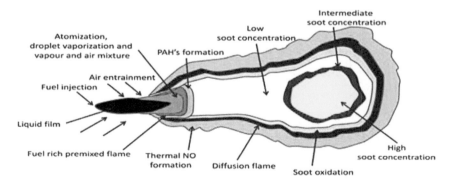

FIGURE 1.25   Illustration of fuel jet sprays during combustion.

FIGURE 1.26   CI engine combustion chambers.

These CI combustion chambers may be classified as either direct-injection (open-chamber) or divided-chamber (IDI) CI geometries. Particular power applications will dictate which of these CI combustion chambers is more suitable for use. In automotive applications, where a high power-to-weight requirement exists, for example, a high-speed divided-chamber engine is usually needed, whereas in marine and industrial power applications, where weight and size are less critical than fuel economy, large slow-speed open-chamber engines are more often used.

The open combustion chamber uses space above the piston crown and head for combustion control. Fuel-injected directly into this single-chamber combustor requires a high-pressure injection process to ensure proper atomization and penetration of fuel spray throughout the chamber volume. Air motion is relatively nonturbulent, so that intake valve location and piston crown geometries are varied in an attempt to introduce motion and air rotation within a chamber. Chamber geometry is frequently shaped to conform to the fuel spray pattern, with a fuel injector generally positioned close to the piston centerline. Open combustion chamber geometry may have less influence on combustion than air motion. A squish, or reduced area, configuration allows the volume between piston crown and clearance volume at the edge of a piston when at TDC to be quite small.

This geometry is an attempt to cause an increased inward motion of a charge at the end of the compression stroke. An open-chamber CI engine requires considerable excess air but, in general, is easy to start. Smaller surface-to-volume ratios for an open chamber result in a lower heat loss than for other geometries. This design is utilized most often in large, low-speed diesel engines that burn diesel fuels having longer ignition delays, i.e., lower cetane numbers. Open-chamber CI engines generally require CRs in a range 12:1–16:1 and have better fuel consumption, higher pressure rise, and less throttling losses of all the CI chambers. High-speed open-chamber CI engines have a tendency to develop diesel knock.

The divided combustion chamber diesel uses two-chamber combustion in an effort to circumvent problems encountered with open-chamber fuel injection designs. A variety of configurations and combustion methods have been developed, including pre-combustion chambers, swirl or turbulent chambers, and air or energy cell chambers. Divided chambers are most frequently used with small high-speed engine applications in an effort to help shorten the time needed for combustion. The additional chamber volume can be located in the piston crown, cylinder wall, or head. Secondary chamber volumes can be as much as 50% of clearance volume.

Pre-combustion chamber CI engines are IDI configurations consisting of the main chamber and a separate fuel ignition volume. Air enters the main and prechamber volumes during a compression stroke, but fuel is injected only into the smaller uncooled prechamber. Deep fuel spray penetration is not necessary, allowing a lower injection pressure and shorter ignition delay. Initial ignition occurs in the prechamber and produces a relatively rich homogeneous and turbulent preconditioned charge with jet spray expansion of reactive gases through orifice openings into the main chamber for complete reaction. Prechamber designs do not depend on air motion within their secondary chamber to burn diesel fuels.

A swirl, or turbulent chamber, is another IDI divided-chamber design. Air enters both chambers during compression, and fuel is injected only into the secondary volume, much like the prechamber. However, significant air motion and turbulence are purposely generated within the swirl chamber during compression. Turbulent-chamber diesels reduce stringent fine spray injection atomization requirements needed for open chambers and allow larger fuel droplets to be burned. The boundary distinguishing a pre-chamber from a turbulent-chamber design is somewhat arbitrary, but, in general, pre-chamber volumes are relatively smaller and burn only a minor portion of the charge there in, whereas swirl-chamber volumes are larger, with a major portion of the burn occurring therein.

The air, or energy, cell is a divided combustion chamber configuration but, in this instance, fuel is injected into the main chamber. The air cell is located directly across from an injector, and fuel is sprayed across the main chamber and enters into an energy cell prior to ignition. This method offers benefits of both the open and turbulent geometries. Air-cell engines rely on their small antechamber and the ensuing air motion and charge interaction to reduce the injection pressure required for an open-chamber design. The piston motion during one mechanical cycle causes a turbulent flow to occur between the antechamber and main chamber. The rapid pressure rise and peak pressure in the figure-eight-shaped main is gas dynamically and thermally controlled by the energy cell. Burn in an air-cell engine occurs with slow initial combustion of injected fuel, followed by a rapid secondary burn resulting from flow jetting back into the main chamber during an expansion stroke. Since the energy cell is not cooled, the design regenerates a charge by returning some energy released by combustion to the charge at a later stage in the burn process. Energy-cell CI engines are generally characterized by larger heat losses to the wall, starting difficulties, and lower thermal efficiency but higher IMEP. Air- or energy-cell chambers are easier to start than the swirl or turbulent chamber since fuel is injected into the main chamber.

Divided-chamber combustors have larger surface-to-volume ratios, more fluid motion, greater heat transfer coefficients, and, therefore, higher heat loss and lower charge temperatures relative to comparable open-chamber geometries. IDI CI engines usually require CRs of 18:1–24:1 to ensure reliable ignition, but they are often smoother-running engines. Additional general characteristics of a divided- chamber IDI CI engine include higher VE, lower peak pressure, lower IMEP, lower thermal efficiency, and higher specific fuel consumption. Two-chamber CI configurations are not as compatible with two-stroke or large displacement volume diesels.

CI combustion is slower than SI combustion, which means that maximum engine speeds are lower for diesel engines. Power output, therefore, cannot be raised by increasing engine speed, so that CI engine output is usually increased by turbocharging, a technique that can also produce improvements in fuel economy. Turbocharging may mean that a reduction in CR is required in order to maintain the peak pressure and temperature of the basic engine design. In particular, two-stroke diesel engine designs are often turbocharged.

The higher thermal efficiency and greater fuel economy of diesel engines compared to SI engines are chiefly a result of the higher CRs required for diesels. Toa lesser degree, diesel performance gains are due to the simplicity of the autoignition process, lower pumping losses as a result of removing the throttle valve, and overall lean AF mixtures required for combustion.

Diesel engines also have the advantage of being able to maintain a higher sustained torque than a spark-ignition engine. Diesel engines control load by simply regulating the amount of fuel injected into the combustion air. This approach eliminates throttling as required for SI engine power control. Because of the much localized fuel-rich nature of CI heat release, an engine operating at constant speed, and hence constant airflow rate, can meet a wide range of power demand by simply controlling the rate and amount of fuel injected. Overall rich or stoichiometric conditions are necessary for high power output. In this instance, the rich combustion and high carbon-to-hydrogen ratio of diesel fuels produce carbon atoms and generate black smoke, an objectionable pollutant. Idle engine operation means lower peak reaction temperatures, less air turbulence, incomplete combustion, and the formation of white smoke.

Several types of smoke can be identified with specific CI engine operating conditions. Cold-starting, idle, and low-load engine operation produce white smoke, liquid particulates consisting of unburned, partially burned, or cracked fuel with a small fraction of lubricating oil. Black-gray smoke occurs at maximum loads and rich combustion and consists chiefly of solid carbon particulates resulting from incomplete combustion. Carbon is opaque and will therefore yield black smoke, whereas liquid hydrocarbon droplets have high optical transmissions and, hence, make smoke appear grayer. The intensity of smoke generated by CI engines is influenced by fuel cetane number, injection rate, cutoff, and atomization. Particulate emissions may also include fuel additives and blue smoke, a common characteristic of lubricating oil aerosols. Smoke can be reduced by burning supplementary fuel additives, de-rating (reducing the maximum fuel flow), avoiding engine overload, afterburning the exhaust gases, and proper maintenance.

Variations in CI fuel composition, as specified by an ultimate analysis, will not affect the heat of combustion and heat release as much as it influences the formation of certain undesirable by-products. These by-products can severely impact the status and lifetime of diesel engine components. For example, carbon residue is a prime source of carbon deposits on diesel engine parts and is usually limited to less than 0.01%. Sulfur content in diesel fuels most frequently must be kept to 0.5–1.5% to limit the formation of corrosive combustion products. Ash and sediment, abrasive constituents of fuel mixtures, should be kept to 0.01–0.05% to protect injection systems and piston-cylinder wear. Ash tends to be concentrated when refining heavy residual fuel oil.

1.18.2   SALIENT FEATURES OF FUEL COMBUSTION IN CI ENGINE

As described above, essentially for a CI engine a low self-ignition temperature is preferred as there is no external source of ignition. The fuel and air enters separately. Moreover, the CI engine operates with lean mixtures. The air during compression attains pressure and temperature conditions that are sufficient enough to ignite the fuel that will be injected at the end of compression stroke. The CI engine is quality governed type where in the amount of air is unthrottled and the regulation of speed or load will be done by varying the amount of fuel injection. The way the air-fuel mix in a CI engine combustion chamber dictates the completeness of combustion process. The fuel will be undergoing delay both in physical and chemical forms. The fuel has got to be atomized properly as its viscosity is higher than gasoline fuel. The role of fuel injection system (fuel pump and injector) plays a vital role.

Before the fuel take part in combustion, the following events will take place:

- atomization;
- penetration;
- dispersion;
- evaporation;
- ignition; and
- combustion.

The fuel droplet once it overcomes delay, will take part in combustion suddenly as if the fuel and air have well mixed just similar to SI engine combustion and the phase is called pre-mixed mode and later, the fuel will mix with remaining part of air and undergoes diffusion or mixed controlled combustion.

Even after, if any leftover fuel is available then it burns mostly without contributing to power generation.

Therefore, the diesel fuel which most commonly employed in CI engine undergoes 4-stages as against to 3 phases of combustion in a SI engine.

The delay period plays a crucial role in such a way that fuel before it takes part in combustion, fuel will be injected continuously and as and when conducive environment prevails, the accumulated fuel burns instantaneously with fast rise in peak heat or peak pressures.

The trends of pressures and heat release rates upon combustion of a typical diesel fuel are depicted in Figures 1.27 and 1.28, respectively.

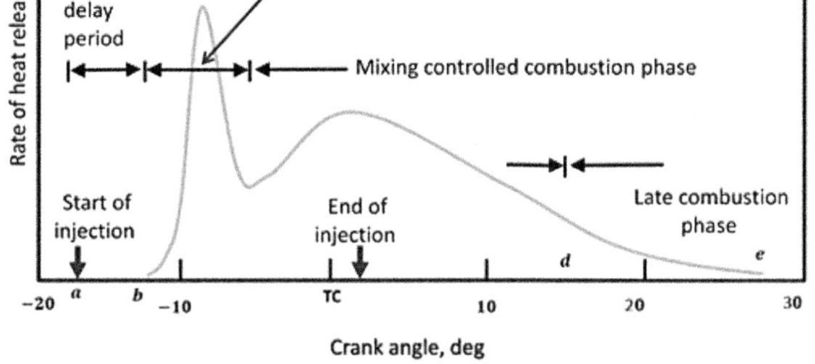

Four stages of combustion in CI engines

FIGURE 1.27   Heat release rate diagram: CI engine.

The ignition quality of diesel fuel is given in terms of cetane number for which the reference fuels are n-cetane and hepta methyl nonane or alpha methyl naphthalene.

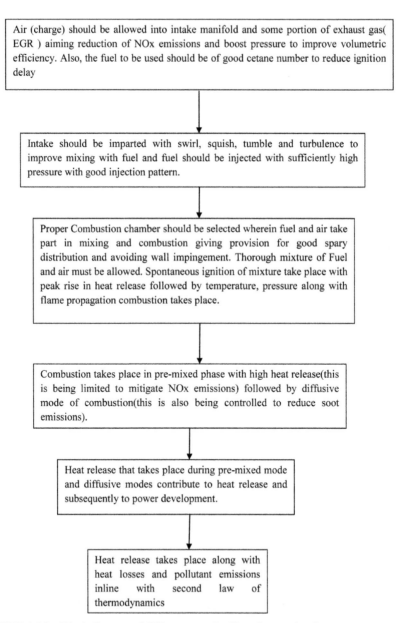

FIGURE 1.28   Block diagram of different steps in CI engine combustion.

KEYWORDS

- **compression ratios**
- **external combustion engines**
- **gasoline direct injection**
- **internal combustion engine**
- **spark-ignition**
- **tetraethyl lead**

REFERENCES

Ferguson, C. R., & Allan, T. T. (2001). Kirkpatrick, Second Edition, *Applied Thermo Sciences-Internal Combustion Engines,* John Wiley & Sons, Inc.

Heywood, J. B. (1988). *Internal Combustion Engines Fundamentals,* McGraw Hill.

Obert, E. F. (1973). *Internal Combustion Engines and Air Pollution,* Harper and Row.

Pulkrabek, W. W. (2004). *Engineering Fundamentals of the Internal Combustion Engine,* Pearson Prentice Hall.

Sean, B. (2010). *Modern Diesel Technology Diesel Engines,* Delmar, Cengage Learning.

Van Wylen Gordon, J., & Sonntag, R. E. (2008). *Fundamentals of Thermodynamics,* John Wiley & Sons, Inc.

CHAPTER 2

Formation Mechanism of Pollutant Emissions

2.1 INTRODUCTION

The air is a source of life on the planet. The air differentiates one planet to other planets. Without air, no life being exists on the earth. Such air is also responsible for Earth's gravity. At sea level, temperature, and pressure, respectively, are the basis for the existence of air in balance. These conditions of standard dry air mentioned in terms of as 15°C and 1.01325 bars. The composition of standard air, by mole or volume, such dry air contains 78.08% nitrogen, 20.95% oxygen, 0.93% argon, 0.0314% carbon dioxide, and trace amounts of other gases, by volume as depicted in Figure 2.1.

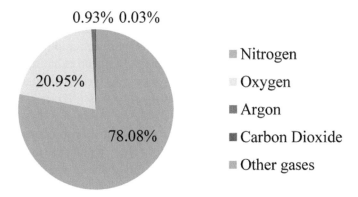

FIGURE 2.1   Typical composition of clean air under STP conditions.

The values given may vary a moderately little with location or time. Air, is composed of about 1% moisture and that's why it is not dry all the time. It is the moisture which grossly depends on region, season, and time and makes place with more humid or less humid regions. In some places, the air composed in the form of water vapor or liquid droplets or ice crystals.

Clean air when it is contaminated or mixed with undesirable species, it is treated as pollutant air. Contamination is not actually present in atmosphere but due to various reasons, the unwanted gases are formed and would mix with clean air. The pollutant air, in addition to the components mentioned above, thus contains the species of incomplete products of combustion releasing out of burning of hydrocarbons, and CO_2, even though some traces are CO_2 would be present in air but of late it is increasing in content due to indiscriminate use of hydrocarbon fuels. The species that are constituting pollutant air, or pollutant emissions (harmful emissions-more precisely), interferes with human health and cause negative environmental impacts. The greenhouse gases, which are responsible trapping hot gases, are Green House Gases (GHGs). Therefore, the air includes air pollutant emissions and greenhouse gas (GHG) emissions. The species in dry air when extremely high in concentrations or exceed permissible levels (higher-than-normal concentrations in the atmosphere) and become toxic and subsequently become sources of global warming as GHGs. The GHGs are resulting in possible catastrophic consequences on planet Earth. The GHGs are causing alarmingly ill effects on the environment and health of Earth planet.

Impact could be the consequences of pollutant input on the environment, organisms or goods. The complex processes of emission transport because pollutant input readings to fluctuate daily and seasonally. There is a great need to mitigate pollutant levels by adopting control methods identifying point source of pollutants of air. The atmospheric air is getting polluted not only by automobiles, power plants but also every source employing fuels for various purposes. Therefore, there are emission limits on exhaust pollutants, e.g., from vehicles or household heating systems, as well as air quality limits on carbon monoxide, sulfur dioxide, nitrogen oxides (NOx), particulates, lead ozone.

The exhaust pollutants produced during combustion processes can be divided into harmless exhaust gas components, which are unavoidable natural products of combustion, and harmful exhaust gas components, which may or may not be subject to limits. The emissions which are provided with limits are called regulated emissions and otherwise unregulated emissions.

Both natural and anthropogenic origins are responsible for formation of emissions and degrading the quality of air. These origins include emissions from fertilizers, pesticides, volcanoes, waste gases from industries, burning of fossil fuels, etc. These emissions are becoming hazardous and interfering with clean air are responsible for affect life of majority of people, living being sand plant life in a region.

Air pollutants comprise primary and secondary air pollutants. Pollutant emissions such as particulate matter (PM), sulfur dioxide (SO_2), nitric oxides (NO_x), hydrocarbon (HC), volatile organic compounds (VOCs), carbon monoxide (CO), ammonia (NH_3), etc. all constitute primary Air pollutants. Secondary air pollutants are produced by the chemical reactions of two or more primary pollutants or by reactions with normal atmospheric constituents. Ground-level ozone, formaldehyde, smog, and acid mist constitute secondary air pollutants. The reactions such as coagulation and condensation or photochemical reactions are the causes for formation of secondary air pollutants are formed through complex physical and/or chemical reactions, e.g., the pollutants such as CO, unburned or partially burned hydrocarbons, SOx; Nox, etc. are emitted in the form of gases and mostly are transparent. The atmospheric air is also being polluted heavily by particles in addition to the above listed gaseous pollutants. There is a concern mounting on the levels of particulates or particulate matter. The major sources of particulates are diesel engines, forest fires, wood-burning stoves, earthmoving machinery, industrial burning, etc. The particulate matter is a mixture of solid particles and liquid droplets suspended in the air. Examples of PM are smoke, fumes, and haze. Most gaseous air emissions such as NOx, SO_2, HC, VOCs, and ammonia (NH_3) are converted into PM.

The PM exists either in suspension, fluid or in solid-state, has a varied composition and size, and is commonly called aerosols. Particulate matter such as PM_{10}, $PM_{2.5}$, PM_1, and $PM_{0.1}$, defined as the fraction of particles with an aerodynamic diameter smaller than respectively 10, 2.5, 1, and 0.1 μm. The major concern about this matter is that it interferes with visibility effects and may clog nostrils of humans and is more of concern for small children, old age people, and animals.

Emissions are not only confined to exhaust sources but from evaporation. As the evaporative emissions would be from leakage past the container lids, spillages, etc. These evaporative emissions are volatile organic compounds (VOCs), contain mostly fuel borne chemicals that made of carbon and/or hydrogen and evaporate easily. Ground-level ozone is mainly caused by and is a common air pollutant, and a proven public health hazard.

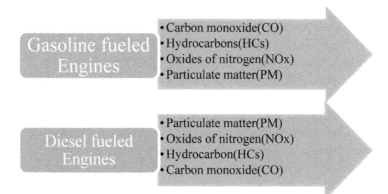

FIGURE 2.2   Constituents of pollutant emissions in fossil fuelled engines.

If each kg of hydrocarbon fuel (typically Octane or iso-Octane, a reference fuel for SI engines) when completely burnt produces majorly 3.1 kg of CO_2 and 1.3 kg of H_2O. Most of the undesirable exhaust emissions are produced in minute quantities (parts per million), and these are: oxides of nitrogen, generally termed NOx, unburnt hydrocarbons (HCs), carbon monoxide (CO), carbon dioxide (CO_2), lead salts, poly-aromatics, soot, aldehydes, ketones, and nitro olefins. Some of these are formed either due to excess amount of air, high-temperature combustion or deficient air and low-temperature combustion. Out of the mentioned pollutants only the first three are of major significance in the quantities produced. However, concentrations, in general, could become heavier as increasing numbers of vehicles come onto roads. Calculation exhaust emissions concentrations of a typical engine with probable 11 species forming out of combustion and calculated percentages of species by writing FORTRAN code.

By the end of the 1980s, CO_2 was also beginning to cause concern, not because it is toxic but because it was suspected of facilitating the penetration of our atmosphere by ultraviolet rays emitted by the sun.

Among different species of exhaust gases of both engines-CO, CO2, SOx, NOx, HCs are gaseous components. The diesel engine emits PM emissions, a carbonaceous product which is visible smoke(commonly known) and can be collected over a filter paper or in general soot/PM can be collected. Therefore, it should be dealt in a different manner. Sometimes not properly tuned SI engines and Gasoline Direct Injection engines are emitting PM emissions. Of late the PM10 and PM2.5 have become order of the day in the discussion where cities where large number of diesel fuel operated engines are plying.

2.1.1 UNREGULATED EMISSIONS

The emissions that are not currently covered by legislation are unregulated, i.e., all other compounds emitted from the exhaust except: carbon monoxide, total hydrocarbons (HC), oxides of nitrogen (NOx) and particulate matter (PM). However, the unregulated portion of the exhaust is still of interest for various reasons. For example, trials are on in few nations to collect data on CO_2 emissions from vehicles with a view to the possibility of these being legislated in the future. In addition, much attention has been given over recent years to the potential carcinogenicity/mutagenicity of diesel particulates.

Some emissions have become less of a concern. Formation of sulfur oxides across diesel oxidation catalysts (DOCs) is directly related to the sulfur content of the fuel and can account for a significant proportion of total particulate matter. Nations like the USA, Japan, Europe have also started reducing the sulfur content of fuels drastically and introduced diesel fuel called ultra-low sulfur diesel, from 0.05% wt. diesel fuel to < 0.005% wt. sulfur diesel fuel for certain urban applications. Adoption of Low sulfur fuels will also help to maintain catalyst performance over the lifetime of the vehicle without poisoning the catalytic converter.

Among the carcinogenic hydrocarbons of concern are, polycyclic aromatic compounds (PAH) are 3-, 4-, 5- or 6-ring compounds. Naphthalene and phenanthrene, the smaller PAHs are present in the vapor phase as well as the particulate phase. A potent carcinogen, a 5-ring compound of the particulate PAHs (i.e., those having 4 or more fused rings.) is Benzo(a) pyrene. The major source of PAHs is diesel engines and the carcinogenicity of diesel particulates is due to the presence of PAH. By the addition of an oxidation catalyst, the harmful levels of PAHs can be reduced. The degree of control depends on lighter and heavier compounds of PAHs-the lighter compounds be effectively controlled (~ 80%), the heavier compounds (~ 30%), or in some cases, very little conversion. Filtration can be done for PAH compounds and be collected on a filter paper. PAHs after extracting into a suitable solvent, the sample is then cleaned and analyzed with HPLC of with fluorescence detection is used.

PAH, compounds are attached to one or more of the rings of nitro group. Nitrated polycyclic aromatic hydrocarbons (NPAHs) are formed by the reaction of PAHs with nitrating agents in exhaust gases, such as nitric acid. Some NPAHs are known to be the potent carcinogens, and 1,8-dinitropyrene, which is commonly detected in diesel exhaust, is reported to be the most mutagenic compound known. The NPAHs are responsible for a significant proportion of the direct-acting Amines activity of diesel particulate.

In summary, the possible vehicle exhaust species of concern upon combustion of hydrocarbon fuels are Hydrocarbons included both Saturated and unsaturated hydrocarbons (includes C_2H_2, C_2H_4, C_3H_3, Aromatics: benzene, toluene, benzpyrene), Oxygenated hydrocarbons-Alcohols: CH_3OH, C_2H_5OH, Aldehydes of formaldehyde and acetaldehyde), Ketones, Ethers: CH_3-CH_2-O-CH_2-CH_3. Also, Particulate matter – Soot, soluble organic fraction (SOF), sulfides, and water vapor.

The sources of the compounds listed in the preceding subsection are as follows: Intake air compounds: N_2, O_2, CO_2, and H_2O.

And, unburned fuel- MTBE, ETBE additives, Products due to complete combustion-CO_2, H_2O and incomplete combustion, Fuel contaminants: Sulfides and Generated from after-treatment: N_2O, NH_3.

2.1.2   MITIGATION OF POLLUTANT EMISSIONS

The problem of engine exhaust emissions was started gaining attention after the observance of Photochemical Smog in early 1960s. This was recognized as major threat to vehicle traffic, health of life on the planet and eco-system that comprises environment. Also, subsequently with further exploitation fossil fuels and its heavy dependence to power automobiles and other prime movers, the problem of global warming has accelerated the measures to contain the issue of pollution caused by combustion of hydrocarbon fuels and other fuels.

Photochemical smog (also known as photosynthetic and photoelectric smog) is produced mainly by vehicle tailpipe gases. The photochemical smog can be seen as yellowish/light brown haze. The effects of photochemical smog are just as serious as those produced by sulfurous smog; include crop damage, eye irritation, and breathing problems in animals, plant leaves, and humans.

Photochemical smog is produced in two stages. First, hydrocarbons (HC) and NOx from vehicle tailpipes react with sunlight, produces ozone, then begins to react with hydrocarbons to produce smog. Formation of ozone at ground level is a major cause of concern as it is highly toxic in low concentrations, and the smog that results from exposure to light is a major problem in the US.

2.1.3   GLOBAL WARMING

Carbon dioxide (CO_2) is classified as a GHG. It is not by definition a harmful emission; after all, humans expel it from lungs every time they exhale. However, carbon dioxide emissions are associated with global warming and much legislation wants to regulate. This is a problem. Because both gasoline and diesel fuel are composed of approximately 85% carbon, when legislators talk about reducing (or taxing) CO_2 emissions.

Combustion is a chemical reaction. When diesel fuel is burned, it uses whatever oxygen is available in ground-level air. This oxygen is known as the reactant. When oxygen reacts with diesel fuel, it is oxidized. This oxidation reaction forms: H_2O (water), CO_2 (carbon dioxide). Neither the water nor the carbon dioxide is a harmful result of the combustion process. For this reason, if these two gases were the only products of burning fuel in an engine cylinder, the term complete combustion could be used to describe what has happened. Harmful emissions are what are produced when the combustion of an HC fuel is not "complete."

2.2 POLLUTANT EMISSIONS FROM COMBUSTION ENGINES

Though there are various ways by which the harmful air emissions are formed (air pollutants), the concern is majorly of pollutant emissions formed through the use of fuels from vehicles/prime movers powered internal combustion engines (ICEs), upon combustion. The engines that are powering the largely used automobiles are fuelled by hydrocarbons obtained by fossil resources-petroleum crude. Upon destructive distillation of crude, fuels are produced, are mainly, the hydrocarbon fuels, represented by the formula, CxHy fuel, consisting only of C and H atoms. The primary objective of using engine is met by allowing refined petroleum fuels (i.e., hydrocarbons) to mix with oxidizer (preferably atmospheric air).

The emissions could be: (i) evaporative emissions, (ii) exhaust or combustion generated emissions or tail pipe emissions, and (iii) blow-by gases, of these, as shown in Figure 2.1, the tail pipe or exhaust emissions is of concern as it is causing disturbance or imbalance in the composition of the clean air (Figure 2.3).

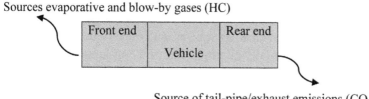

Sources evaporative and blow-by gases (HC)

Front end | Rear end
Vehicle

Source of tail-pipe/exhaust emissions (CO,HC,NOx)

FIGURE 2.3 Sources of emissions from a typical automobile engine.

The combustion or liberation of heat is done by allowing mixing of fuel and oxidizer inside the cylinder. Eventually chemical combination of these reactants takes place with exothermic type of reactions (helps in liberation of heat and consequent rise in temperature about to the level of adiabatic flame temperature (AFT). Depending on the load/speed conditions, different amounts of fuel/oxidizer are allowed to undergo combustion. If fuel undergoes complete combustion, all carbon will be converted to carbon dioxide (CO_2), whole hydrogen will be resulting in (H_2O) and nitrogen (N_2), as nitrogen being inert under conditions of stoichiometric combustion of fuel (Figure 2.4). However, in real combustion conditions, not all the time the fuel be combining with stoichiometric air (i.e., theoretical amount of air or chemically correct amount of air) and thus the products would be carbon monoxide (CO), unburned hydrocarbons (HC), hydrogen (H_2), NOx, and particulates also appear in addition to the components CO_2, H_2O and N_2 and few traces of OH, N, O, NO, NO_2, particulates, unburnt fuel, etc., and thus about eleven species are normally assumed to form upon combustion (Borman). Therefore, the exhaust gas composition would be complex concentrations of various species. The species of products would contain few radicals and the formation mechanism of all the exhaust species are different and pose high degree of difficulty in control mechanism.

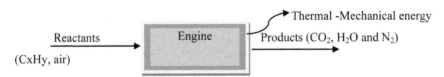

FIGURE 2.4   Simple combustion reaction under stoichiometric condition.

Since, majority of automotive fuels are derived from fossil hydrocarbons and in all the given conditions, the CO_2 and H_2O are the major products combustion and could not be eliminated its formation. Any process that would alter the totally these products lead to formation of incomplete products of oxidation and other products thereof. Since such species are being left to the atmosphere, it is mixing with the supposedly clean air and thus altering the concentration of clean air species with obnoxious products of tail pipe emissions. It cannot totally eliminate harmful emission products while burning fossil fuels but what is of major concern is their concentrations are increasing and exceeding the permissible limits prescribed by EPA or other controlling norms and thus pollutant emissions being the major cause of concern. Also, some fuel components are highly volatile and thus are leaving into the atmospheric air.

The automotive population has increased with population and is synonymous with energy/fuel consumption and so also with pollutant emissions that are harmful products and detrimental to human health, vegetation, and animals (all life existing on mother planet), CO_2, which is partially responsible for the greenhouse effect, is not viewed as a pollutant, since it does not pose a direct health hazard and appears as the final product of every complete oxidation of a hydrocarbon. A reduction of CO_2 in the exhaust gas is thus only to be achieved through a reduction in fuel consumption or through an altered fuel having a smaller amount of carbon with reference to its heating value.

As the engines could, be either gasoline or diesel-fuelled prime movers, the major species of concern are CO, HC, NOx, and PM. The major reason being insufficient amount of air (oxidizer) available for combustion and the concentration of species would vary with the amount of air available during combustion. Figure 2.5 represents variation of pollutant emission species with percentage of air of typical SI engine fuelled by gasoline.

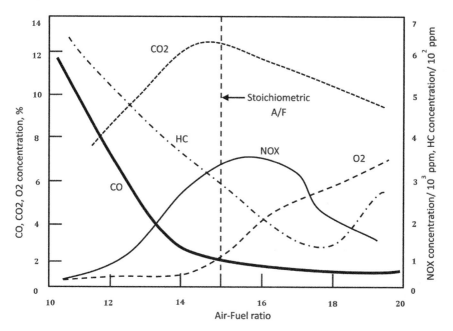

FIGURE 2.5   Variation of concentration of pollutant emissions, of a SI engine, with air-fuel ratio.

It can be seen that under fuel rich conditions (deficient air or oxygen) fuel cannot undergo complete oxidation and result in both HC and CO emissions and thereby increased levels of these emissions. Also, due to poor combustion and misfires, HC emissions increase at very lean mixtures. Thus lean mixtures with high in-cylinder temperature condition favor formation of lower HC and CO emissions. However, the NOx emission formation behaves in a different way. The generation of NOx emissions is a function of the combustion temperature, being greatest very near to stoichiometric conditions when temperatures are the highest. Peak NOx emissions occur at slightly lean conditions, where the combustion temperature is high and there is an excess of oxygen to react with the nitrogen.

> Thus, it can be said here that HC and CO emissions are forming due to insufficient oxidizer, insufficient time where as NOx emissions form due to presence of lean mixtures and high in-cylinder temperatures.

At this stage, a distinction is to be made between the concepts of complete and incomplete combustion as well as between perfect and imperfect combustion. Also, it should be understood that complete combustion and efficient combustion are two different terms although looks alike.

The combustion reaction drives toward completeness means products of hydrocarbon fuel combustion lead to only CO_2, H_2O, and N_2 whereas efficient reaction pertains to happening of complete combustion in short time interval.

Also, CO, and HC rise as products of incomplete combustion in a rich mixture (AF < 1.0) whereas NOx formation is most rapid at high temperatures with sufficient levels of oxygen (AF~ 1.1). With a lean mixture (AF > 1.2), the combustion temperature drops, so that NOx emissions fall off and HC emissions increase.

It is mandatory to understand clearly the mechanism of formation mechanism of incomplete products of combustion in order to evolve proper methods to mitigate the pollutant emissions. Thus, it is of more concern about these emissions that hamper the health of vegetation, humans, and animals.

Since air is essential for live beings especially humans and if it is imbalanced with pollutant emissions formed and emitted by automobiles and other hydrocarbon fuel consuming sources, interest is increasing more on how these pollutant emissions can be contained to provide clean air and reduce atmospheric pollution.

2.3 FORMATION MECHANISM OF TAIL-PIPE EMISSIONS OF SI ENGINE

2.3.1 *FORMATION OF CARBON MONOXIDE*

Almost all fossil fuel run prime movers use petroleum-derived fuels and even alternative fuels (AFs) that contain hydrocarbons, inherently undergo combustion process for release of energy. In such a case, the hydrogen will convert to H_2O and carbon to CO_2 upon complete combustion or oxidation when it meets stoichiometric amount of air. However, not all the time fuel (Hydrocarbons-HCs) reacts with stoichiometric air but do undergo oxidation either with deficient air or excess air depending on load/speed variations. The major by-products of incomplete combustion, or engine pollutants, associated with SI engine fuel chemistry are carbon monoxide, oxides of nitrogen and hydrocarbons, as shown in Figure 2.4.

In fact, the carbon monoxide has some heating value and oxidizes in high temperature conditions with further release of heat.

Therefore, the fuel upon combustion necessarily emits partially oxidized compounds such as carbon monoxide or hydrocarbons. Also, equilibrium thermo-chemistry is independent of time, but the process of combustion in an SI engine is, of necessity, time-bounded. For complete understanding of emission's formation, simple thermo-chemical calculations is not sufficient but rather requires a thorough knowledge of both fuel and engine characteristics.

The conditions that promote for formation of CO and HCs (unburned hydrocarbons) are almost similar. The fuel that does not undergo complete combustion may thus leads to incomplete combustion and is responsible for release of incomplete products of combustion. Incomplete combustion is a natural consequence of SI engine operation and results from the unique nature of each particular fuel-engine interface; i.e., the quality of fuel used, the specific geometry of an engine's combustion chamber, and the particular means of ignition and burning utilized to produce power.

The incomplete products of combustion under certain circumstances would be hazardous to humans and may lethal even and to life existing on the earth. Carbon monoxide is toxic because it is absorbed by the red corpuscles of the blood, stopping/inhibiting absorption of the oxygen necessary for sustaining life. Inhale of excess amounts of CO will reduce oxygen carrying capacity of blood.

The following are reasons for incomplete combustion:

1. Deficient amount of air (non-stoichiometric) or fuel rich mixtures;
2. Insufficient time for combustion-high speed operation;
3. Dissociation of CO_2 into CO and O_2 at elevated temperatures;
4. Dilution of fuel with burnt gases when charge is mixed with recycled gases.

Carbon monoxide, an intermediate product of combustion of any hydrocarbon fuel, is generally fuel-rich SI engine combustion product produced at both full-load and idle operation. The major source of CO is a result of chemical kinetics within the bulk gas; however, CO is also produced by partial oxidation of UHC during the exhaust stroke as well as dissociation of CO_2 produced during combustion. The rapid drop in gas temperature during expansion will freeze CO concentrations at levels different from those predicted on the basis of equal temperatures and equilibrium composition calculations. Control of CO is chiefly by improved Fuel-Air management and lean burn, such as with a stratified engine. CO could also be reduced with the use oxidation catalyst enabled catalytic converters, will be discussed in late chapters. Carbon monoxide formation increases steeply with decreasing air-fuel ratio (AF), as not enough oxygen is available to completely oxidize the mixture.

For conversion of chemical energy to thermal energy, one depends heavily on hydrocarbon fuels. All carbon under the conditions of complete combustion will be converted to carbon dioxide but due local prevailing conditions of conversion device, sometimes partial oxidation happens resulting in more popular carbon compound-carbon monoxide-CO. Thus, CO arises as an intermediate product of oxidation. Under stoichiometric ($\phi = 1.0$) and hyper-stoichiometric ($\phi > 1.0$) conditions, CO can theoretically be completely oxidized to CO_2. In case of local air deficiency ($\phi < 1.0$), CO generally remains intact as a product of incomplete combustion.

The main reactions in the oxidation of CO are the reactions with a hydroxyl radical and a hydro-peroxyl radical.

$$CO + OH \rightarrow CO_2 + H \tag{1}$$

$$CO + HO_2 \rightarrow CO_2 + OH \tag{2}$$

Since reaction (1) is dominant under most conditions, CO oxidation is heavily dependent on the concentration of OH radicals. The reaction rate of reaction (1) is much slower than that of the reaction between OH radicals and hydrocarbons. For this reason, CO oxidation is usually inhibited until the fuel molecules and hydrocarbon intermediate species are oxidized. During rich combustion ($\phi < 1.0$), CO oxidation progresses, due to a lack of O_2, in competition with H_2 oxidation.

$$H_2 + OH \rightarrow H_2O + H \tag{3}$$

As opposed to the kinetically controlled reaction (1), reaction (3) is practically in equilibrium at higher temperatures. With an increasing air ratio and increasing temperature, the deviation of the kinetics from the HC equilibrium is reduced, and as a result, the CO concentration is reduced with increasing air ratio l.

With stoichiometric combustion ($\phi = 1.0$), reactions (1) and (2) can be described with very good approximation as a gross reaction via the water-gas reaction which in this case proceeds near equilibrium, because the surplus concentrations of the chain propagators H and OH are very large.

$$CO + H_2O \rightarrow CO_2 + H_2 \tag{4}$$

In the super-stoichiometric range ($\phi > 1.0$), CO oxidation no longer progresses in competition with H_2 oxidation and is dominated again by reaction (2.1). During lean combustion ($\phi > 1.4$), increased CO develops again due to the lower temperatures and incomplete combustion in the area near the wall of the combustion chamber. Generally, CO oxidation is highly contingent on temperature, so that reaction (2.1) becomes increasingly slow during expansion as well. The CO concentration in the exhaust gas thus corresponds approximately to the equilibrium concentration at 1700 K.

In the combustion of hydrocarbons, CO arises as an intermediate product of oxidation. Under stoichiometric ($\phi < 1.0$) and hyperstoichiometric ($\phi > 1.0$) conditions, CO can theoretically be completely oxidized to CO_2. In the case of local air deficiency ($\phi < 1.0$), CO generally remains intact as a product of incomplete combustion.

The decisive reactions in the oxidation of CO are the reactions with a hydroxyl radical and a hydro-peroxyl radical.

Thus, it can be summarized here that the following are major reasons for formation of CO:

1. Fuel bound CO-fuel consisting of higher carbon.
2. Non-stoichiometric amount of oxygen during combustion reactions-presence of rich mixtures and too lean mixture will reduce combustion temperatures as fuel cannot completely oxidize.

3. Too cool combustion chamber walls-over cooling of cylinder.
4. Very high speed operation of engine-In sufficient time for mixing of carbon present in fuel with air.
5. Existence too high temperatures and pressures during the combustion process, significant quantities of CO form even when there is sufficient oxygen for complete combustion to occur. This is due to dissociation of the CO_2 molecules into CO and O_2 because of the high temperature molecular vibration.

2.3.2   FORMATION OF HYDROCARBONS

The hydrocarbon-based fuels are prone for formation of unburned hydrocarbons due to lower amounts of oxidizer availability and lower in-cylinder temperatures, with consequent formation of multiple species of hydrocarbons. Also, release of hydrocarbons also takes place due to high volatility nature of few hydrocarbon fuels or volatile organic compounds.

Thus, hydrocarbon emissions occur not only in the vehicle exhaust, but also in the engine crankcase, the fuel system, and from atmospheric venting of vapors during fuel distribution, dispensing, and spillage.

Therefore, following are species of HCs normally form:

1. Volatile organic compounds (VOCs).
2. Unburnt or partially burnt HCs called combustion generated pollutants.

The hydrocarbon fuels are comprised of major species of about 10–20 and to the tune of 100–200 minor species. The mixture of air-fuel is homogeneous, in conventional SI engines, there exists few unburnt/ partially burnt HCs formed in the event of improper combustion and are of concern as many unburnt of partially burned HCs are pollutants emissions. In conventional SI engines, the most important sources of HC emissions are; Flame extinguishing within crevices as a result of flame quench, adsorption, and desorption of the fuel in the oil film on the cylinder liner. Adsorption and desorption of the fuel in deposits on the combustion chamber walls with lubricating oil or cool combustion chamber walls, could be major sources of HC emissions. As discussed above, in addition to carbon monoxide, similar conditions also prompt for formation of partially burnt hydrocarbons or unburned hydrocarbons or unburned fuel. However, there are specific reasons for formation of hydrocarbons upon combustion.

Hydrocarbons are grouped into two types-total hydrocarbons (THC) and non-methane hydrocarbons (NMHC). Methane does not react with atmosphere even though it is one of the GHGs (Ferguson). In SI engine, about 9% fuel supplied is left unburned during normal combustion process. Turns (2000) and Kalyand Puri (2007) analyzed the possible pathways by which species HC emissions are left into exhaust-crevices, oil layers, deposits, liquid fuel and flame quench, and flow past exhaust valves, as represented (Figure 2.6).

Hydrocarbons have harmful effects on environment and human health. With other pollutant emissions, they play a significant role in the formation of ground-level ozone. Automobile Vehicles are responsible for about 50%of the emissions that form ground-level ozone. Hydrocarbons are toxic with the potential to respiratory tract irritation and cause cancer.

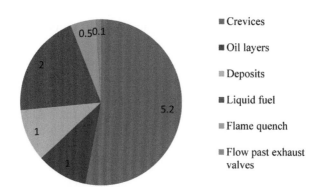

FIGURE 2.6   Sources of SI engine HC emissions (Ferguson).

The lighter hydrocarbons (C_1–C_{12}) which are present in the gaseous phase are important in gasoline exhaust but tend to be present in insignificant quantities in diesel exhaust. However, the heavier hydrocarbons (C_{14} and above)), which mainly exist in the vapor phase, are more commonly found in diesel exhaust. Typically occurring chemical types are: some PAHs, substituted naphthalenes, heavier aliphatics, alkylbenzenes, and some sulfur- or updated oxygen-containing compounds such as naphthaldehydes. Vapor-phase hydrocarbons are sampled via tubes containing an adsorbent material. They can then be analyzed via gas chromatography (GC)/mass spectrometry (GC/MS).

The following are the causes of HC emissions:

1. Incomplete Combustion: The conditions known to researchers and well established for incomplete combustions are (a) insufficient oxidizer, (b) insufficient time, (c) flame quenching, and (d) presence of residual gases (diluents). Whatever may be the conducive conditions are provided for combustion to take there always some unburnt or partially burnt HC compounds will be formed in the exhaust. Mostly under rich mixture conditions, the fuel by not finding sufficient oxygen to react with, incomplete mixing of the air and fuel results in some fuel particles left unburnt. These will lead to formation of unburnt and partially burnt hydrocarbons. Flame quenching at the walls leaves a small volume of unreacted air-and-fuel mixture. The thickness of this unburned layer is of the order of tenths of an mm.

2. Flame Quenching: It is another cause for high exhaust HC levels, however, is a relatively minor mechanism along the surfaces. The combustible mixture, near the wall that does not originally get burned as the flame front passes, will burn later in the combustion process as additional mixing occur due to swirl and turbulence. Another cause of flame quenching is the expansion which occurs during combustion and power stroke. As the piston moves away from TDC, expansion of the gases lowers both temperature and pressure within the cylinder. This slows combustion and finally quenches the flame somewhere late in the power stroke. This leaves some fuel particles unreacted. The hydrocarbons during expansion are later oxidized when they diffuse into the burned gases and do not contribute significantly to the tail pipe hydrocarbon emissions.

During idling or low conditions due to poor combustion, high exhaust residual gases release with likelihood of quenching in expansion. In order to mix recycled gases (high levels of EGR) for reducing NOx emissions may eventually result in HC emissions. Efforts are also done to employ a second spark plug is added to an engine combustion chamber, to reduce HC emissions. By starting combustion at two points, the flame travel distance and total reaction time are both reduced and less expansion quenching results.

3. Air-Fuel Ratio (AF): Influence of AF is strong function on the formation of HC emission and its levels as shown in Figure 2.5. With a fuel-rich mixture, there is not enough oxygen to react with all the carbon, resulting in high levels of HC and CO in the exhaust products. This is particularly true in engine startup, when the air-fuel mixture is purposely made very rich to overcome residuals and heat losses. It is also true to a lesser extent during rapid acceleration under load. If AF is too lean, it results in poor combustion, again promoting HC emissions. Under too lean condition misfire takes place. This occurs more often as AF is made leaner. Approximately, one misfire out of 1000 cycles gives exhausts HC emissions of 1 gm/kg of fuel used.

4. Crevice Regions: Very narrow regions in the cylinder and piston rings are called crevices, the ring gap being a large percent of crevice volume. During the compression stroke and early part of the combustion process, air, and fuel are compressed and trapped into the crevice volume of the combustion chamber at high pressure. As much as 3–5% of the fuel in the combustion chamber could be forced into the crevice regions. Reverse blow-by occurs later in the cycle during the expansion stroke when the pressure in the cylinder is reduced below crevice volume pressure. Fuel and air flow back into the combustion chamber, where most of the mixture is consumed in the flame regions. However, by the time the last elements of reverse blow-by flow occur, flame reaction has been quenched and unreacted fuel particles remain in the exhaust. Also, location of the spark plug relative to the top compression ring gap affect the amount of HC in engine exhaust.

HC in the exhaust increases with increase in distance of spark plug from the ring gap since more fuel will be forced into the gap before the flame front reaches. Crevice volume around the piston rings is greatest when the engine is cold, due to the differences in thermal expansion of the various materials.

5. Flow Through Exhaust Valve: As pressure increases during compression and combustion, some air-fuel is forced into the crevice volume around the edges of the exhaust valve and between the valve and valve seat. A small amount of mixture leaks past the valve into the exhaust manifold. The flow partially burnt gases and some of products of combustions get recessed through exhaust valve due to pressure differential created between high pressure of exiting gasses and atmosphere. When the exhaust valve opens, the air-fuel mixture present in crevice volume gets carried into the exhaust manifold, and there is a momentary peak in HC concentration at the start of blow-down.

6. Valve Overlap: During this period, both the exhaust and intake valves are open. The valve overlap is helpful in scavenging operation, however too high valve overlap, short-circuits the incoming charge into exhaust along with expelling gases. A well-designed engine minimizes this flow, but a small amount can occur. Too high HC emissions are observed during idling period with valve overlap, which is a worst condition.

7. Combustion Chamber Walls: A very thin layer of lubricating oil is formed on the cylinder walls of an engine to provide lubrication between them and the moving piston. During the intake and compression strokes, the incoming air and fuel comes in contact with this oil film. In much the same way as wall deposits, this oil film absorbs and desorbs gas particles, depending on gas pressure. During compression and combustion, when cylinder pressure is high, gas particles, including fuel vapor, are absorbed into the oil film. When pressure is later reduced during expansion and blow down, the absorption capability of the oil is reduced and fuel particles are desorbed back into the cylinder. Some of this fuel ends up in the exhaust (Pulkrabek).

Gas particles, including those of fuel vapor, are absorbed by the deposits on the walls of the combustion chamber. The amount of absorption is a function of gas pressure, so the maximum occurs during compression and combustion. Later in the cycle, when the exhaust valve opens with reduced cylinder pressure, absorption capacity of the deposit material is lowered and gas particles are desorbed back into the cylinder.

These particles, including some HC, are then expelled from the cylinder during the exhaust stroke. This problem is greater in engines with higher compression ratios (CR) due to the higher pressures. Clean combustion chamber walls with minimum deposits will reduce HC emissions in the exhaust. Most gasoline blends include additives to reduce deposit buildup in engines. Older engines will typically have a greater amount of wall deposit build-up and a corresponding increase of HC emissions. This is due both to age and to fewer swirls that was generally found in earlier engine design. High swirl helps to keep wall deposits to a minimum.

Often, as an engine ages, clearance between the pistons and cylinder walls increases due to wear. This increases oil consumption and contributes to an increase in HC emissions in three ways: increased crevice volume, increased absorption-desorption of fuel in the thicker oil film on cylinder walls, and there is more oil burned in the combustion process.

Oil consumption also increases as the piston rings and cylinder walls wear. In older engines, oil being burned in the combustion chamber is a major source of HC emissions. HC emissions will increase as oil consumption increases. Further, as oil consumption going up with piston rings wear, blow-by, and reverse blow-by also increase. The increase in HC emissions is therefore both from combustion of oil and from the added crevice volume flow.

2.3.3 FORMATION OF OXIDES OF NITROGEN

Generally, hydrocarbon fuels are used for conversion of chemical energy to mechanical energy in a prime mover (e.g., automobile vehicle). The reaction of these fuels with atmospheric air is aimed to undergo exothermic reactions (Figure 2.2) with release of large amount heat and consequent rise in high temperatures. As discussed in Chapter 1, the larger proportion of air contains diatomic nitrogen. The atmospheric nitrogen exists as a stable diatomic molecule at relatively low temperatures (when compared to high in-cylinder temperatures), and only very small trace amounts of oxides of nitrogen are found. However, at the very high temperatures that occur in the combustion chamber of an engine, diatomic nitrogen (N_2) breaks down/ dissociates to monatomic nitrogen (N) which is highly reactive:

$$N_2 \sim 2\,N \tag{5}$$

Due to exothermic reactions that form during combustion, high temperatures are realized. Under the conditions of high in-cylinder temperatures, the molecular nitrogen dissociates and combines with atomic oxygen to form complex compounds of oxides of nitrogen cyclically. The NOx (NOx) represent the seven oxides of nitrogen such as nitric oxide (NO), nitrogen dioxide (NO_2), nitrous oxide (N_2O), dinitrogen dioxide (N_2O_2), dinitrogen trioxide (N_2O_3), dinitrogen tetroxide (N_2O_4), and dinitrogen pentoxide (N_2O_5). The major contributors to air pollution due to oxides of nitrogen for the purpose of emission calculations, the NOx which are of concern for photochemical smog are often referred only to NO and NO_2 by the norms of Environmental Protection Agency (EPA). The formation of NOx is highly dependent on temperature, with a much more significant amount of N generated in the 2500–3000 K temperature range that can exist in an engine. Other gases that are stable at low temperatures but become reactive and contribute to the formation of NOx at high temperatures include oxygen and water vapor, which break down as follows:

$$O_2 \sim 2\,O \tag{6}$$

$$H_2O \sim OH + H \tag{7}$$

The higher the combustion reaction temperature, the more diatomic nitrogen, N_2, will dissociate to monatomic nitrogen, N, and the more NOx will be formed. Thus, due to the reaction of atomic oxygen and nitrogen, the combination forms oxides of nitrogen, formed during the entire combustion process with conducive conditions of high temperatures and lean mixtures. At low temperatures very little NOx is created. Although, maximum flame temperature will occur at a stoichiometric AF.

The following are major causes of formation of oxides of nitrogen (NOx):

1. High in-cylinder temperatures during combustion.
2. Presence of large amount of air during combustion.

The toxicity of hydrocarbons and oxides of nitrogen, on the other hand, arises indirectly as a result of photochemical reactions between the two in sunlight, leading to the production of other chemicals. Unburnt hydrocarbons can come from evaporation from the carburetor float chamber and fuel tank vent as well as from inefficient combustion due in different instances to faulty ignition, inadequate turbulence, poor carburation, an over-rich mixture, or long flame paths from the point of ignition. The relationships between emissions and the AF are illustrated in Figure 2.1. Other factors are overcooling, large quench areas in the combustion chamber, and the unavoidable presence of a quench layer of gas a few hundredths of a millimeter thick clinging to the walls of the combustion chamber, and quenching in crevices such as the clearance between the top land of the piston and the cylinder bore.

> *The deficient amount of oxygen and low temperatures are responsible for formation mechanism of HC, CO emissions. However, oxides of nitrogen form mainly due to prevalence of high temperature and excess oxygen (favorable conditions) which would help in dissociation of molecular nitrogen of atmospheric air and there by formation of cyclic compounds oxides of nitrogen (NOx).*
> *Therefore, it can be inferred that causes of formation mechanism of HC, CO, and (NOx) respectively are opposite to each other.*

Referring to the Figure 2.5, at the equivalence ratio (∅~0.95), the flame temperature is still very high, and in addition, there is an excess of oxygen that can combine with the nitrogen to form various oxides, which favorable conditions. In addition to temperature, the formation of NOx depends on pressure, AF, and combustion time within the cylinder, chemical reactions not being instantaneous. In SI engines, the highest concentration of NOx is formed around the spark plug, where the highest temperatures occur. Because they generally have higher CRs and higher temperatures and pressure, CI engines with divided combustion chambers and IDI tend to generate higher levels of NOx. Under advanced spark timing conditions, even higher concentrations of NOx are formed.

Photochemical Smog is serious problem in many large cities when automotive population density is high, NOx is one of the primary causes of photochemical smog. Smog is formed by the photochemical reaction of automobile exhaust and atmospheric air in the presence of sunlight. NO_2 decomposes into NO and monatomic oxygen:

$$NO_2 + hv \text{ (photon) light} \sim NO + O + smog$$

Zeldovich extensively studied the chemical kinetic mechanism behind formation of cyclic compounds of nitrogen and oxygen and proposed the popular Zeldovich mechanism, with temperature dependence.

In this case, so-called thermal NO, which is formed among the combustion products at high temperatures according to the mechanism from atmospheric nitrogen, so-called prompt NO, which develops in the flame front via the Fenimore mechanism from atmospheric nitrogen, NO formed from the N_2O mechanism, and finally so-called fuel NO, which is produced by nitrogen portions in the fuel. The importance of the various formation mechanisms is highly dependent under engine conditions on the operational conditions. At low temperatures, reached for example in exhaust gas recycling, the prompt NO path increases in importance under fuel-rich conditions, while the N_2O path becomes more important under nitrogen-rich conditions.

NOx is a very undesirable emission, and regulations that restrict the allowable amount continue to become more stringent. NOx when released out reacts in the atmosphere to form ozone and is one of the major causes of photochemical smog. NOx is created mostly from nitrogen in the air. Nitrogen can also be found in fuel blends, which may contain trace amounts of NH_3, NC, and HCN, but this would contribute only to a minor degree. There are a number of possible reactions that form NO, all of which are probably occurring during the combustion process and immediately after. These include but are not limited to:

1. Thermal NO: Its formation occurs "behind" the flame front in the hot burned gases. It was first described by Heywood (1988). The two-step mechanism formulated by Zeldovich was later extended by Heywood (1988). This extended Zeldovich mechanism consists of the three elementary reactions;

$$O + N_2 \rightarrow \{k_1\} \text{ NO} + N$$
$$N + O_2 \rightarrow \{k_2\} \text{ NO} + O$$
$$N + OH \rightarrow \{k_3\} \text{ NO} + H$$

It can be seen that dissociation of oxygen (O_2) is quicker and relatively low temperatures the first reaction and thus triggered by atomic oxygen in splitting molecular nitrogen and forming NO. The dissociations are endothermic absorbing heat released due to combustion.

NO_2 is formed out of NO with the following two reactions:

$$NO + H_2O \rightarrow NO_2 \rightarrow H_2$$
$$NO + O_2 \rightarrow NO_2 + O$$

The k_s are called specific reaction rate coefficients and whose temperature dependent experimental determination forms an important aspect of Zeldovich mechanism in forward and backward reactions.

If the instantaneous NO concentration is below the equilibrium concentration at the corresponding temperature, as is the case throughout much of the combustion process, the forward reaction has a decisive influence on the total conversion. Only when the momentary NO is above the equilibrium concentration of the corresponding temperature is the total conversion substantially determined by the reverse reaction. In the engine, however, this situation appears at best towards the end of the expansion stroke, when temperature is already quite low.

The first reaction possesses a high activation energy because of the stable N_2 triple bond and thus proceeds sufficiently fast only at high temperatures; hence the designation "thermal." It is therefore the rate-limiting step. The above numerical values show that at 1800 K, the first reaction progresses at around seven to eight decimal powers slower than the second and third.

In a typical combustion process, the residence time is shorter than the characteristic time. As a result, NO does not reach its equilibrium concentration. Therefore, it is better for the actual NO concentration in the flame to be determined using this equation.

2. Prompt NO: The formation of prompt NO in the flame front itself is much more complicated than thermal NO formation, because this process is closely related to the formation of the CH radical, which can react in many ways. "Prompt NO" formation was first described by Turns (2000). According to Fenimore, the decisive reaction is that of CH with N_2 to form HCN (hydrocyanic acid), which rapidly reacts further to form NO:

$$CH + N_2 \rightarrow k_{ff} HCN + N$$

Acetylene (C_2H_2), as a precursor of the CH radical, is only formed under fuel-rich conditions in the flame front, hence the concept "prompt NO." Because of the relatively low activation energy of the reaction, prompt NO formation proceeds already at temperatures of about 1,000 K. Subsequently, HCN reacts using various paths to form NCO and NH:

3. NO Formed via N$_2$O: This reaction mechanism is only significant when lean air-fuel mixtures repress the formation of CH, and thus little prompt NO is formed, and furthermore if low temperatures stifle the formation of thermal NO. N$_2$O (nitrous oxide or laughing gas) is formed analogously to the first, rate limiting reaction of the Zeldovich mechanism:

$$N_2 + O + M \rightarrow N_2O + M \qquad (8)$$

The reaction takes place however with a collision partner M, which is not changed by the reaction and considerably lowers the activation energy in comparison with reaction (14). NO formation then results trough oxidation of N$_2$O.

$$N_2O + O \rightarrow NO + NO \qquad (9)$$

Because N$_2$O is only formed in a three-way collision reaction, this reaction path progresses preferably at high pressures. Low temperatures hardly slow down this reaction. NO formed via N$_2$O is the essential NO source in lean premixed combustion in gas turbines. This mechanism is however also observable in SI engine lean combustion. It is therefore important in the case of lean combustion in SI engines as well as modern diesel-engine combustion processes with high point pressures. It is presumably also the essential NO formation mechanism in lean HCCI combustion.

4. Fuel Nitrogen: The conversion of nitrogen bound in the fuel to NOx does not play a role in engine combustion, because fuels for ICEs contain negligible amounts of nitrogen. It can, however, play a role in the case of certain low-quality heavy oils. The formation of NO via fuel nitrogen is also of importance in the combustion of coal, since even "clean" coal contains about 1% of nitrogen.

The reaction progresses via HCN and ammonia (Miller and Bowman, 1989). The further conversion of HCN takes place in accordance with the mechanisms indicated in Eqs. (25)–(27).

Most fuels contain nitrogen element, the NO originated from this part of nitrogen is referred to as fuel NO. The amount of nitrogen in fuel varies with the fuel type. As summarized in Depending on the refinery process, some natural gases contain virtually no nitrogen but others have quite a lot of nitrogen in form of N$_2$. Unlike fossil fuels, biomass is characterized with high nitrogen. Although the amount of nitrogen in fuel is relatively small, the fuel nitrogen is much more reactive compared to the nitrogen present in the combustion air.

Consequently, the formation of fuel NO from a nitrogen-rich fuel is higher than that from a nitrogen-lean fuel. Sometimes, as much as 80% of the NO in the flue gas of a coal-firing furnace is produced from fuel nitrogen. The fuel NO is sensitive to stoichiometry rather than the temperature because it forms readily at quite low temperatures. For a simple comparison purpose, among the relative importance of thermal NO, prompt NO and fuel NO formation at different temperatures, overall NO emissions increase with temperature, mainly due to the increase of thermal NO.

2.3.4   PARTICULATE MATTER

The unburned lubricating oil and tetraethyl lead (TEL) in the exhaust, from ash-forming fuel and oil additives are the important sources particulate matter emissions from four-stroke spark-ignition engines. The levels of PM emissions are low when compared to an equivalent diesel engine; the PM emissions are of concern. They can be formed mostly due to poor maintenance or engine wear lead to high oil consumption, or when oil is mixed with the fuel, as in two-stroke engines. Nowadays, gasoline direct injection engines (GDI) have become significant source of PM emissions. The PM emissions from diesel engines are regulated but these emissions from spark-ignition engines are not yet regulated.

2.3.5   FORMATION OF OTHER EMISSIONS

Other than emission species described above, prominent pollutant being aldehydes – either formaldehyde or acetaldehyde. These emissions are generally being high when alcohols are added either to improve octane rating or to fuels as blends or directly as neat fuels (in Brazil-as majority of vehicles run on alcohols in fuel flexible vehicles).

Aldehydes are part of unregulated emissions, are oxygenates with the general formula R-CHO where R is an organic side chain. Aldehydes are formed when combustion is incomplete; their formation is favored when the environment is oxygen-rich. These compounds can be irritants or odorous; they may also be toxic and can contribute towards photochemical smog. Formaldehyde, for example, is both toxic and a highly active ozone-precursor.

Aldehydes levels can be significant in diesel exhaust, and may become of more concern in the future as oxygenate additives are increasingly used in diesel fuels. Formaldehyde and acetaldehyde are typically the dominant compounds and can be present at levels equivalent to or in excess of those seen from gasoline exhaust. The characteristic odor which has been associated with diesel engines in the past can be partly attributed to the presence of aldehydes. However, these compounds can be effectively controlled by oxidation catalysts, where typical average conversion levels would be around 70%. Aldehydes are usually analyzed as their hydrazones via high-performance liquid chromatography (HPLC). The method is highly sensitive; compounds are measured by ultraviolet (UV) detection at optimum wavelength. Since sulfur is one of the constituent of petroleum fuels and would cause corrosion effects for components and forms oxides of sulfur and sulfurous and sulfuric acid when combined with water vapor.

Many countries have laws restricting the amount of sulfur allowed in fuel, and these are continuously being made more stringent. During the 1990s, the United States reduced acceptable levels in diesel fuel from 0.05% by weight to 0.01%. The amount of sulfur in natural gas can range from little (sweet) to large (sour) amounts. This can be a major mission's problem when this fuel is used in an IC engine or any other combustion system.

2.3.5.1 EVAPORATIVE AND REFUELING EMISSIONS

Gasoline being a volatile liquid fuel, a significant amount of unburned hydrocarbons is formed as evaporative emissions from Gasoline-fueled vehicles of its fuel system. Over a period of extensive research, four main sources of evaporative emissions have been identified: breathing (diurnal) losses from fuel tanks caused by the expansion and contraction of gas in the tank with changes in air temperature; hot-soak emissions from the fuel system when a warm engine is turned off; running losses from the fuel system during vehicle operation; and resting losses from permeation of plastic and rubber materials in the fuel system. By venting the fuel tank and the carburetor to a canister containing activated carbon, evaporative emissions can be minimized. When the engine is not running, volatile emissions from the fuel system are absorbed by activated carbon material. When the engine is running, intake air is drawn through the canister, purging it of hydrocarbons, which then form part of the fuel mixture fed to the engine. Refueling vapor emissions can also be controlled using a larger canister by capturing them through the refueling nozzle and conducting them back to the service station tank.

2.3.5.2 TOXIC POLLUTANTS

In earlier days, TEL, used as an octane enhancer in leaded gasoline. This has become an out dated technique as it started emitting unburnt lead as such from engines. Prior to incorporation of catalytic converters, lead compounds used to toxic components, However, in modern engines, benzene, 1,3-butadiene, and aldehydes have become toxic compounds. Aldehydes are intermediate products of hydrocarbon combustion. Aldehydes are highly reactive, form as intermediate products during combustion. Aldehydes may form due to quenching of hydrocarbon fuels and also due to partially reacted mixture specifically in contact with cool wall surfaces. Aldehydes primarily be both formaldehyde and acetaldehyde are irritants and suspected carcinogens. They form in large quantities when ethanol and methanol as fuels undergo the primary oxidation reactions and hence these compounds are often found in significant concentrations in the exhaust.

2.3.5.3 CRANKCASE EMISSIONS

Blow-by of burnt or compressed gas past the piston rings are the major sources of unburned hydrocarbons from engines and majorly all reciprocating engines experience leakage consists of unburned or partly burned air-fuel mixture and contains unburned hydrocarbons., Generally to remove blow-by gases, a vent is provided from the engine to the atmosphere, by means of a positive crankcase ventilation (PCV) valve. As far as the case of older vehicles is concerned, hydrocarbon emission levels from the crankcase were about half the level of uncontrolled exhaust emissions. In the PCV system, unburned hydrocarbons in the blow-by gas are recycled to the engine and burned. The PCV also help to prevent condensation of fuel and water in the blow-by gases, thus reducing oil contamination and increasing engine life. Recycling of burnt gases cause fouling of intake manifold and can be eliminated using a suitable and effective gasoline detergent.

2.3.5.4 REFUELING EMISSIONS

Refueling emissions occur when the gasoline liquid fuel is filled at the gasoline pumping stations. Evaporative and refueling emissions are strongly affected by fuel volatility. Diurnal emissions also vary depending on the daily temperature range and the amount of vapor space in the fuel tank.

Due to the conduction of heat from the warm engine to the carburetor as it is exposed to the atmosphere, hot-soak emissions results. The hot-soak losses are greatly reduced in sealed fuel systems, such as those used with fuel injection systems. Such systems may have higher running losses, however, due to the recirculation of hot fuel from the engine back to the fuel tank.

2.4   COMPRESSION-IGNITION (CI) ENGINE EMISSIONS

2.4.1   FORMATION OF DIESEL ENGINE EMISSIONS

Diesel fuel is much more suitable for compression-ignition engines (CI engines) due to inherent fuel properties and it undergoes mostly the hetero-geneous mode of combustion without the aid of any external source of igni-tion. Due to these differences, the diesel emissions are different from those of spark ignited engines. Charge for the diesel engines is only air. The CI engines work with overall lean equivalence ratios (~20:1 at part-load to 80:1 at full load, 20:1 is leaner than stoichiometric AF). During the compression stroke, a diesel engine compresses only air. The process of compression heats the air to about 700–900°C (depending on CR), which is well above the self-ignition temperature of diesel fuel. When the piston is nearing the top of the compression stroke, liquid fuel is injected into the combustion chamber under high pressure and high velocity through a number of small orifices in the tip of the injection nozzle, the periphery of each jet mixes with the hot and dense air already available. After a brief period known as the ignition delay, this fuel-air mixture ignites.

Diesel emissions include also pollutants originate from various non-ideal processes during combustion, such as incomplete combustion of fuel, reac-tions between mixture components under high temperature and pressure burning of engine lubricating oil and oil additives as well as combustion of non-hydrocarbon components of diesel fuel, such as sulfur compounds and fuel additives. The pollutant components of diesel engine include unburned hydrocarbons (HC), carbon monoxide (CO), NOx, or particulate matter (PM). Total concentration of pollutants in diesel exhaust gases typically amounts to some tenths of 1%.

Generally, the diesel fuel, in a typical CI engine, undergoes four phases: ignition delay, premixed mode, mixing-controlled or diffusion combustion and late burning. The essence of pre-mixed mode of combustion of diesel fuel is that a fuel burns very rapidly and gives rise very high rate of pressure and temperature rise, as it overcomes the ignition delay period. The subsequent rate of burning is controlled by the rate at which the remaining fuel and air can mix. Combustion always occurs at the interface between the air and the fuel. Most of the fuel injected is burned in this mixing-controlled combustion phase, except under very light loads, when the premixed combustion phase dominates. Diesel engine emits about 1/5th of SI engine HC emissions.

Typically, diesel engine exhaust consists of particulate matter, oxides of nitrogen, hydrocarbons, and carbon monoxide. Diesel engine is more prone for formation of solid soot or smoke/particulate matter due to its way combustion.

In modern diesel engines, fuel is directly injected under high pressure, usually shortly before the top-dead center, into the combustion chamber. The fluid fuel entering the combustion chamber is atomized into small droplets, vaporized, and is mixed with air, resulting in a heterogeneous mixture of fuel and air. Combustion is initiated by the high temperatures and pressures by an auto-ignition process, hence the alternative name compression-ignition (CI) engine. In the conventional diesel combustion process, usually there is only a very short amount of time available for mixture formation. A fast injection and an as good as possible atomization of the fuel are therefore the prerequisites for a fast and intensive mixing of fuel and air.

Therefore, load of the engine is controlled by the amount of injected fuel and combustion start is controlled by the start of injection. Diesel engines are usually operated with a globally lean AF, but direct injection (DI) leads to different mixture areas ranging from very lean through stoichiometric to very rich mixture ratios.

Diesel engines normally run at a constant speed and the air intake remains unchanged and fuel is required to be regulated to vary the load/speed and thus the diesel engines are regarded as quality governed engines. As long as air is unthrottled, the fuel undergoes near-complete combustion which normally happens during 25–75% of load operation or part load condition.

The polycyclic aromatic hydrocarbons, or "PAHs," which are of importance in diesel engine soot formation, in addition to few formaldehyde, benzene, and polynuclear aromatic hydrocarbons, as well as other unidentified mutagenic compounds, included among the unburned hydrocarbons in typical diesel engine exhaust. The mutagenic (and presumed carcinogenic) activity is reduced with the techniques that are meant for control of diesel hydrocarbon and particulate matter emissions. Except at high loads, unburned hydrocarbons and CO are not a serious problem except. However, the two most troublesome emissions from diesel engines are soot (particulates) and oxides of nitrogen.

Although DI Diesel engines are becoming popular because of their fuel economy as well as low exhaust emissions, they emit a higher amount of visible exhaust termed as smoke. The characterization of this smoke has remained a challenge in engine development and modeling work. Diesel and spark-ignition engines produce more or less similar emission species. On the other hand, owing to the low volatility of diesel fuel relative to that of gasoline and the fact that carburetors are not employed, evaporative emissions are not so significant.

Sulfur, which plays a major part in the production of particulates and smoke emissions, is present in larger proportions in diesel fuel than in petrol. This is one of the reasons why the combustion of diesel fuel produces between 5 and 10 times more solid particles than that of petrol.

Because diesel power output is governed by regulating the supply of fuel without throttling the air supply, there is excess air and therefore virtually zero CO in the exhaust under normal cruising conditions. As a diesel engine is opened up towards maximum power and torque, NOx output increases because of the higher combustion temperatures and pressures. At the same time HC, CO, and sooty particulates also increase.

However, because of the relatively low volatility of diesel fuel and the extremely short time available for evaporation, HC, and CO emissions are more likely to form under cold conditions. In the latter circumstances, under idling or light load conditions, the situation is aggravated by the fact that the minute volumes of fuel injected per stroke are not as well atomized as when larger volumes are delivered through the injector holes.

Hydrocarbons can arise from fuel absorbed in deposits and oil layers or from very lean mixtures. At light loads, the fuel is less apt to impinge on surfaces; but because of poor fuel distribution, large amounts of excess air, and low exhaust temperature, lean fuel-air mixture regions may survive to escape into exhaust. White smoke is often observed at low load conditions and is really a fuel particulate fog. The more typical black smoke sometimes observed during periods of rapid load increase, or for older engines at higher loads, is made up primarily of carbon particles. Diesels provide reduced hydrocarbons in automobiles even though they have no catalyst system. If the diesel starts promptly, which is not the case in cold weather conditions, and thus starting aids such as glow plugs are often used in automotive applications. The problem of NO production in NA diesel engines arises from the early rapid burning, which produces very-high temperature products (ref. Chapter 1) (Figure 2.7).

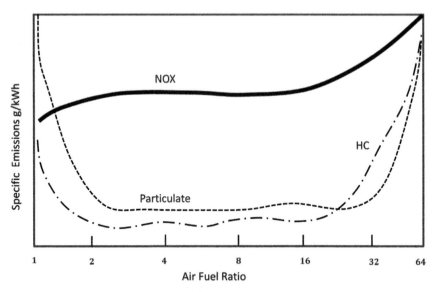

FIGURE 2.7   Variation diesel engine exhaust species with air-fuel ratio.
Source: Ref.: Air Pollution from Motor Vehicles.

Figure 2.6 represents the typical relationship between diesel engine exhaust emissions with AF. Under part-load (corresponding to very light load), at very high AFs (lean mixtures)-left side of figure, the temperature in the cylinder after combustion is too low to burn out residual hydrocarbons, so emissions of gaseous hydrocarbons and particulate SOF are high. The major part of PM emissions is soot (carbonaceous matter engulfed with sulfates etc.) is a result of fuel-rich mixtures available under high load operation, so under rich mixture conditions or at lower AFs, less oxygen is available for soot oxidation, so soot emissions increase. As long as excess AF) remains above 1.6, this increase is relatively gradual. Soot and visible smoke emissions show a strong non-linear increase below the smoke limit (at about $\phi> = 1.5$).

In naturally aspirated engines (those without a supercharger/turbocharger), the amount of air in the cylinder is independent of the power output. The diesel engines are also specified based on smoke emission such as power output at higher loads for these engines is normally smoke-limited, that is, limited by the amount of fuel that can be injected without exceeding the smoke limit. Therefore, maximum fuel settings on naturally aspirated engines represent a compromise between smoke emissions and power output. Where diesel smoke is regulated, this compromise results in smoke opacity below the regulated limit.

2.4.2 UNBURNED HYDROCARBONS EMISSIONS

2.4.2.1 SOURCES OF HC EMISSIONS-UNBURNT HYDROCARBONS

The unpleasant smell of a diesel engine exhaust is due to hydrocarbons (HCs) which form through the lubricating oil and improper mixing of fuel-air mixtures in fuel rich zone. This happens predominantly in the high load regime. There are three main reasons for this. First, at low temperatures and light loads. Also, such mixtures subsequently fail to burn. Secondly, because of the low volatility of diesel fuel relative to petrol, and the short period of time available for it to evaporate before combustion begins, HCs are generated during starting and warming up from cold. In these circumstances, fuel droplets, together with water vapor produced by the burning of the hydrogen content of the remainder of the fuel, issue from the cold exhaust pipe in the form of what is generally termed white smoke, but which is in fact largely a mixture of fuel and water vapors. At about 10% load and rated speed, both HC and CO output are especially sensitive to fuel quality and, in particular, cetane number.

Thirdly, after cold starting and during warm-up, a higher than normal proportion of the injected fuel, failing to evaporate, is deposited on the combustion chamber walls. This further reduces the rate of evaporation of the fuel, so that it fails to be ignited before the contents of the chamber have been cooled, by expansion of the gases, to a level such that ignition can no longer occur. Similarly, the cooling effect of the expansion stroke when the engine is operating at or near full load can quench combustion in fuel-rich zones of the mixture. This is the fourth potential cause of HC emissions.

Unburnt HCs tend to become a problem also at maximum power output, owing to the difficulty under these conditions of providing enough oxygen to burn all the fuel. As fuel delivery is increased, a critical limit is reached above which first the CO and then the HC output rise steeply. Injection systems are normally set so that fuelling does not rise up to this limit, though the CO can be removed subsequently by a catalytic converter in the exhaust system.

A small amount of liquid fuel is more likely to get trapped on the tip of the fuel injector nozzle even when injection stops. This small volume of the fuel is called sac volume. The sac volume of fuel is surrounded by a fuel-rich environment and therefore it evaporates little slowly resulting in HC emissions in the exhaust. Sac volume is also the fuel contained in the volume between the pintle needle seat and the spray hole or holes. After the injector needle has seated and combustion has ceased, some of the trapped fuel may evaporate into the cylinder. Finally, the crevice areas, for example between the piston and cylinder walls above the top ring, also contain unburnt or quenched fractions of semi-burnt mixture. Expanding under the influence of the high temperatures due to combustion and falling pressures during the expansion stroke, and forced out by the motions of the piston and rings, these vapors and gases find their way into the exhaust.

The air-fuel mixture in a CI engine is heterogeneous with fuel still being added during combustion. It causes local spots to range from very rich to very lean and many flame fronts exist at the same time unlike the homogeneous air-fuel mixture of an SI engine that essentially has one flame front. Incomplete combustion may be caused by under mixing, in fuel-rich zones, some fuel particles do not find enough oxygen to react with, and in fuel-lean zones, some local spots will be too lean for combustion to take place properly. With over mixing, some fuel particles may be mixed with burned gases and it will, therefore, lead to incomplete combustion.

Typically DI diesel fuel undergoes heterogeneous combustion whereas SI engine fuel mostly gasoline takes part under homogeneous condition except in lean stratified intake-manifold fuel injection or HCCI combustion processes. Thus, diesel engines emit the largest part of unburned hydrocarbons already in the cold start and warm-up phases where the temperature in the combustion chamber is relatively low, where little re-oxidation occurs.

2.4.2.2   HYDROCARBONS AND RELATED EMISSIONS

The levels of CO and HC emissions are low in diesel engines due to its lean mixture working conditions compared to gasoline engines. The incomplete combustion of the fuel or the presence lubricating oil in exhaust is the main reason for diesel engine's HC emissions. However, the HC emissions of diesels are easy for control even fairer that gasoline engines. By correctly deciding the mixture strength requirement and allowing thorough mixture formation with good fuel injection equipment will matching of the air-fuel mixing process and injection equipment, the HC emissions can be controlled. Further, mechanical developments have reduced oil- derived hydrocarbon levels by reducing the amount of oil entering the cylinder.

Moreover, by properly reducing the nozzle sac/hole volume and not eliminating secondary injections, the HC emissions are further reduced. The HC emissions are normally formed at lower loads (as well as the unburned-fuel portions of the particulate SOF). The major source of light-load hydrocarbon emissions is excessive air-fuel mixing, which results in air-fuel mixture(s) that are too lean to burn.

Fuel deposits on the combustion chamber walls or in combustion chamber crevices by the injection process, fuel retained in the orifices of the injector which vaporizes late in the combustion cycle, partly reacted mixture quenched by too-rapid mixing with air, and vaporized lubricating oil can be cited as few more sources of HC emissions. Along with these HCs by following the same phenomena, aldehydes, and ketones and however small may be carbon monoxide are formed by following the same processes.

The unregulated pollutants from diesel engines include aldehydes and ketones polycyclic and nitro-polycyclic aromatic compounds (PAH and nitro-PAH), and traces of hydrogen cyanide, cyanogens, and ammonia. The products in the diesel exhaust such as polycyclic aromatic hydrocarbons and their nitro-derivatives are of serious concern as these are causes of mutagens and suspected carcinogens. PAH and acetylene appear to form during the combustion process, possibly via the same polymerization reaction that produces soot (Williams et al., 1987).

Indeed, the soot particle itself can be viewed as a very large PAH molecule. PAH and nitro-PAH can be reduced by decreasing the aromatic content of the fuel. Diesel engines emit significantly more aldehydes and ketones than comparable gasoline engines. Aldehyde emissions are largely responsible for diesel odor and cause irritation of the nasal passages and eyes.

These emissions are strongly related to total hydrocarbon emissions and can be controlled by the same techniques. Sulfur dioxide emissions are the result of naturally occurring sulfur in the fuel and can only be reduced by lowering the fuel sulfur content. Hydrogen cyanide, cyanogens, and ammonia are present only in small quantities in diesel exhaust and are unlikely to become regulated.

The diesel fuel undergoes diffusive mode of combustion in heterogeneous unsteady and conditions. The diesel engine is a self-ignition engine in which fuel and air are mixed inside the engine. By virtue of high temperatures owing to relatively high CRs, the air required for combustion is highly compressed inside the combustion chamber. The fuels for CI engines are chosen in such a way that has low-self ignition /auto-ignition temperature and this generates high temperatures during compression, are sufficient for the diesel fuel to ignite spontaneously when it is injected into the cylinder. Most commonly used fuel in CI engine is diesel fuel (which has about 12–22 carbon atoms in its molecular structure). Subsequently, the diesel engine uses heat to release the chemical energy contained in the diesel fuel and to convert it into mechanical force (Bosch, 2005).

Carbon and hydrogen construct the origin of diesel fuel like most fossil fuels. For ideal thermodynamic equilibrium, the complete combustion of diesel fuel would only generate CO_2 and H_2O in combustion chambers of engine (Prasad and Bella, 2010). However, many reasons (the AF, ignition timing, turbulence in the combustion chamber, combustion form, air-fuel concentration, combustion temperature, etc.) make this out of question, and a number of harmful products are generated during combustion. The most significant harmful products are CO, HC, NOx, and PM and these species with approximate weightage in typical diesel combustion are shown in Figure 2.8 Apart from the mentioned causes, flame quenching, crevice volume, oil-film, and deposits on the cylinder wall, misfiring, etc. also contribute significantly for formation of HC emissions.

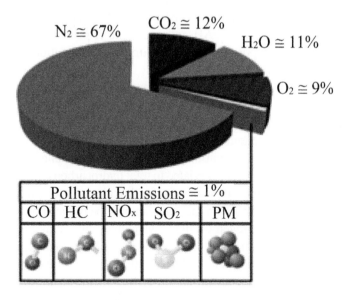

FIGURE 2.8   Typical compositions of diesel exhaust gas.
Source: Reşitoğlu, et al. (2015). Open access.

2.4.3   CARBON MONOXIDE (CO)

All carbon should burn to carbon dioxide in ideal combustion condition, however, due to varying operating conditions, carbon cannot completely oxidizes and leaves small traces of partially oxidized component called carbon monoxide. The conditions that are favorable for formation of CO is insufficient oxidizer and insufficient time and results in incomplete combustion and his concentration is largely dependent on air/fuel mixture and it is highest where the excess-air factor (λ) is less than 1.0 that is classified as rich mixture (Wu et al., 2004). Carbon monoxide results from the incomplete combustion where the oxidation process does not occur completely.

Diesel engines are lean combustion engines which have a consistently high AF ($\lambda > 1$). So, the formation of CO is minimal in diesel engines. Nevertheless, CO is produced if the droplets in a diesel engine are too large or if insufficient turbulence or swirl is created in the combustion chamber (Demers and Walters, 1999). It can be caused especially at the time of starting and instantaneous acceleration of engine where the rich mixtures are required. In the rich mixtures, due to air deficiency and reactant concentration, all the carbon cannot convert to CO_2 and be formed CO concentration. Although CO is produced during operation in rich mixtures, a small portion of CO is also emitted under lean conditions because of chemical kinetic effects (Faiz et al., 1996).

Carbon monoxide is an odorless and colorless gas. It is a toxic gas. In humans, CO in the air is inhaled by the lungs and transmitted into the bloodstream. It binds to hemoglobin and inhibits its capacity to transfer oxygen. Depending on CO concentration in the air, as thus leading to asphyxiation, this can affect the function of different organs, resulting in impaired concentration, slow reflexes, and confusion (Raub, 1999; Kampa and Castanas, 2008; Walsh, 2011; Strauss et al., 2004).

2.4.4 PARTICULATE MATTER (PM)

Diesel particulates consist mainly of combustion generated carbonaceous material (soot) on which some organic compounds has been absorbed. Most particulates are generated in the fuel rich zones within the cylinder during combustion due to incomplete combustion of fuel hydrocarbons. Formation of particulate matter results in poor economy.

> Engine that pollutes more also consumes more and vice versa.

Anything in the exhaust gas that is in the solid state can be classified as particulate matter (PM). Carbon soot and dust form particulate matter.

Incomplete combustion is the main culprit in the formation of PM. These minute particulate compounds can cause respiratory problems in human and animal life. The EPA classifies particulates into two general categories:

1. Inhalable particulates that range between 2.5 and 10 microns in diameter;
2. Respirable particulates that tend to be 2.5 microns or less in diameter.

The spark-ignition engines are generally categorized that they operate close to stoichiometric AF, and particulates or soot emissions from these engines are insignificant. Few particulates are being observed in GDI engines. Emission formation in both SI engines and CI engines is a strong function fuel-air ratio (FA). During combustion of the rich fuel-air mixtures, a carbonaceous particulate matter commonly called smoke or soot and is observed with the release of black smoke in the exhaust indicates high concentration of soot in the exhaust gases.

The main four phases that are responsible for soot formation viz.:

1. Particle formation and nucleation;
2. Particle surface growth;
3. Coagulation and agglomeration; and
4. Oxidation.

Even though the diesel engines operate under relatively lean mixtures (compared to its counterpart-SI engine), still smoke formation is a major concern in diesel engine. This emphasizes that pollutant formation is strongly dependent on the FA distribution in the spray, as shown below. Majority of the diesel engines work for full load operation and due to quality type governing, the load/speed is regulated through injection more fuel. This develops fuel rich zones and fined deficient oxidizer and is responsible for particulate formation. The three distinct regions can be seen with varying in fuel-air equivalence ratios of a typical diesel fuel spray, as shown in Figure 2.9. It represents a scoop like conical spray during its release. The core of the spray even exceeds (fuel-rich zones) flammability limits of a typical diesel fuel. The figure also depicts the distribution of fuel-air zone along and around diesel spray.

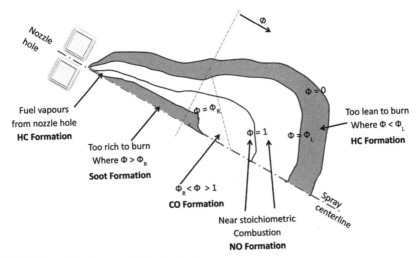

FIGURE 2.9   Distribution of diesel emissions along spray jet.

The third phase of diesel fuel-diffusion mode, with fuel rich mixtures, is responsible for formation of particulate matter or soot or solid carbon of very small.

Apart from particulate (soot) formation, the diesel fuel is prone for formation of large amounts of oxides of nitrogen (NOx).

The NOx emissions form mainly due to prevalence of high temperatures (Figure 2.10).

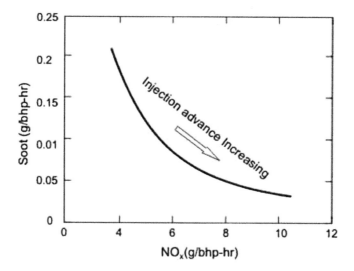

FIGURE 2.10   NOx – PM emission trade-off.

   Unburned HC and other oxygenated hydrocarbons like aldehydes and PAH originate in the region where due to excessive dilution with air the mixture is too lean at the spray boundaries. Towards the end of combustion, fuel in the nozzle sac and orifices gets vaporized, enters the combustion chamber, and contributes to HC emissions. Also, CO is formed in fuel rich mixtures forms due to partial oxidation of flammable region (Figures 2.11–2.13). Soot formation is minimal in case of alcohol fuels followed by paraffins, olefins and highest with aromatics.

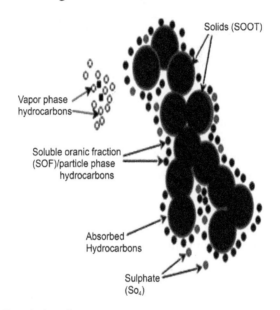

FIGURE 2.11   Description of soot.
Source: Adapted from Pulkrabek, 2004.

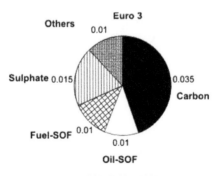

PM=0.08 g/kWh

FIGURE 2.12   Typical constituents of PM emission.

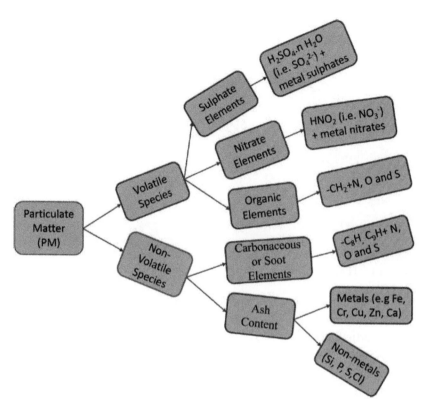

FIGURE 2.13 Conceptual model of typical PM composition.
Source: Reprinted from Raza, et al., 2018. Open access.

In diesel-fuelled engines, the soot particles form primarily from the carbon. The C/O and H/C ratios of molecular structure of diesel fuel is indicative of soot formation. This is particularly happens with the carbon atoms containing 12 to 22 and H/C ratio of about 2. Also, combustion environment conditions of 1000–2800 K and 50–100 bar makes more conducive for diesel fuel from soot and these soot particles typically of few hundred diameters.

The typical particulate matter was comprised of primary particles (Diameter: 22.6 ± 6.0 nm approximately) with the classic onion-skin structure consisting of concentrically wound nano crystalline graphite (domain sizes between 2 and 3 nm) that is often reported for exhaust particulate, carbon black and macroscopic graphite. Raman spectroscopy, which is very effective at detecting the amorphous forms of sp_3 and sp_2 hybridized carbon (often referred to just as disordered carbon) indicated that the soot was not entirely crystalline, but this was also the case for commercial graphite. These details are taken through Electron microscopy of the diesel soot (Figure 14).

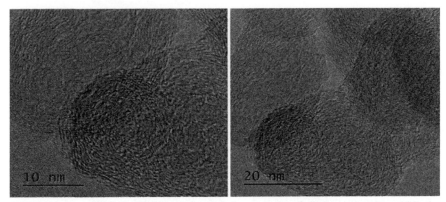

FIGURE 2.14   TEM of diesel soot, showing the presence of distinct spherules within the primary particles.
Source: Reprinted from Ramdas, 2015. With permission. Open access.

Diesel particulate matter has two main components:

1. Dry Soot or Solid Carbon Material: Dry soot is mainly the carbonaceous fraction of the particulate and its typical chemical formulae are C_8H, C_9H, and $C_{10}H$. About 5 to 10% by mass oxygen and 0.5% nitrogen are also present. Typical empirical formula of dry soot would be $CH_{0.11}O_{0.06}5N_{0.005}$. Dry soot results from several processes like pyrolysis, dehydrogenation, and condensation of fuel molecules.

Particulate matter emissions from diesel engines are considerably higher (six to ten times) than from gasoline engines (PM emissions form mostly in gasoline injected engines). Size of diesel particulate matters are typically spheres around 15–40 nm in diameter, and approximately more than 90%of PM is smaller than 1 μm in diameter. Many factors majorly include the combustion and expansion process, fuel quality (sulfur and ash content), lubrication oil quality, and consumption, combustion temperature, exhaust gas cooling (Burtscher, 2005) influence the formation process of PM emissions.

There are three main components of diesel particle emissions: soot, soluble organic fraction (SOF), and inorganic fraction (IF). Soot, seen as black smoke, is a major constituent of the total PM emissions with more than 50%. Next to soot is SOF, which consists of heavy hydrocarbons adsorbed or condensed on the soot. SOF is derived partly from the lubricating oil, partly from unburned fuel, and partly from compounds formed during combustion.

Under light engine loads when exhaust temperatures are low SOF values are too high (Sarvi et al., 2011; Tighe et al., 2012; Metts et al., 2005; Stanmore et al., 2001; Sharma et al., 2005).

Visible black smoke emissions from diesel engine result on account of high concentration of soot in the exhaust gas. All the diesel engine design and operating variables that affect soot formation and oxidation also influence black smoke intensity. Smoke emissions increase with increase in engine load due to overall richer FAs and hence, the rated engine power was specified based on the maximum permitted smoke density to curb black smoke emissions during engine operation. The rated power was also known as 'smoke limited power.' Poor control of fuel injection rate during acceleration also increases smoke. As the diesel particulate matter (PM) mainly consists of soot and the adsorbed unburned hydrocarbons (SOF) on soot core, the PM content has been related to exhaust soot content and HC concentration. A typical constituent of species of PM emission is depicted below:

Particulate matter emissions in the exhaust gas are resulted from combustion process. They may be originated from the agglomeration of very small particles of partly burned fuel, partly burned lube oil, ash content of fuel oil, and cylinder lube oil or sulfates and water (Demers and Walters, 1999; Maricq, 2007). Most particulate matters are resulted from incomplete combustion of the hydrocarbons in the fuel and lube oil.

Many researchers have performed studies to detect the impact of PM emissions on environment and human health. In these researches, it is stated that inhaling of these particles may cause to important health problems such as premature death, asthma, lung cancer, and other cardiovascular issues. These emissions contribute to pollution of air, water, and soil; soiling of buildings; reductions in visibility; impact agriculture productivity; global climate change (Englert, 2004; OECD, 2002; Michael and Kleinman, 2000).

KEYWORDS

- **carbon monoxide**
- **greenhouse gas**
- **hydrocarbon**
- **particulate matter**
- **polycyclic aromatic compounds**
- **volatile organic compounds**

REFERENCES

Annamalai, K., & Ishwar, P. K. (2007). *Combustion Science and Engineering,* CRC Press–Taylor & Francis Group.

Burtscher, H. (2005), Physical characterization of particulate emissions from diesel engines: a review. *Aerosol. Sci., 36,* 896–932.

Demers, D., & Walters, G. (1999). Guide to exhaust emission control options. *BAeSAME,* Bristol.

Englert, N. (2004). Fine particles and human health—a review of epidemiological studies. *Toxicol Lett 149,* 235–242.

Faiz, A., Weaver, C. S., & Walsh, M. P. (1996). Air Pollution from Motor Vehicles Standards and Technologies for Controlling Emissions. The World Bank Washington, D.C.

John, H. B. (1988). *Internal Combustion Engines Fundamentals,* McGraw Hill.

Kampa, M., & Castanas, E. (2008). Human health effects of air pollution. *Environ. Pollut., 151,* 362–367.

Maricq, M. M. (2007). Chemical characterization of particulate emissions from diesel engines: a review. *Aerosol Sci. 38,* 1079–1118.

Metts, T. A., Batterman, S. A., Fernandes, G. I., & Kalliokosi, P. (2005). Ozone removal by diesel particulate matter. *Atmos Environ. 39,* 3343–3354.

Michael, R. A., & Kleinman, M. T. (2000). Incidence and apparent health significance of brief airborne particle excursions. *Aerosol. Sci. Technol., 32,* 93–105.

Miller, J. A., & Craig, T. (1989). *Bowman Progress in Energy and Combustion Science, 15*(4), Pages 287–338.

Organization for Economic Co-Operation and Development (OECD) (2002). Strategies to reduce green house gas emissions from road transport: analytical methods. OECD, Paris.

Pulkrabek, W. W. (2004). *Engineering Fundamentals of the Internal Combustion Engine,* Pearson Prentice Hall.

Prasad, R. & Bella, V. R. (2010). A Review on Diesel Soot Emission, its Effect and Control. *Bulletin of Chemical Reaction Engineering and Catalysis, 5*(2).

Ramdas, R., Nowicka, E., Jenkins, R., Sellick, D., Davies, C., and Golunski, S. (2015). Using real particulate matter to evaluate combustion catalysts for direct regeneration of diesel soot filters, *Applied Catalysis B: Environmental, 176&177,* 436–443. http://creativecommons. org/licenses/by/4.0/).

Raub, J. A. (1999). Health effects of exposure to ambient carbon monoxide. Chemosphere: global change. *Science 1,* 331–351.

Raza, M., Chen, L., Leach, F., and Ding, S. (2018). A review of particulate number (PN) emissions from gasoline direct injection (GDI) engines and their control techniques. *Energies, 11,* 1417. doi: 10.3390/en11061417. https://creativecommons.org/licenses/by/4.0/

Resitoglu, I. A., Altinisik, K., & Keskin, A. (2015). The pollutant emissions from diesel-engine vehicles and exhaust aftertreatment systems. *Clean Techn Environ Policy, 17,* 15–27.

Robert Bosch (2005). *Bosch Automotive Handbook,* Bentley Publishers.

Sarvi, A., Lyyranen, J., Jokiniemi, J., & Zevenhoven, R. (2011). Particulate emissions from large-scale medium-speed diesel engines: 1. Particle size distribution. *Fuel Process Technol., 92,* 1855–1861.

Sharma, M., Agarwal, A. K., & Bharathi, K. V. (2005). Characterization of exhaust particulates from diesel engine. *Atmos Environ.* 39, 3023–3028.

Stanmore, B. R., Brilhac, J. F., & Gilot, P. (2001). The oxidation of soot: a review of experiments, mechanisms and models. *Carbon 39,* 2247–2268.

Strauss, S., Wasil, J. R., & Earnest, G. S. (2004). Carbon monoxide emissions from marine outboard engines. *Society of Automotive Engineers,* 2004–32–0011.

Tighe, C. J., Twigg, M. V., Hayhurst, A. N., & Dennis, J. S. (2012). The kinetics of oxidation of diesel soots by NO_2. *Combust Flame 159,* 77–90.

Turns, S. R. (2000). *An Introduction to Combustion: Concepts and Applications,* McGraw Hill.

Walsh, M. P. (2011). Mobile source related air pollution: effects on health and the environment. *Encyclopedia of Environ Health, 3,* 803–809.

William, F. A. (1987). *Combustion Theory,* Taylor & Francis Group.

Zhang, W.-G., Xing-Cai, L., Jian-Guang, Y., Zhen, H. (2004). Effect of cetane number improver on heat release rate and emissions of high speed diesel engine fueled with ethanol–diesel blend fuel. *Fuel 83,* 2013–2020.

CHAPTER 3

Emissions Measurement Methods and Instrumentation

When you can measure what you are speaking about, and express it in numbers, you know something about it.

—Lord Kelvin

3.1 INTRODUCTION: EMISSION MEASUREMENT METHODS

Internal combustion engines (ICEs) are the main prime movers for automobiles used in transportation. These engines are combustion engines or transportation engines that have become the source and major cause of concern at ground level for the contribution to atmospheric pollution (Springer and Patterson, 1973). Combustion reactions are exothermic reaction associated with the high heat release with consequent rise in high temperatures and these reactions are very complicated and rapid chemical reactions, and the products of combustion are exhaust species of interest which needs control when exceed certain prescribed set norms. The species highly dependent on pressure and temperature. The way fuel and air mix homogeneously dictates the quality of combustion and combustion, in turn, depends on the equivalence ratio or quality of mixture. The thorough is the mixture formation the completeness is the combustion reaction, i.e., any hydrocarbon fuel that participates in combustion form carbon dioxide and water vapor as the products of complete combustion. Under complete combustion conditions, when stoichiometric amount of air is supplied, the temperature rise will high enough and under adiabatic condition with no work transfer, the temperature attained is theoretically maximum, called adiabatic flame temperature (AFT). In practice, not all the time stoichiometric air is supplied and hence there are no chances of complete combustion to take place. Sometimes it is required to supply excess amount of air and in other cases lesser amount of air as required by application. It is guaranteed that with excess oxygen or air there will not be any carbon monoxide present there exists the presence of some carbon monoxide and unburned oxygen in the products. This can be due to incomplete mixing, insufficient time, etc. for complete combustion, and other factors. The formation mechanism of various species of exhaust is described in Chapter 2.

If fuel and air react as per chemically balance reaction (or with 100% theoretical air), the products of combustion include CO_2, H_2O, and N_2 only. However, this doesn't happen in practical situations and if the amount of air supplied is less than the theoretical amount of air, CO_2 and CO are both present in the products depending on the percentage of completion of reaction and also contains some unburned fuel in the products. Through the help of computation study, Borman et al. (1998) observed that about 11 species are normally be seen in the exhaust.

By the application of principles of conservation of mass to the chemical equation, for a given fuel, assuming complete combustion, the respective amounts of the products can be estimated. Therefore, it is practically difficult to estimate accurately the concentrations of exhaust species in case of incomplete combustion. Moreover, instead of simply mentioning the different species, it is essential to quantify them as the concentration of these gases will vary depending on the geometrical and operating conditions of engines. These gases also vary with the type of fuel and engine ignition system also. Hence, the means of knowing the constituents of precise species and their concentrations are done through exhaust gas sampling and measurement.

The quantification has become mandatory with the environment (earth's atmosphere) is getting degraded as per EPA/clean air act. Moreover, with promulgation of periodic stringent emission legislation, it has become essential for both the research community and vehicle manufacturing units to maintain standards of emissions strictly adhere to. Therefore, it is mandatory to measure and monitor the exhaust gas pollutant concentrations.

Sampling and dilution systems of exhaust gases and its measurement comprises of exhaust gas composition measuring equipment. The determination of environmentally relevant mass emissions requires the determination of their concentrations as well as the engine's exhaust gas volumes.

3.2   EXHAUST GAS MEASUREMENT METHODS

The main concern is measurement of emissions (especially pollutant emissions) emitted from automotive engines.

The exhaust of vehicular engines contain gaseous species are measured using:

1. non-separation methods;
2. separation methods.

However, the diesel engine exhaust contains smoke/soot/Particulate Matter is measured using Smoke Meter or dilution tunnels (Figure 3.1).

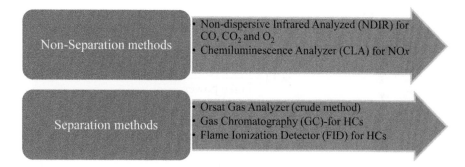

| Non-Separation methods | • Non-dispersive Infrared Analyzed (NDIR) for CO, CO_2 and O_2
• Chemiluminescence Analyzer (CLA) for NOx |

| Separation methods | • Orsat Gas Analyzer (crude method)
• Gas Chromatography (GC)-for HCs
• Flame Ionization Detector (FID) for HCs |

FIGURE 3.1 Exhaust gas measurement methods.

By using measurements, mole fractions of the gaseous products of combustion can be obtained and the values are generally expressed on a "dry" basis.

Excluding water vapor, even though its proportion can be significant, mole fractions are obtained for all gaseous products, a feature of dry product analysis. Since water is formed when hydrocarbon fuels are burned. The dew point temperature plays a significant role when the exhaust products of combustion are cooled at constant mixture pressure, where water vapor begins to condense. The information about dew point temperature is important in order to mitigate or control corrosion as water deposited on ducts, mufflers, and other metal parts can cause corrosion.

3.3   SPECIFYING CONCENTRATION OF GASES

The proportion a specific gaseous pollutant (the levels of pollutant) can be given either be given specified on volume basis or on mass basis. If specified on volume basis, the units of are either ppmv (parts per million by volume) or ppbv (part per billion per volume).

$$1 \text{ ppmv}_{\text{pollutant gas}} = \frac{\text{Volume of a pollutant gas}}{\text{Total volume of gas mixture} \times 10^6}$$

Also, 1 ppmv is equal to 0.0001% by volume.

This type of specification is referred to as mole fraction of a gas species multiplied by either 10^2 or 10^3. More commonly, the unit of a pollutant gas is expressed on mass basis, the units either mg/m^3 of $\mu g/m^3$.

However, this sort of specification causes confusion and nowadays, the concentration of exhaust gas levels is measured in terms of specific units or normalized units.

Therefore, the specific emission of a pollutant viz.; CO, NOx, or HCs are expressed as:

$$sCO = \frac{m_{CO}}{\text{brake power}}$$

$$sNOx = \frac{m_{NOx}}{\text{brake power}}$$

$$sHC = \frac{m_{HC}}{\text{break power}}$$

The units of specific pollutants are $\mu g/kW\text{-}h$ or $mg/kW\text{-}h$ or $g/kW\text{-}h$.

However, neither mole fraction units nor specific units (normalized units) give information on amount of fuel used in emitting a certain pollutant. In this regard, the pollutant levels are now being measure in Emission Index. This is defined as a ratio of mass flow rate of pollutant to the mass flow rate of fuel used.

Emission Index of Carbon monoxide or $EI_{CO} =$

$$\frac{\text{mass flow rate of CO}}{\text{mass flow rate of fuel}} = mCO\left(\frac{g}{s}\right) / mfuel\left(\frac{kg}{s}\right)$$

Emission Index of Carbon monoxide or $EI_{NOx} =$

$$\frac{\text{mass flow rate of CO}}{\text{mass flow rate of fuel}} = mNOx\left(\frac{g}{s}\right) / m_{fuel}\left(\frac{kg}{s}\right)$$

Emission Index of Carbon monoxide or $EI_{HC} =$

$$\frac{\text{mass flow rate of CO}}{\text{mass flow rate of fuel}} = mHC\left(\frac{g}{s}\right) / m_{fuel}\left(\frac{kg}{s}\right)$$

This type of specification gives clear indication of the amount of pollutant produced per unit amount of fuel consumed (Heywood, 1988; Turns, 2003).

3.4 EXHAUST MEASUREMENT INSTRUMENTATION

The following sections deal with the principle, working of instrumentation used for measurement of CO, HC, NOx, and PM emissions. Among these well-known pollutants CO, HC, and NOx are gaseous phase emissions whereas PM are non-gaseous emission.

3.4.1 PRINCIPLE OF OPERATION OF NDIR ANALYZER

The NDIR analyzers are used for measuring the concentrations of carbon monoxide and carbon dioxide. The device is based on the principle that the infrared (IR) energy of a particular wavelength, peculiar to a certain gas. The IR energy of other wavelengths will be transmitted by that gas. A method in which broadband radiation is used for sampling gaseous pollutant, is specific to NDIR method, one of the most popular methods in exhaust gas analysis. In the IR part of the electromagnetic spectrum, a variety of the pollutant gases get absorbed. The amount of absorption depends on the product of gas concentration level and path length over which the electromagnetic radiation travels through the gas. The oxides of carbon-CO and CO_2 absorb about 3% in the wavelength range of 4.0–5.0 μm and about 10% radiation in the wavelength range of 4.2–4.3 μm respectively for the same path length concentration product. The hydrocarbons (HCs) absorb about 8% of radiation in the 3.3–3.5 μm wavelength band under the same condition. It can be seen that CO, CO_2, and HCs absorb in the narrow range. Nitric oxide (NO) has also a weak absorption band, allowing it to be analyzed by NDIR but lack of sensitivity and interference by water vapor do not give high accuracy with low concentrations.

Since hydrocarbons are of very wide range, it is difficult to detect which specific hydrocarbon absorbs the infrared radiation; hence, NDIR is normally not used to precisely account for the level of a specific hydrocarbon.

Normally exhaust sample of engines are a mixture of various gases, measured based on absorption of radiation of suitable wavelength. Therefore, these gases need not require separation for the purpose of measurement.

3.4.2   *WORKING OF NDIR ANALYZER*

Depending on the specific gas in IR region working of non-dispersive infrared (NDIR) technique Analyzer relies upon the energy absorption characteristics. In a simple NDIR instrument, IR energy passes through two identical tubes and falls on a detector. The first tube is the reference cell and is filled with a non-absorbing gas such as nitrogen. The second tube is the measurement cell and contains the gas sample to be analyzed.

A characteristic feature of NDIR analyzer is that it does not disperse the light emitted from an IR source i.e., splitting up of the component wavelengths by means of a prism or grating. In NDIR, by a band-pass filter, a broadband of light is produced which is chosen to coincide with an absorption peak of the pollutant molecule.

The schematic of a typical NDIR analyzer is shown in Figure 3.2. It contains two separate cells through one of which IR radiation passes, a reference cell of clean dry air, and a sample gas passes through another cell. A **"microphone"** type of detector is used in the unit. It consists of two chambers separated by a thin metal diaphragm filled with gas of the species being measured.

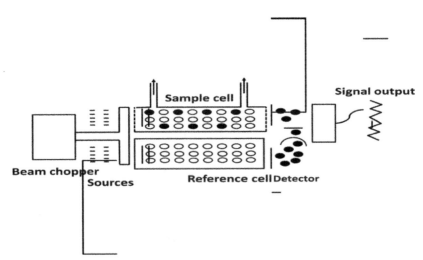

FIGURE 3.2   Typical layout of a NDIR analyzer.

The kinetic energy of the molecules increases when they absorb the IR radiation causing the pressure in each chamber to increase. If absorbing molecules are present in the sample cell, the amount of energy reaching that side of the detector will reduces. Hence, it results in a displacement of the diaphragm due to development of a pressure differential between the two chambers.

As the instrument electronics senses the signal results a change in capacitance, the instrument also includes a beam chopper. For easy detection and amplification, an alternating signal in the detector is provided.

A limitation of this type of analyzers is that the interference with the other gases being of the in the region of IR may try to influence the accuracy of actual gas being sampled. For example, the value of CO_2 would interfere with CO detection. By the use of a desiccant, water vapor can be readily removed from the sample using a silica gel.

The detector measures the intensity of two different wavelengths, one at the sample gas absorption wavelength and the other is at reference gas absorption wavelength when the IR Source continuously sends an IR waves through the gas tubes.

As the reference, gas generally contains nitrogen (an inert gas to certain temperature level) and detector receives 100% signal. If a target gas is (such as CO or CO_2) present in sample gas means the received signal will be attenuated at the detector side. The detector measures these two signals and their difference is proportional to the amount of absorbing a target gas in the sample cell. i.e., CO_2 gas. So finally, the CO_2 gas concentration is measured with the difference in absorption of IR radiation in the sample and reference cells. CO_2 gas concentration measuring unit is ppm (Richad Atkins-Engine testing).

One tube which is regarded as reference tube is normally filled with dry air. The sample cells may be divided by windows into various lengths to give different ranges of sensitivity, the unused sample cells generally being flushed with a non-infrared absorbing gas such as oxygen or nitrogen, or gas free of the component being measured, e.g., fresh air for carbon monoxide analyzers. The radiation from the source is interrupted by a rotating two-bladed shutter driven by a synchronous motor. Thus chopping is necessary to avoid spurious signals which would otherwise be caused by slow thermal changes, and it provides an alternating signal which is more convenient for amplification. An electronic bridge which detects the diaphragm movement is fed from a.r.f. oscillator. Matched coils in series with the detector capacities on either side of the diaphragm form two arms of a bridge. When the detector diaphragm is displaced, it increases one capacity and reduces the other, thus unbalancing the bridge.

The unbalanced signal is then recorded directly in parts per million or percent of the sampled gas. To change the range it is possible to switch the sample to different cells or, alternatively, to change the sensitivity of the electronic circuiting. To zero an analyzer a non-infrared-absorbing gas, e.g., dry air, is passed through the instrument. For other points on the scale calibrating gases with known concentration of carbon dioxide, carbon monoxide, and nitric oxide are passed through the analyzer.

3.4.3   THE BEER-LAMBERT LAW FOR NDIR GAS ABSORPTION

According to this law, the IR intensity will reduce on the active detector as and when an IR-absorbent gas enters the cell and the relationship known as the Beer-Lambert Law is denoted as:

$$I = I_0 \exp(-klc) \tag{1}$$

where, I = absorbed Radiation energy; I_0 = incident radiation energy; k = gas characteristic absorption constant, m^2/gmol; c = gas concentration, gmol/m^3; l = gas column length, m.

From this law, it is clear that the gas concentration **c** can be measured.

3.4.4   ADVANTAGES OF NDIR TECHNIQUE

The NDIR method is versatile in its use due to its benefits:

- Hazardous chemical environment does not affect its functioning;
- Not affected by poisoning effects unlike other sensors;
- Elimination cross-sensitivity due to nondetection of Hydrogen;
- Anaerobic conditions does not affect gas analysis;
- No sensor burn-out or deterioration upon exposure to high gas concentrations;
- Stable and long-term operation needs minimum recalibration;
- High stability even in after longer storage times;
- Economical than catalytic sensors.

3.5   MEASUREMENT BY CHEMILUMINESCENCE ANALYZER FOR OXIDES OF NITROGEN (NOX)

3.5.1   PRINCIPLE OF CHEMILUMINESCENCE ANALYZER

The oxidizer for combustion most of the hydrocarbon fuels is atmospheric air. Major portion of air contains Nitrogen (N_2). Under the conditions of combustion, high temperatures are prevailed and such conditions are conducive for dissociation nitrogen. This makes formation of complex compounds of oxides of nitrogen (NOx) with oxygen also getting dissociated. The NOx emissions are highly poisonous and require continuous monitoring and measurement. NOx refers majorly to a combination of both NO and NO_2. If NOx is subjected to high temperature of about 1000°C, NO_2 breaks down to NO. The sample gas will thus contain only NO.

If NO is allowed to react with ozone, the following exothermic reaction takes place:

$$NO + O_3 \rightarrow NO_2 + O_2$$

$$NO_2^* \rightarrow NO_2 + h\nu(photon)$$

NO_2 formed through above reaction will be in the excited state and hence emits radiation. This is called chemiluminescence.

The emitted radiation is proportional to the concentration of NO_2 and hence provides a method of measuring NOx level.

3.5.2 WORKING OF CHEMILUMINESCENCE ANALYZER

A chemiluminescence detector (CLD) is the industry standard method of measuring c concentration. The reaction between NO and O_3 (ozone) emits light. This reaction is the basis for the CLD in which the photons produced are detected by a photo multiplier tube (PMT). The CLD output voltage is proportional to NO concentration. The light-producing reaction is very rapid so careful sample handling is important in a very rapid response instrument. Measurement is based on the chemiluminescence produced by blending NO and ozone O_3. A chemical reaction converts NO and O_3 into NO_2. Since ozone is required to allow the reaction to take place, a major component in the working of CLA or CLD is ozone generator; an ozone generator in the analyzer itself generates the required ozone O_3 from oxygen O_2, which may be pure oxygen, synthetic air, or ambient air depending on the type of analyzer. A schematic of a CLA is shown in Figure 3.3.

A CLD is only able to measure NO. Therefore, every NO_2 molecule is converted into NO before the CLD detector so that NO_x ($NO_x = NO + NO_2$) can be measured. An NO_2/NO converter performs this conversion much like a catalytic converter.

Since NO_2 can also react with water, any water condensation must be prevented at least up to the NO_2/NO converter. Otherwise, the NO_2 being measured would be lost and aggressive acids would form. Therefore, the complete sample gas path and the analyzer are usually heated. Older analyzers often use an unheated CLD detector. Then, the sample has to pass through the NO_2/NO converter first (to avoid NO_2 to condensed water) and then proceed to a gas drier (to remove water).

NOx quenching occurs when $NO_2{}^*$ molecules collide with other suitable molecules before they have released light. Then, the energy is not released as light but to the other molecules. Thus, less light is generated and the measured value is too small. In exhaust gas, such molecules are primarily H_2O and CO_2. The more molecules are in the measuring cell, the more often such collisions occur, and, thus, the greater the quenching is. Hence, most CLD analyzers are operated in a vacuum (at approximately 20–40 mbar absolute). This significantly reduces the number of molecules and thus the potential for quenching. In addition to analyzer calibration and linearization, two diagnostic tests are particularly important for CLD. The H_2O and CO_2 quench test measures the degree of quenching and the NO_2/NO converter test measures convertibility and efficiency. Typically, the efficiency ought to be over 90%.

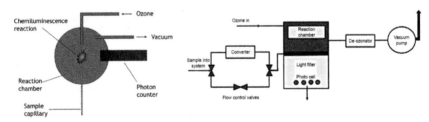

FIGURE 3.3   Layout of a chemiluminescent analyzer.

3.6   MEASUREMENT OF HYDROCARBONS

Hydrocarbons are measured with a flame ionization detector (FID). Exhaust gas contains a multitude of hydrocarbon compounds. Normally, it is important to measure their cumulative rather than their individual components. The FID measuring principle measures most different hydrocarbon compounds, thus allowing a cumulative result.

3.6.1   FID MEASURING PRINCIPLE

Hydrogen has a very high heating value and when hydrogen flame is used as a heating source for the sample exhaust gases, the gases get ionized. This instrument is mainly used to measure hydrocarbon concentrations in the gas sample. A pure hydrogen-air flame produces very little ionization; a large amount of ionization is produced when such a flame is passed with few hydrocarbon molecules into the flame zone. This is the principle behind flame ionization detection. The carbon atoms present in the hydrocarbon molecule is proportional to ionization. A flame generated from a constant flow of synthetic air and a gas mixture of hydrogen and helium burns inside the measuring cell. The flame burns in an electrical field between the cathode and anode. This flame is blended with a constant sample gas flow. Hydrocarbon molecules are cracked and ionized in the process. A weak current produced between the cathode and anode when ions generated are transported in the flame zone (measurement signal).

3.6.2   WORKING OF FID

A diagrammatic description of FID is shown in Figure 3.4.

The FID comprises a burner, an ignitor, an ion collector, and electronic circuitry. Hydrogen, or a mixture of hydrogen and nitrogen, enters one leg of the capillary tube and the sample enters by another leg. The length and bore of the capillary tubes are selected to control the flow rates. The mixture of gases then flows up the burner tube to a mixing chamber where a fixed proportion of air is admitted. This mixture is ignited and a diffusion flame stands at the exit to the burner tube.

An electrostatic field is produced around the flame by a polarizing battery. This causes the electrons to flow to the burner jet and the ions to the collector. The collector and the burner form part of an electrical circuit. The flow of ions to the collector and the electrons to the burner complete the electrical circuit. The D.C. level of the current flow is recorded on a meter. This is calibrated directly in amount of hydrocarbons. To calibrate, the detector samples of known hydrocarbons are measured and the meter reading noted.

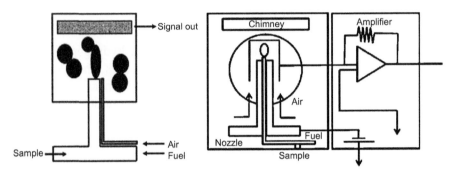

FIGURE 3.4   Schematic of flame ionization detector (FID).

When measuring hydrocarbons in diesel engine exhaust gas, the entire exhaust gas sample must be heated to 190°C from the sampling point to the FID since diesel engine exhaust gas contains hydrocarbons that would already condense below this temperature. Without heating, the condensed hydrocarbons would not be measured and the gas passages would also be contaminated. This is called HC hang-up.

Since the FID measures the sum of all organically bound hydrocarbon atoms in the hydrocarbons, calibration is performed with a component that is representative of the exhaust, e.g., C_6H_{14} or C_3H_9. The magnitude of the ion flow is an indicator of the HC concentration in the exhaust.

3.7   OXYGEN ANALYZERS

3.7.1   WORKING PRINCIPLE

Oxygen analyzer measures residual oxygen contents in the exhaust. The measurement principle is based on the paramagnetic property of oxygen. Permanent magnets in the measuring chamber that is passed by the measuring gas create a strongly non-homogeneous magnetic field. If oxygen is present in the measuring gas, the oxygen molecules are drawn into the non-homogeneous magnetic field due to their paramagnetic behavior. The highly non-homogeneous character of the magnetic field creates oxygen particulate pressures of different magnitudes. These particulate pressures are higher at locations showing high field strength than at locations of low field strength. The pressure differential between the partial pressure of the measuring gas and the partial pressure of a reference gas indicates the oxygen concentration in the exhaust gas.

At least three types of oxygen analyzers are in common use, as follows:

1. Those that work using the property of paramagnetism;
2. Those that use oxygen as part of the electrolyte in an electrical cell; and
3. Those where the output of a fuel cell will vary, depending on the level of oxygen that enters the analyzer.

3.7.2 WORKING OF OXYGEN ANALYZERS

Paramagnetic analyzers have an advantage over the other two types because the sensing element does not have to be replaced regularly (as in the case of the fuel cell type) or serviced (as in the case of the electrolyte type). Figure 3.5 gives an example of a paramagnetic O_2 analyzer. Oxygen is strongly paramagnetic, which means that the oxygen molecules will tend to align themselves in a way that adds to the strength of a magnetic field. In the design of one popular instrument, the exhaust gas is passed through a chamber shaped similarly to a dumbbell. The aligning force within the oxygen molecules is measured, and this force, therefore, is proportional to the amount of oxygen in the exhaust gas.

Another popular method is to create a magnetic field around the outlet of a small pipe. As oxygen collects around the end of the pipe, it restricts the amount of nitrogen that will flow through the pipe. The change in flow of the nitrogen, therefore, is proportional to the amount of oxygen in the exhaust gas. Most gas analyzers measure the concentration of a gas by comparing the gas with two known reference points. The reference points are known as the zero and span values. The analyzer is calibrated before each measurement by passing a zero gas (usually nitrogen or air) through the analyzer and setting the response of the analyzer to zero. A gas containing a known concentration of the gas to be measured (span gas) then is passed through the analyzer, and the sensitivity or gain of the analyzer is adjusted to match the value of the span gas (Figure 3.5).

3.8 GAS CHROMATOGRAPHY (GC)

The FID provides with a measurement of total or gross hydrocarbon concentration, it doesn't give information on individual hydrocarbons. Gas chromatography (GC) is used to identify the hydrocarbon components. Generally, it is used in combination with FID to improve accuracy of measurement. GC is a well-known chemical analysis tool for physically identifying the components in a chemical mixture. A number of chromatographic methods are used. The important section of the instrument is the column. It is a packed tube containing a solid or a solid-liquid which have certain absorbing properties. The gas to be sampled is injected into an inert gas stream called the carrier gas (or solvent) (Figure 3.6).

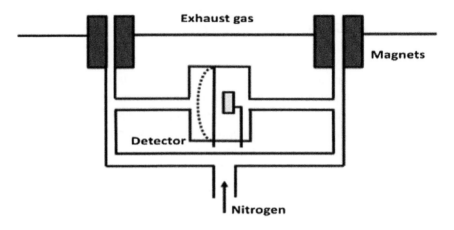

FIGURE 3.5   Paramagnetic O$_2$ analyzer.

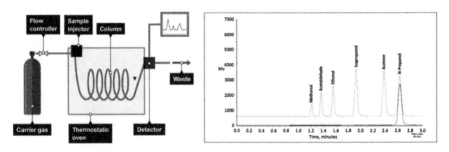

FIGURE 3.6   (a) Layout of a gas chromatograph, (b) typical gas chromatogram.

The carrier gas is the moving phase and the solid or solid-liquid is the stationary phase. As the gas flows down the column, chemical components in the moving phase migrate into the stationary phase and then back again into the moving phase. The rate of migration into and out of the stationary phase depends on the absorbing properties of the components and the stationary phase. Some components absorb more strongly than others, so that there will be a tendency for some of the components in the carrier gas to lag behind others as the gas flows down the tube. If the column is long enough the sample will separate into discrete zones of each component in the mixture with sharp discontinuities at the zone interfaces. The gas leaving the column therefore contains slugs of the constituent substances in the original mixture.

On leaving the column, the gas enters a detector (Figure 3.6). The detector serves two functions: the first to measure the time a component is in the column called the elution time, and the second to measure the quantity of the component in the original sample. For hydrocarbon constituents we can use a FID as the detector. If the output from the detector is connected to a time base recorder, then as the carrier gas passes through the detector the product of the output signal times time will be directly related to the quantity of that component currently being detected.

The signal will reach a maximum for each component and then decrease. A record of the type shown in Figure 3.6(b) will be obtained. In this chromatogram, the first component is methane or methanol as may be the case, so that the quantity of methane/methanol is almost proportional to the output. There are small bump in the record followed by another sharp peak corresponding to hydrocarbon components. To calibrate a chromatograph, fixed quantities of known substances are injected into the sample gas and the elution time is noted as well as the output calibration. Thus, the hydrocarbon can be identified by the elution time and quantitatively measured.

3.9   ORSAT APPARATUS

The analysis of exhaust gases using Orsat Apparatus is a conventional and crude method. It is primarily used for exhaust gas analysis. Orsat Analyzer determines the volume fraction of individual gases in a mixture by selective absorption of the constituents by reagents and the consequent reduction in volume of a sample gas mixture. Reagents that are used are:

- For absorption of CO_2 it is 40% solution of potassium hydroxide.
- Mixture of pyrogallic acid (1,2,3-Trihydroxybenzene) and solution of KOH is used for absorption of O_2.
- Alternately Chromos chloride (Chromium dichloride, $CrCl_2$) for absorption of oxygen.
- Alternately copper mono chloride (CuCl) may be used as the reagent for absorption of CO.

The apparatus works on the basis of chemical titration and if handled correctly it can give excellent results but cannot measure extremely small quantities or concentrations. In this method of analysis, 100 c.c. of exhaust gas is withdrawn from the pipe directly into a water-jacketed measuring burette. The sample gas is then successively passed into a solution of caustic potash to absorb carbon dioxide, a solution of pyrogallic acid in caustic potash and water to absorb oxygen and finally, a solution of cuprous chloride to absorb carbon monoxide. At each step, the reduction in volume of the 100 c.c. sample corresponds to the percentage of the constituent in the exhaust gas. Accurate exhaust gas analysis will enable the airflow to an engine to be calculated from the fuel flow (Figure 3.7).

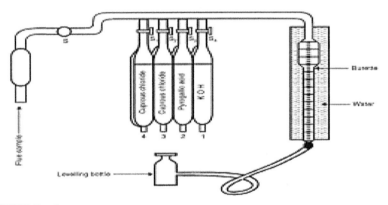

FIGURE 3.7   Orsat apparatus.

3.10   MEASUREMENT OF DIESEL ENGINE EXHAUST SMOKE AND PARTICULATE MATTER

3.10.1   *PRINCIPLE OF OPERATION OF SMOKE METERS USED IN STATIONARY DIESEL ENGINES*

One of the important and visible parts of diesel engine exhaust is smoke. The soot or smoke number is measured by a diesel smoke analyzer. The smoke is measured in both stationary and automotive type diesel engines. The smoke doesn't carry a molecular formula as it can be a partially burned carbon engulfed with other compounds such sulfates. The smoke from diesel-fuelled engines is measured connecting the engine to an engine dynamometer. The smoke analyzer measures the soot content in the exhaust of diesel engines in a photoelectrical process using the filter paper method. During this measurement, an extraction sensor extracts a specified exhaust flow from the engine and draws it in across a special filter paper having a predefined surface. The soot particulates in the exhaust cause the filter paper to assume a grayish color. This grey value is detected by a reflection photometer and is correlated with a smoke scale rating. A 100% reflection corresponds to a smoke number of zero while a 0% reflection corresponds to a smoke number that often shows the basic design of such a system.

Smoke is the emission of particles that are visible to the naked eye. There are three types of smoke:

1. **Black Smoke:** It is caused by the incomplete combustion of the fuel;
2. **White Smoke:** It is caused by the emission of atomized fuel that has not ignited; and
3. **Blue Smoke:** It is caused by lubricating oil that has passed from the engine and into the exhaust pipe.

Two techniques for the measurement of smoke are in common use:

1. Filter darkening type; and
2. Light extinction type.

The first technique uses a pump to draw exhaust gas through a filter paper. This is popularly called Bosch type smoke meter (Figure 3.8(a)). The blackness of the paper then is given a value by measuring the amount of light that is reflected back from the blackened area.

In the second technique, Hartridge type smoke meter. In this type, light is passed through the exhaust gas, and the amount of light that is able to reach a detector is measured. In the latter (light extinction technique), light can be passed either across the exhaust pipe (across the duct method) or along the exhaust pipe. The filter paper technique measures only the carbon particles that are deposited on the paper; it cannot detect blue or white smoke. The light extinction technique measures all smoke but cannot determine the color of the smoke. The main advantage of the light extinction technique is that the smoke emission can be measured continuously. With the filter paper technique, only snapshot or steady-state measurements are possible. When taking exhaust gas samples for analysis, it is recommended that a special probe design (Figure 3.8(b)) is adopted to ensure that the relative gas velocity in the main exhaust pipe does not impinge on the sample rate of the relative analyzer.

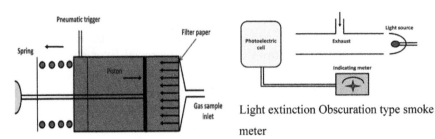

Light extinction Obscuration type smoke meter

FIGURE 3.8 (a) Bosch hand pump, (b) Hatridge type meter.

3.10.2   *WORKING OF SMOKE METERS USING THE FILTER PAPER TECHNIQUE*

The simplest smoke meters consist of a calibrated chamber, a spring-loaded piston assembly, a trigger mechanism (pneumatic), and a filter holder. The chamber is evacuated by displacing the piston against the resistance of the spring. The piston is held in the evacuated position by the trigger mechanism. A filter paper is inserted in the holder and clamped using the knurled nut. When the engine is running at the required condition, applying a pneumatic signal to the trigger mechanism triggers a reading. The piston is displaced by the spring, and a calibrated volume of exhaust gas is drawn through the filter paper. The smoke level in the exhaust gas is determined by measuring the light that is reflected from the blackened filter paper.

The following are the main problems encountered with these types of smoke meters:

1. Premature blackening of the filter paper from deposits remaining on the holder.
2. Condensation on the filter paper.
3. Blocking of the sample probe.
4. Heat damage to the sample line.
5. Inaccurate calibration of the sample volume.
6. Inaccurate calibration of the evaluation unit.

The amount of light that fails to reach the detector depends on the following:

1. The number of smoke particles;
2. The size of the smoke particles;

3. The light absorption characteristic of the particles;
4. The length of the path of the light beam.

For a particular smoke meter, the length of the path of the light beam is constant. Therefore, changes to the number, size, or absorption characteristics of the smoke particles contained in the exhaust gas will alter the amount of light that reaches the detector. The change in the amount of light that reaches the detector will cause a change in the electrical output of the detector. To achieve accurate measurements, it is important to keep the lenses of the light emitter and detector clean. This can be difficult when dirty exhaust gas is being passed continually over the lenses. A curtain of filtered air often is blown across the lenses to prevent the exhaust gas from coming into contact with the lenses and depositing particles on them. If an air curtain is used, care must be taken to ensure that only the minimum amount of air is introduced to the exhaust gas. Otherwise, the exhaust gas will be diluted by the air and will cause an inaccurate measurement. This is especially important at low exhaust gas flow rates.

The following are the main sources of problems with these smoke meters:

1. Contamination of the lenses due to loss of purge air;
2. Contamination of the lenses due to water or oil in the purge air;
3. Poor optical alignment of the emitter and detector;
4. Mechanical or heat damage to the emitter, the detector, and the electrical leads and connections.

The main advantage of the across-the-duct smoke meter is the simplicity of the installation. The absence of the need to install an exhaust gas sampling system greatly reduces the chances of errors due to hang-up, condensation, and sample system blockage. Care must be taken to present the exhaust gas to the meter correctly. The meter should not be installed close to any bends or sudden changes in exhaust pipe cross-sectional areas. One possible disadvantage of across-the-duct smoke meters is that the relatively short length of the path can cause reduced sensitivity to exhaust gas with low smoke content.

3.10.3   PARTICULATE MATTER

Production of Particulates: Of all the emissions from the diesel engine, particulates are perhaps the most problematic. They have been identified as an irritant and possible carcinogen in a number of environmental health studies and are, therefore, subjected to very tight legislative limits. Currently, global emissions legislation considers the total mass of PM being emitted by a vehicle. Although a good starting point, this approach can give a false picture as two engines producing the same total mass of PM may, in fact, have completely different emission constituents and, hence, a markedly different reading on air quality. Typical composition of diesel exhausts PM.

Carbon, 31%; Unknown, 8%; Sulphate and water, 14%; unburned fuel, 7%; unburned oil, 40%.

There are a number of concerns over possible adverse health effects of diesel particles. Potentially the most serious arises from published research in which mortality and morbidity have been shown to correlate with the concentrations of fine particles equal to or smaller than 2.5 microns in diameter (termed PM2.5). It is likely that these extremely small particulate sizes are the most detrimental to health as they can travel deep into the respiratory system, can become embedded more easily, and have a larger surface area-to-volume ratio so that they can carry more potentially carcinogenic chemicals.

3.10.4   AUTOMOTIVE VEHICLE EXHAUST GAS SAMPLING AND CONDITIONING

3.10.4.1   INTRODUCTION

The diesel fuelled vehicle smoke or particular measurement can be done by mounting a vehicle on chassis dynamometer. For the purpose of exhaust gas measurement, the exhaust gas needs to be collected. In many instances, sampling is done from the diluted exhaust. There are two systems namely CVS system and partial flow dilution system through either of which exhaust gas dilution can be performed in a full flow dilution tunnel, normally used for regulatory testing. The dilution step is very important to obtain representative measurement; sample losses or changes in sample properties can cause significant errors.

Exhaust gases need to be sampled from the exhaust system before subjecting it to measurement. The sampling system acts via media between the exhaust and measurement systems. Sampling helps in pre-conditioning the exhaust to meet the requirements of the particular measurement instruments—such as temperature, concentration, or the presence of moisture and volatiles without effecting physical and/or chemical changes of the measured species. Most measurement instruments are designed to operate close to ambient temperature. Thus, the hot exhaust gases are required to cool sufficiently.

For regulatory purposes, the U.S. federal, the United Nations Economic Commission for Europe (ECE), and the Japanese test procedures are the most preferred test procedures for measuring vehicle emissions. Even though these belong to three different places, there are common elements. The important feature of three test procedures is that they differ in respective vehicle driving cycles used for testing on chassis dynamometers as far as light-duty vehicles (LDV) are concerned and the heavy-duty vehicles are performed on engine dynamometers. For driving cycles, constant volume sampling (CVS) is followed.

The vehicle emission tests are given on pollutant emissions produced per kilometer traveled by vehicles of a given class. Computer models estimate vehicle emission factors as functions of speed, ambient temperature, vehicle technology, and other variables.

ECE and the FTP-75 tests are the methods for which standardized measurement methods used with normally CVS method. An essential aspect of CVS is to dilute all of the engine's exhaust gas (full flow) and maintain a constant volumetric flow of diluted exhaust gas (exhaust gas and dilution air). With atmospheric air, the exhaust gas is diluted and is collected in sampling bags. A blower feeds the thinned exhaust at a constant displacement.

The exhaust gas mass is determined using blower speed and temperature of the gas mixture. The mean volumetric air-to-exhaust ratio must be not less than 8:1 so that condensation can be avoided. A metering pump extracts the samples into the bag. This process is carried out according to a predetermined sequence. At the same time, part of the flow of ambient air required for dilution is sampled in bags. This is necessary to allow the ambient air to be checked for pollution. The bag contents are then analyzed. The pollutant mass of the individual components is computed from the mass of the supplied gas and the pollutant concentration in the exhaust. The pollutant mass determined in this process may be referenced to the respective test or the test duration. The emission values can then be indicated in grams of pollutant per test or in grams of pollutant per mile.

3.10.5 EXHAUST GAS DILUTED SAMPLING

As the exhaust contains volatile material (water, sulfur compounds, VOCs.), cooling may lead to super saturation of many of these species, resulting in condensation and nucleation. To avoid this, the exhaust gas has to be either (1) diluted to a level where super saturation can be avoided or (2) the volatile material has to be removed from the exhaust. Dilution—which is also an effective means to obtain the required temperature reduction—is the preferred method in most exhaust gas measurements. Dilution may be also required to lower the concentrations to match the ranges of the instruments used.

The system of measurement is illustrated in Figure 3.9.

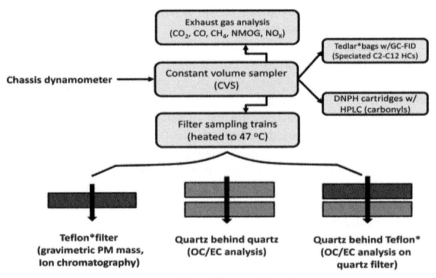

FIGURE 3.9 Constant volume sampling of diesel vehicle exhaust.

Since the diesel engine contains soot/particulate matter, the sampling for gaseous and particulate emissions is different. Enough care will be exercised for sampling gaseous emissions for the reason that no chemical reactions take place during collection process train, and hence is done in an inert atmosphere for arresting gases to contact with other foreign reactive substances. For sample, gases to flow or pass through Teflon and stainless steel tubes are used. In CVS systems, the issues of nucleation, condensation, and residence time are not important for collection/sampling of gaseous pollutant emissions. Eventually at certain temperatures, if condensation is anticipated, then water traps are required to use prior to measuring.

However, nucleation and condensation phenomena substantially interfere with the particles during sampling of particulate matter. Hence, conditions under which particulate matter sampling has to be carried out may carefully be handled to avoid nucleation/condensation to the extent possible or be kept under control.

In PM, sampling one should take care of coagulation which is an issue of concern. The residence time or sampling time is another point while sampling PM as it tends stick particle to particle and hence the particle size distribution to continuously change towards larger particles. Hence, the time collection should be as small as possible. Therefore, immediate strong dilution is a powerful way to reduce the influence of coagulation.

To satisfy the requirements of a given emission analyzer and/or measurement procedure, the dilution may need conditioning. These days as per European particle number (PN) measurement developed under the particle measurement program (PMP), there needs to a requirement for the removal of volatile particles.

3.10.6   SAMPLING OF RAW EXHAUST GAS

For simplicity and where accurate results are not required to the third decimal, undiluted or raw gas sampling is carried out. During simple sampling the effects high concentration of moisture, particulates, and high temperature may interfere with results. Such sampling is allowed in some regulatory and for mainly used for non-regulatory and/or field testing.

For handling high concentrations of undiluted exhaust samples, at elevated temperatures to avoid condensation, in case of an undiluted sample, simple instruments such as smoke meters or opacity meters can be employed.

A flow condition called isokinetic sampling is adopted for taking a sample from a gas flow (diluted or raw), for keeping the flow velocity in the sampling line same as in the flow. Otherwise, in the sampling line, aerodynamic effects would lead to depletion or an enrichment of coarse particles. The importance of this effect increases with increasing particle size. For diesel emission measurements, errors due to non-isokinetic sampling are not taken into account.

3.10.7   SMOKE OPACITY

Smoke opacity instruments measure optical properties of diesel smoke, providing an indirect way of measuring of diesel particulate emissions. There are two groups of instruments: opacity meters, which evaluate smoke in the exhaust gas, and smoke number meters, which optically evaluate soot collected on paper filters. Correlations have been developed to estimate PM mass emissions based on opacity measurement. Second generation opacity meters based on laser light scattering are much more sensitive and appear to hold promise for application to newer engines with much lower particulate emissions.

Smoke and smoke opacity meters are instruments measuring the optical properties of diesel exhaust. These instruments have been designed to quantify the visible black smoke emission utilizing such physical phenomena as the extinction of a light beam by scattering and absorption. In general, smoke, and opacity meters are much simpler (some of them very simple) and less costly in comparison to most other instruments used for PM measurement. They are often used to evaluate smoke emissions in locations outside the laboratory, such as in maintenance shops or in the field. In fact, the smoke opacity measurement is the only relatively low-cost and widely available method to measure a PM-related emission parameter in the field.

While the demand for advanced diesel engines is growing, it is essential to take care of the following issues in the metering of diesel emissions:

- Resolution;
- Cross sensitivity to nitrogen dioxide;
- Insensitivity to small particles.

Multiple light path systems with mirrors have been used to improve the sensitivity of diesel smoke density measuring units. Laser light scattering based opacity meters called generation II have shown to be promising and sensitive to measurements with newer engines, i.e., particulate filters incorporated diesel engines. Numerous correlations have been developed to correlate Smoke opacity values with PM measurement parameters but proved to be incorrect as sulfates, HCs, water vapor or physical conditions affect opacity readings and led no accurate correlation.

3.10.8   *PARTICULATE MATTER MEASUREMENTS*

Particulate matter is a complex exhaust constituent which can be characterized by several parameters, including particle mass, number, size distribution, surface, etc. Traditionally, emission compliance testing has required gravimetric determination of diesel particulate mass emissions. However, new European regulations additionally require measurement of particle number emissions. Instruments utilizing collecting or *in-situ* measurement techniques are used for the analysis of various particle parameters.

When compared to gaseous pollutants, the structure of diesel particulate matter (PM) is not well-defined substance due to its complex mixture with varying chemical composition and physical properties. Elemental carbon (EC), a variety of organic species of different volatility, sulfur compounds and metal oxides are the major components of diesel PM emissions. Of these, semi-volatile species may be in the gas phase or condensed on particles, and thus function of the temperature.

Type of fuel, engine technology, operating conditions, and exhaust aftertreatment vary the composition of diesel PM emissions. The size and composition of particulate matter would also vary with time and undergo transformation once exited into the atmosphere or during sampling line or in the measuring apparatus. The variation in PM can be attributed to particle coagulation, evaporation, and/or condensation of volatile compounds. Particulate transformation occurring in the sampling/measurement equipment may result in the formation of artifacts, i.e., compounds that are not present in either the undiluted engine exhaust gas or in the atmosphere around diesel vehicles. Therefore, there is no single absolute measure of diesel PM emissions because of the complexity, which is not present in gaseous emissions.

Nowadays, the measurement method of particulate matter has been changed, as the size of the particle is becoming smaller and smaller due to advance CRDI systems and it is based on the gravimetric quantification of the mass of particulates collected on the filters. Hence, there are challenges in evaluation of its contribution to the airborne concentrations. The focus is on the newly evolving instrumentation and instrument-based emission measurement methods for measuring the exhaust gases like carbon monoxide, carbon dioxide, nitrogen oxides (NOx), and particulate matter.

In practice, the automotive Diesel engine emission measurements are performed on an engine or vehicle dynamometer, over a standardized emission test cycle. The test cycles could be vehicle driving cycles (such as EUDC, NEDC, IDC, MIDC, etc.). The test cycles are different for different countries. The outcome of emission test cycles is aimed at to comply with EURO-Emission norms. The vehicle driving emission test cycles are sequential operating conditions simulated in the laboratory mimicking real-life operation. For the sampling, the CVS method is used in which exhaust gases are mostly diluted with air.

Emissions can be characterized, regulated, or controlled only if they can be accurately measured. In recent times, due to increased awareness with regards to hazardous affects due to very small size PM levels, a wide range of measurement techniques of different levels of sophistication, equipment cost and accuracy are being developed. High precision, high accuracy enabled standardized instrumentation with well-developed techniques are used for regulatory purposes to obtain comparable results.

In the late 1990s, high accuracy on-board emission analyzers called as portable emission measurement systems (PEMS), entirely sophisticated vehicle emission test equipment was installed on a vehicle to measure real-time emissions with a laboratory level of accuracy. PEMS systems include laboratory-class analyzers packaged in one or more portable units and a simplified sampling system. These analyzers provide results as both raw gas concentrations and mass-based emission factors, in g/km or g/kWh.

PEMS can be used for compliance testing of new and aging vehicles, development of mobile source emission models, engine control strategies, aftertreatment systems, and OBD systems. Emission measurement equipment is also useful in vehicle inspection and maintenance (I&M) programs, where vehicles must pass a periodic emission check. There is a growing trend to increasingly rely on vehicle's OBD information for I&M purposes in lieu of emission testing.

A very advanced after-treatment device called diesel particulate filter or a trap is being developed and used that removes exclusively solid fraction effectively (by more than 99%, if a wall-flow filter is used), while the volatile fraction passes through the filter in the gas phase. In modern diesels, before the PM emissions exited into atmosphere, it is made mandatory to include a diesel particulate filter or trap that traps, in EU nations.

In a particulate filter about two orders of magnitude of the solid fraction is collected in accumulation mode. After this mode, volatile species makes a nucleation mode with ultrafine particles. The method of measurement makes lot of difference in fraction of volatile material- a part of PM strongly depends on the on temperature and dilution. This condensation of volatiles is not important for older vehicles as-is for newer vehicles. Condensation of liquids on agglomerates may also change their structure by capillary forces, leading to more compact structures.

3.10.9   MEASUREMENT OF 'WET' AND 'DRY' PARTICLES

Accurate and repeated measurement of PM parameters – mass, number or size distributions with the presence of volatile material will be difficult due to the nucleation and condensation phenomena during sampling. Diesel particle measurement methods can be classified into two categories on how the volatiles:

- **Dry Measurement:** At the wish of operator, volatile material from the exhaust gas sample can be eliminated before making ready instrument for measurement. Nucleation can also be avoided by choosing a dilution ratio which is high enough to avoid the supersaturation required for nucleation.
- **Wet Measurement:** Also, one could also cool the undiluted exhaust prior to the measurement to enforce nucleation/condensation of all material that can condense. This 'wet' measurement would present a worst-case measurement, including all material which potentially can form particles.

Majority of PM measurement provides results somewhere in between these two extremes. The conventional gravimetric analysis, done at a moderate dilution and temperature, includes a significant part of the semi-volatile material and can be classified as a relatively 'wet' measurement. The European particle number measurement, on the other hand, represents a 'dry' measurement—only solid particles are measured and nucleation is prevented by adequate pre-treatment.

The exhaust of engines consists of both gaseous emissions and particulate matter. The later is more prevalent in compression-ignition engines (CI engines). Nowadays, the particulates are also present gasoline direct-injected (GDI) engines. The concentrations of gaseous components or species proportions are measured with gas analyzers while particulates are measured with special traps. Selective measuring processes tailored to suit the component to be measured are used for gaseous components. Solid matter, e.g., particulates, is detected gravimetrically on the basis of filter loading or of the discoloration of the filter paper.

To ensure that correct measurement technology is applied that also meets the legally specified boundary conditions, various physical measurement principles have been introduced.

3.10.9.1 *EMISSIONS' MEASUREMENT AND TESTING PROCEDURES*

Due to differences among models of vehicles, operating conditions, and other factors that cause wide variations emissions even though similar tests are conducted. To remove aberrations, there is a need to have an emission regulations test that is consistent and replicable. This procedure should as well be used for in-use conditions or severe enough to ensure effective emission control system performance under all conditions.

The European test procedure and emissions standards are used in the European Union. In India, in addition, EDC/EUDC, includes an Indian driving cycle or its modified version. These tests are performed on a chassis dynamometer following a specific driving cycle.

The CVS method is used for checking and measuring emissions. Basically, these driving cycle tests are designed to replicate the normal road/conditions that an operator experiences. These conditions include principal concern of control programs, in stop-and-go driving conditions typical of urban areas. The layout of a typical emissions testing laboratory is shown in Figure 3.10.

CVS Sampling System

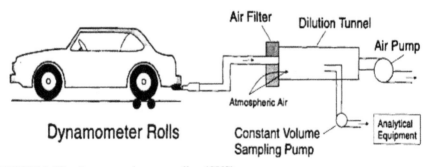

FIGURE 3.10   Constant volume sampling (CVS).

For testing in-use vehicles with engine dynamometers, to avoid removal of engine from bonnet, the whole vehicle is placed on a chassis dynamometer and subjected for emissions measurement. The levels of in transient tests, Particulate matter and hydrocarbon emissions in transient tests are observed to be higher than in steady-state tests. In transient conditions, diesel particulate matter and hydrocarbon emissions are particularly sensitive to the test cycle used.

However, emissions of NOx show a better correlation between transient and steady-state tests. As vehicles operate in a variety of speed-load conditions, it is important that testing procedures reflect these conditions. For such conditions to include it is sufficient to subject the vehicle for a limited number of operating conditions and hence be insufficient for assessing control regulations. Therefore, well developed and programmed electronic engine control systems are required to subject to. However, this situation poses difficulty in a transient test cycle. For overcoming this difficulty, the European Union has developed an appropriate transient test cycle.

3.10.10   PARTICLE NUMBER COUNTING

With the more and more stringent norms are being promulgated, it has become to reduce PM as it not only a pollutant but interferes with visibility. Particle number (PN) is a well-established metric in all of the engine development process and it is mandatory for certification of Euro 6 passenger cars, light commercial vehicles (diesel and gasoline) and Euro VI for Heavy Duty Engines. European Union is also put restrictions on the PM emissions of gasoline direct injection (GDI) vehicles.

Since only very sensitive instruments are able to capture the particulate emissions of modern combustion engines. The future homologation procedures include modified EPA particulate measurement (2007) as well as particle number counting.

First, coarse particles that stem from reintroduced wall deposits rather than directly from combustion are separated. Second, the exhaust gas is diluted and subsequently heated to 400°C. Third, dilution is performed once more before the particle number counter (PNC) to cool the exhaust gas and lower the particle number further.

A particle number is obtained in a condensation particle counter (CPC) and volatile nanoparticles are eliminated.

Thus, only non-volatile particles, i.e., chiefly soot particles, are counted. This requirement is rooted in two factors.

On the one hand, non-volatile particles are toxicologically more relevant to human health. On the other hand, reproducible measurement of volatile particle emissions has proven to be extremely difficult. This is not a problem with the measurement per se since volatile particles can be counted just like solid particles. However, the formation of homogeneously condensed hydrocarbons and sulfates after particulate filters is extremely sensitive to the slightest changes in engine or exhaust gas conditioning.

The conditioned exhaust gas may also be analyzed for such particle properties as size distribution, active surface area, etc. (not required by PMP). Although not explicitly required by PMP, CPC are the most common and most sensitive systems for particle number counting in the range of sub-microns to a few nanometers (Figure 3.11).

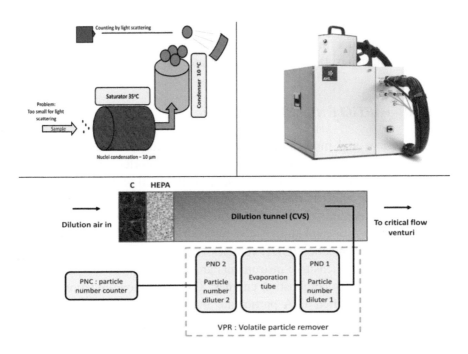

FIGURE 3.11 Condensation particle counter (CPC).
Source: https://www.avl.com/-/apcplus-avl-particle-counter.

3.10.11   PARTICLE NUMBER MEASURING INSTRUMENT

The numbers of diesel powered vehicles are increasing day by day with the advancement in the diesel engine and aftertreatment technology. Also, diesel power train has an edge over gasoline when it comes to fuel economy. However, these vehicles emit more amount of PM from their engine during combustion compared to gasoline engines. Engine PM is a complex mix of solid, semi-solid, gaseous, and liquid hydrocarbons, metals, EC, sulfates, and nitrates that range in size from a few nanometers to several microns in aerodynamic diameter. For most of the last forty years, transportation-related PM emission regulations have focused on reducing the mass of PM emitted. However, a growing body of evidence suggests that mass alone is not a sufficient measure of exposure to PM and its associated health risks. In recent years, researchers have studied the impacts of ultrafine particles

(UFPs), with aerodynamic diameters of less than 100 nanometers (nm). Although UFPs are not a major factor in mass-based PM measurements, they are the dominant contributor to the overall number of particles.

A CPCs from AVL was used during all the tests. This device is as per the recommendation of GRPE PMP program. AVLCPC has got highest linearity independent from particle size with a very low maintenance; switch between high and low dilution without changing the rotating disk. The dilution air used for the primary dilution of the exhaust in the constant volume sampler (CVS) tunnel was first passed through a high-efficiency particulate air (HEPA) filter, charcoal scrubbed, and then passed through a secondary HEPA filter. The volatile particle remover (VPR) provides heated dilution, thermal conditioning of the sample aerosol, and secondary dilution for the cooling and freezing of sample evolution prior to entry into the particle number counter (PNC). In this way, in order to draw a sample from the CVS, the particle-sampling system is required to classify it according to size, to transfer it to a diluter, and to condition the sample so that only solid particles of suitable concentration are measured and sent to the particle counter. Figures 3.12 and 3.13 shows the schematic diagram of the PNC. The VPR consists of a first particle number diluter (PND1), an evaporation tube (ET), and a second particle number diluter (PND2). The PND1 is a rotating-disc diluter with the hot dilution set at 150°C and HEPA-filtered dilution air. After the first diluter, the sample is further divided into two separate flows.

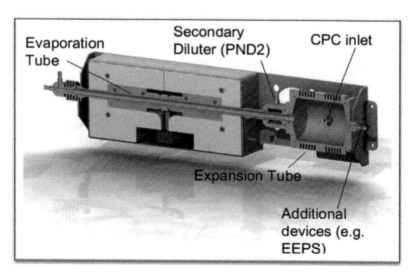

FIGURE 3.12 Particle number counter used during experiments sectional view.

The flow will be conducted to the ET and held at a constant temperature of 300°C. Finally, the sample is diluted again in the PND2 at a dilution ratio of approximately 8.0:1 and transferred to the PNC. The VPR is designed to achieve a greater than 99% reduction of 30 nm tetracontane (C 40) particles and a greater than 80% solid particle penetration at 30 nm, 50 nm, and 100 nm particle diameters. Figure 3.14 shows the internal construction of the particle number counter with actual set up picture in Figure 3.15.

FIGURE 3.13   System layout for Particle number counters on the chassis dynamometer.

3.10.12 FOURIER-TRANSFORM INFRARED (FTIR) SPECTROSCOPY

3.10.12.1 WORKING PRINCIPLE OF FTIR SPECTROSCOPY

A technique used to obtain an infrared spectrum of absorption or emission of any phase- a solid, liquid or gas, is called Fourier-transform infrared spectroscopy (FTIR). The study of interactions between matter and electromagnetic fields in the IR region involves the infrared (IR) spectra. In the infrared spectral region, the EM waves couple with the molecular vibrations. By absorbing IR radiation, molecules are excited to a higher vibrational state. At a certain frequency, the IR frequency when absorbed would actually interact with the molecule. The infrared spectroscopy is a very powerful technique provides accurate information on the chemical composition of the sample such as exhaust gas and both qualitative as well as quantitative analysis can be carried out. Fourier transformation is the principle behind the working of FTIR technique, the output from FTIR is. Interferogram, determined experimentally in FTIR spectroscopy, is representation of wave number against transmittance values. This transformation is carried out automatically and the spectrum is displayed/ infrared spectroscopy results in a positive identification (qualitative analysis). Thus, the basic theory at work is that the bonds between different elements absorb light at different frequencies.

The light is measured using an infrared spectrometer which produces the output of an infrared spectrum. The IR spectrum is a graph of infrared light absorbance by the substance on the vertical axis and the frequency (wavelength) on the horizontal axis.

FTIR spectroscopy is gaining popularity in the measurement of many gaseous components present even larger proportions in the exhaust of engines. It is a portable on-board FTIR analyzer type analyzer mostly used measuring formaldehyde, acetaldehyde, and greenhouse gases (GHGs) (CO_2, methane, and N_2ONO, NO_2), CO, and ammonia. With this analyzer, the results of NOX, CO, and CO_2 obtained on engine dynamometer tests and the compared to have good relation with total emissions per cycle. It was inferred by few researchers that vehicles equipped with modern engine technology, clean-burning fuel, in combination with a three-way catalyst would result in relatively low emissions under a wide variety of operating conditions. It is observed also that except the levels of hydrocarbons, essentially all gaseous pollutants of interest are well measured by on-board FTIR and the instrument proven to be useful.

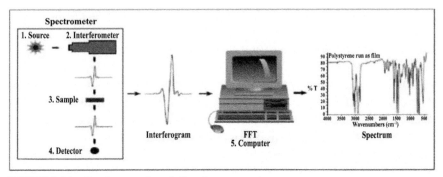

FIGURE 3.14   Fourier-transform infrared spectrometer set-up.

The wide range of wavelengths in the infrared region is measured by FTIR technique. This is accomplished through the application of infrared radiation (IR) to samples of a material. The sample's absorbance of the infrared light's energy at various wavelengths is measured to determine the material's molecular composition and structure.

This technique can also be used for identification of unknown materials, additives within polymers, surface contamination on a material, etc. The results of the tests can help in locating molecular composition and structure of a sample. FTIR technique is too fast to send the signal. A simple device called an interferometer is used to identify samples by producing an optical signal with all the IR frequencies encoded into it.

Signal obtained is decoded by applying a mathematical technique known as Fourier transformation. This computer-generated process then produces a mapping of the spectral information. The resulting graph is the spectrum which is then searched against reference libraries for identification.

The size of samples as small as 20 microns may easily be can be analyzed with FTIR analyzer attached with microscope. FTIR can also measure levels of oxidation in some polymers or degrees of cure in other polymers as well as quantifying contaminants or additives in materials.

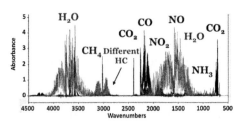

FIGURE 3.15   FTIR spectra.
Source: Adapted from: Vojtíšek-Lom et al. (2018).

3.11   VEHICLE DRIVING CYCLES: INTRODUCTION

To assess the performance of a vehicle behavior, it is subjected to tests on a chassis dynamometer which will help to simulate the road conditions unlike an engine dynamometer. The vehicle performance and emission testing will be on both engine as well as chassis type dynamometers. These tests are useful for certification of vehicles.

It is ensured that the followings things be ready for performing tests:

- Test room or ambient conditions;
- Weight of vehicle mass and road conditions to which vehicles subjected to;
- Testing equipment used for measurement;
- Emission and other analyzers with calibration details;
- Fuels for use;
- Test cycle and gear-shift series;
- Data Logging and Observations.

In real driving situations, a vehicle will be subjected to conditions such as idling, acceleration, constant speed, and deceleration. These varying speed situations are a result of the increasing number of vehicle, traffic congestion problems, road conditions, and traffic signaling are few of the reasons. Also, in some countries, unavoidable traffic variations takes place due to animal movement and unforeseen traffic situations. Moreover, sudden changes in wind/seasonal variations such as rain all leading to variations in vehicle speeds. These issues with serious consequences on emissions and fuel consumption. The condition of a driver and his experience also a matters a lot. Therefore, in order to examine the effects of the driving patterns on fuel consumption and exhaust emissions from cars in any country, a passenger car (both gasoline and diesel versions) vehicle will be subjected to severe tests on a chassis dynamometers resembling speed patters and time. It is common practice to have a driving cycle for each country and hence a driving cycle represents a set of vehicle speed points versus time (Giakoumis, Driving, and Engine Cycles, 2017).

It means to assess the performance and emissions of a vehicle under transient conditions. In a normalized pattern, for comparison of a vehicle from other vehicle, driving cycle is used to assess fuel consumption and pollutants emissions of a vehicle. On a chassis dynamometer, tailpipe emissions of a vehicle are collected and analyzed to assess the emissions rates. Thus, a driving cycle is plot between vehicle speed and time. The driving cycle is not performed for commercial vehicles but engine dynamometers are used for evaluation through a set of engine torque and speed points instead of vehicle speed points.

There are two kinds of driving cycles, the modal cycles as the European standard NEDC, or Japanese 10–15 Mode and the transient cycles as the FTP-75. Main difference is that modal cycles are a compilation of straight acceleration and constant speed periods and are not representative of a real driver behavior, whereas transient cycles involve many speed variations, typical of on-road driving conditions. Vehicle emissions constitute the main source of atmospheric pollution in modern cities. The typical driving profile consists of a complicated series of accelerations, decelerations, and frequent stops and it is simulated by driving cycles on a laboratory chassis dynamometer. The New European Driving Cycle (NEDC) is applied in laboratory test approvals in the EU and is based on traffic data from European capitals (Paris and Rome). The FTP 75 driving cycle and the Japan 10–15 modal cycle are currently used in the United States and Japan, respectively.

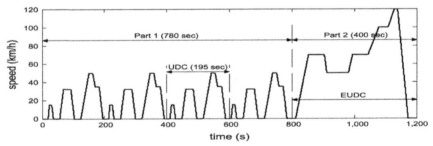

FIGURE 3.16   Typical vehicle driving cycle.

Types of driving cycles:
1. **Transient Driving Cycles:** This involves many changes, representing the constant speed changes typical of on-road driving.
   - American FTP-75 cycle.
   - European Hvzem cycle.
2. **Modal Driving Cycles:** Modal driving cycles involve protracted periods at constant speeds.
   - European NEDC cycle.
3. Japanese 10–15 Cycle;
4. European Driving Cycles:
   - Urban driving cycle;
   - EUDC;

- EUDCL;
- NEDC.

Tzirakis et al. (2006) conducted measurements of fuel consumption and vehicle emissions of three vehicles (Citroen Xsara 1.6L, a Mitsubishi Space Runner 2.0L Turbo and a Chrysler PT Cruiser 2.4L Turbo). For this purpose, they developed an Athens Driving cycle – ADC and compared with European driving cycles, verifying the exceptionality of Athens. Fuel consumption showed an increase for ADC compared to ECE and NEDC in percentages that vary from 9%to about 79% depending on the vehicle.

They were of the opinion that Driving patterns should be constantly updated with new traffic data recordings due to the alterations of the road network, the growing and changing of the car fleet, the traffic adjustments and changing driving behavior.

They should be included in the future or every country in the EU should proceed to the development of individual driving cycles (according to their unique driving conditions) for their own emissions testing and certification procedures.

Vehicle driving cycle has been developed for passenger cars and motor-cycles in Chennai, India. Representative driving cycles for passenger cars and motorcycles reflecting the real-world driving conditions in Chennai, India. To collect second-by-second vehicle speed from a representative set of vehicles for developing the driving cycles, on-board diagnostic (OBD) reader and global positioning system (GPS) receivers were considered. Eleven assessment measures were used in the construction of the driving cycles from the micro-trips. The developed cycles for motorcycles and passenger cars were of 1,448 and 1,065 seconds (peak-hour), which were further compared with existing driving cycles. The study revealed that to develop a driving cycle variables should include for a specific city adequate number of vehicle numbers should be considered. Test routes should consist of urban arterials, highways, and local streets to cover significant volume of traffic along with driving behavior across drivers was not considered. It is concluded that driving cycle should meet the requirement of major cities and it should cover wide variety of vehicle types (Ramadurai and Nagendra, 2016).

1. **Urban Driving Cycle:**
   - The Urban Driving Cycle, also known as ECE R15 cycle, has been first introduced in 1970 as part of ECE vehicle regulations act. The cycle has been designed to represent typical driving conditions of busy European cities, and is characterized by:

      o   Low engine load;
      o   Low exhaust gas temperature;
      o   Maximum speed of 50 km/h.
- The cycle ends on 195 s after a total theoretical distance of 1017 meters, with an average speed of 18.77 km/h.

2. Extra-Urban Driving Cycle:

- The EUDC introduced by ECE R101 in 1990, has been designed to represent more aggressive, high speed driving modes. The maximum speed of the EUDC cycle is 120 km/h.
- Both speed and acceleration higher than ECE 15.
- Total duration is 400 s and theoretical distance is 6955 meters, with an average speed of 62.6 km.

3. New European Driving Cycle (NEDC):

This is a combined cycle consisting of four ECE R15 cycles followed by an EUDC or EUDCL cycle. The following measurements are usually performed in this cycle:

- Urban fuel economy (first 780 seconds).
- Extra-Urban fuel economy (780 to 1180 s).
- Overall fuel economy (complete cycle).
- CO_2 emission (complete cycle).
- Carbon monoxide.
- Unburnt hydrocarbons.
- Nitrogen oxides.

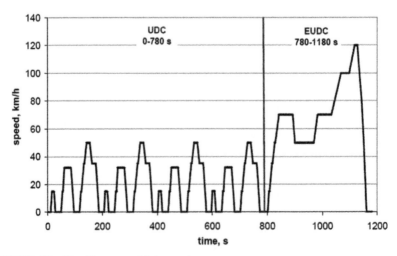

FIGURE 3.17   New European driving cycle.

4.   US Driving Cycles:

- **FTP 72:** The US FTP-72 cycle is also called Urban Dynamom-eter Driving Schedule or LA-4 cycle.

The cycle simulates an urban route of 12.07 km with frequent stops. The maximum speed is 91.25 km/h and the average speed is 31.5 km/h.

The cycle consists of two phases:

    ○   505 s (5.78 km at 41.2 km/h average speed).
    ○   867 s. (6.29 km at 26.12 km/h average speed).

The first phase begins with a cold start. The two phases are separated by stopping the engine for 10 minutes. Weighting factors of 0.43 and 0.57 are applied to the first and second phase, respectively.

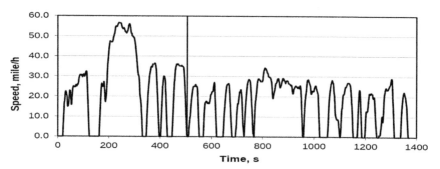

FIGURE 3.18   Federal test procedure cycle.

- **FTP 75:** This cycle is derived from the FTP-72 by adding a third phase of 505 s, identical to the first phase of FTP-72 but with a hot start. The third phase starts after the engine is stopped for 10 minutes.

FTP-75 cycle consists of the following segments:

    ○   Cold start transient phase (ambient temperature 20–30°C), 0–505 s.
    ○   Stabilized phase, 506–1372 s.
    ○   Hot start transient phase, 0–505 s.
    ○   Weighting factors are 0.43 for the cold start phase, 1.0 for the 'stabilized' phase and 0.57 for the hot start phase.

The following are some basic parameters of the cycle:

    ○   Duration: 1877 s.

    O   Distance traveled: 17.77 km.

    O   Average speed: 34.12 km/h.

    O   Maximum speed: 91.25 km/h.

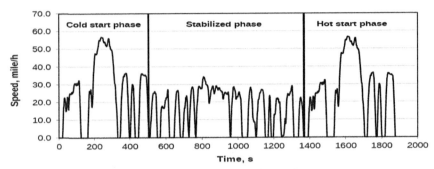

FIGURE 3.19   Federal test procedure cycle under different phases.

- **SFTP-US06:** This was developed to address the shortcomings with the FTP-75 test cycle in the representation of aggressive, high speed and high acceleration driving behavior, rapid speed fluctuations.

The cycle represents a 12.8 km route with an average speed of 77.9 km/h, maximum speed 129.2 km/h, and a duration of 596 seconds.

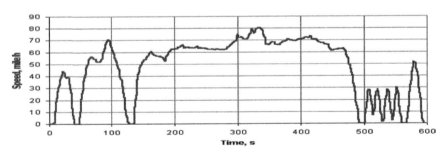

FIGURE 3.20   Supplemental federal test procedure (SFTP).

This has been introduced to represent the engine load and emissions associated with the use of air conditioning units in vehicles certified over the FTP-75 test cycle.

The cycle represents a 5.8 km route with an average speed of 34.8 km/h, maximum speed 88.2 km/h and a duration of 596 seconds.

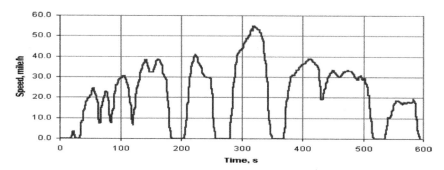

FIGURE 3.21   Supplemental Federal test procedure (SFTP) with FTP.

- **IMDC 240:**
  - This test is a chassis dynamometer schedule used for emission testing of in-use LDVs in inspection & maintenance programs. The test was formulated based on selected segments of the FTP-75 test cycle.
  - It is a short, 240-second test representing a 3.1 km route with an average speed of 47.3 km/h and a maximum speed of 91.2 km/h.

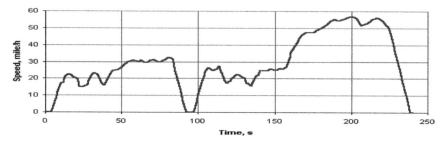

FIGURE 3.22   Light duty FTP cycle.

- IM 240
- LA-92

5. **Indian Driving Cycle:**
   - IDC was formulated around 1985, at ARAI, Pune when the mass emission norms came.
   - Based on the heterogeneous traffic pattern in major cities in India.

○ This is now only followed for two/three-wheelers, which are common modes of transportation in Indian cities.

○ For passenger cars, the European driving cycle, with a modification to provide for a lower maximum speed.

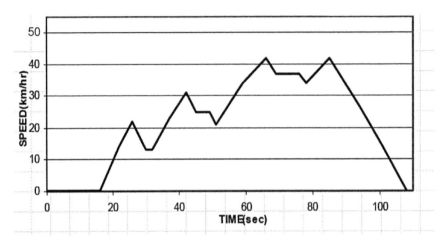

FIGURE 3.23   Indian driving cycle.

3.12   CONSTANT VOLUME SAMPLING (CVS)

- CVS is the accepted method for measuring the mass of pollutants emitted from the exhaust of cars when being emissions certified on a particular legislated drive cycle.
- CVS has to dilute the exhaust gas before it is measured by the analyzers.
- The exhaust gases are drawn into the CVS and into a dilution tunnel where dilution air is mixed with the exhaust gas in order to maintain a constant total flow of exhaust + dilution gas. This constant level of total gas flow rate is maintained by a blower at the other side of the dilution tunnel.
- After the diluted exhaust gas has been mixed then it is taken to Teflon bags via heated sample lines.
- These sample bags are subsequently analyzed and the total mass of pollutants present in the bag is used for certification purposes.

KEYWORDS

- **chemiluminescence detector**
- **gas chromatography**
- **hydrocarbons**
- **inorganic fraction**
- **non-dispersive infrared**
- **particulate matter**

REFERENCES

Arun, N. H. Mahesh, S., Ramadurai, G., & Shiva Nagendra, S. M. (2017/2016). Development of driving cycles for passenger cars and motorcycles in Chennai, India, *Sustainable Cities and Society, 32*.

Borman, G. L., Ragland, K. W. (1998). *Combustion Engineering,* McGraw-Hill International Editions: Mechanical Engineering Series.

Giakoumis, E. G., (2017). *Driving and Engine Cycles,* Springer International Publishing AG.

Heywood, J. B. (1988). *Internal Combustion Engines Fundamentals,* McGraw Hill.

Springer, G. S., & Patterson, D. J. (1973). *Engine Emissions-Pollutant Formation and Measurement.* Plenum Press: London.

Turns, S. R. (2000). *An Introduction to Combustion: Concepts and Applications,* McGraw Hill.

Tzirakis, E., Pitsas, K., Zannikos, F., Stournas, S. (2006). Vehicle emissions and driving cycles: comparison of the Athens driving cycle (ADC) with ECE-15 and European Driving Cycle (EDC). *Global NEST Journal, 8*(3), 282–290.

Vojtíšek-Loma, M., Beráneka, V., Klír, V., Pechoutb, M., and Voříšekc, T. (2018). On-road and laboratory emissions of NO, NO2, NH3, N2O and CH4 from late-model EU light utility vehicles: Comparison of diesel and CNG. Science of The Total Environment Volumes 616–617, March 2018, Pages 774–784.

CHAPTER 4

SI-in-Cylinder Measures

4.1 INTRODUCTION

In Chapter 2, the mechanism of the formation of emissions from both SI and CI engines is described. It is observed that both engines are equally responsible for emitting obnoxious emissions that are harmful to life on the planet Earth. The control of harmful emissions in conventional SI engines fuelled by gasoline/petrol is somewhat straight forward compared to diesel-fuelled CI engine. The main reason that can be attributed to this factor is its homogeneous mode of combustion (pre-mixed mode). The typical pollutant emissions from gasoline fuelled SI engine are carbon monoxide (CO), oxides of nitrogen (NOx), and hydrocarbons (HCs).

The operating period and transient operating conditions play a major role in the formation of pollutant emissions. The running of vehicular engine in initial 60 to 80 seconds have a crucial effect on whether the emission standards will be complied with, as established through the stipulated vehicle driving test cycles, during this period, the engine has to overcome heat losses and idling/residual gas effects. Moreover, the higher frictional power to be overcome and a relatively cool catalytic converter, cold intake pipes and combustion chamber walls, are some of the factors that contribute to sharply increased pollutant emissions, particularly of HCs (partly VOCs) and CO. Precisely timed correlation of parameters such as mixture strength and ignition timing demand rapidly changing operational parameters caused by driver action.

Both engine related parameters (geometrical/operational) and aftertreatment techniques play role in containing harmful emissions. In this direction, some of the engine and combustion variables that affect emissions are: (i) air-fuel ratio (AF); (ii) spark-ignition timing; (iii) turbulence in the combustion chamber; and (iv) exhaust gas recirculation (EGR). The interdependent parameters which would affect these parameters may help in reduction of harmful emissions. The AF tops among these parameters.

By identifying and optimizing certain engine design and operating parameters, effective control strategies, engine-out pollutant emissions can be reduced substantially from uncontrolled levels to marginal levels. This involves trade-offs among engine complexity, fuel economy, power, and emissions. While affecting these parameters, one should achieve lower pollutant without sacrificing the engine performance.

Therefore, the pollutant emissions of spark-ignition engines may be reduced in a number of ways. The approaches can as well be applied to compression-ignition engines (CI engines).

For incorporating variations in design and operating parameters, it is essential to follow the approaches in sequence:

- **Design/Geometrical Parameters:** Changes in the combustion chamber design, sparkplug location, injector position, number of valves, displacement (swept volume), bore-to-stroke ratio, compression ratio (CR), inlet, and exhaust port shape, etc.
- **Operational Parameters:** These include equivalence ratio, ignition timing, injection duration, and injection timing, valve timing, internal/external EGR based on modified engine timing, etc.

4.2   SI ENGINE EMISSION CONTROL TECHNIQUES

As presented in the above section, the tail-pipe or combustion generated pollutant emissions from a typical gasoline fuelled SI can be mitigated by adopting both in-cylinder approaches and exhaust gas aftertreatment methods.

SI engine in-cylinder control strategies can be grouped into two categories; varying engine geometrical parameters and engine operational parameters. Over and above these two, engine-out emissions can be mitigated using aftertreatment techniques (these are also called as engine add-on techniques). In this chapter, only in-cylinder measures are discussed.

4.3   SI ENGINE IN-CYLINDER MEASURES OR TECHNIQUES

In the control of SI engine exhaust emissions, by employing in-cylinder measures, it does not cost extra as in case of aftertreatment devices;

The following may be a few factors to be taken into account in control of emission in both spark-ignition engine:

1. Use of variable CRs;
2. Use of varying spark timings/fuel injection timings;
3. Adoption of fast burn combustion chambers;
4. Use of proper grade fuel to achieve complete combustion without abnormal combustion (knocking/detonation);
5. Use of proper cooling and lubricants;
6. Adoption of supercharging in case of CI engines;
7. Mechanism for proper swirl squishes, tumbles, and tumbles;
8. Use of variable displacement approach;
9. Use of electronic control with feedback systems;
10. Use of exhaust gas recirculation (EGR);
11. Use of multiple or split injection strategy;
12. Use of water-diesel emulsions.

In the present, the SI engine in-cylinder emission control methods will be discussed.

4.4   FEATURES OF ENGINE DESIGN AND OPERATING PARAMETERS

The design parameters of typical spark-ignition engines include cylinder displacement, CR and operating parameters include engine speed and spark ting, etc.

4.4.1   *EFFECT ENGINE DESIGN PARAMETERS*

4.4.1.1   *COMPRESSION RATIO (CR)*

It is understood from thermodynamic analysis of the air-standard cycle or Otto cycle (suitable for SI engine) is a strong function of CR and ratio of specific heats. However, from the practical standpoint of view, it is known that with higher CRs, keeping other parameters unchanged, the engine will detonate, and the engine has to be optimized for a CR, for a given fuel. It is proposed that the highest useful compression ratio is an outcome (HUCR) would be arrived at. Moreover, CR is one of the easiest modifications that can be done to an engine design and is the most sought after approach and was an early method used for reducing the unburnt hydrocarbons (UHC). It is also well confirmed through experiments and is proven to be effective.

The following Figure 4.1 depicts the effect of CR and speed on UHCs. These results pertain to a typical Spark Ignited single-cylinder, overhead camshaft, hemispherical head engine of 73-mm bore X 77-mm stroke dimensions. The engine was tested with and without EGR; the results for the former are shown in Figure 4.1.

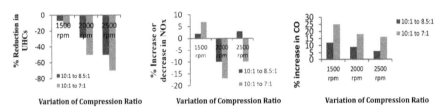

FIGURE 4.1   Variation of engine exhausts emissions with compression ratio.

It can be inferred from Figure 4.1 that by decreasing the CR, there is a drop in unburned hydrocarbons and NOx emissions however with a marginal rise in carbon monoxide emissions. As far as HC emissions are concerned, the main reason is the quenching at the wall surfaces of combustion chamber; by decreasing, the CR the clearance volume largely increases with little increase in surface area, the surface-to-volume ratio (S/V) decreases. Thus, the reduction in an S/V ratio helps in reduction in HC emissions.

To lessen HC emissions, the design of compact combustion chambers are desirable with less clearance volume crevice regions (clearances between the components of the combustion chamber should be close enough). The smaller the specific surface, the smaller is the surface of the cold cylinder walls that may cause the ignition flame to be extinguished. A favorable surface-to-volume ratio, i.e., a relatively small combustion chamber surface, also reduces emissions of unburned hydrocarbons.

Also, with reduced CRs, the exhaust temperatures are increased due to reduced expansions. This helps in exhaust gas oxidation with after reactions and thus further reductions in HC exhaust levels results. However, a higher CR reduces exhaust temperatures due to improved efficiency, in turn, impairs post-reactions of unburned hydrocarbons and CO (Figure 4.2).

Both CO and UHC in the tailpipe or in the exhaust system get oxidized with higher temperature obtained due to a lower expansion ratio. This is maintained in a thermal reactor for reduction of emissions which would get oxidized at temperatures. A further advantage is that the lower end-of-compression temperatures reduce the peak temperatures reached during the subsequent combustion and some reduction in NOx also ensues. However, engine experiences lower efficiency with reduced CR thereby increased fuel consumption and CO_2 takes place.

> *Reduction in compression ratio reduces exhaust HC levels but results in lower thermal efficiency and reduced engine power. Also, a lower compression ratio reduces the NOx levels with consequent reduced maximum cycle temperatures. Therefore to accommodate higher compression ratios, a fuel with high Octane rating is chosen Higher compression rations yields better efficiency with good fuel economy (lower specific fuel consumption). Therefore, for a given fuel, a trade-off is to be arrived at between emissions and compression ratio.*

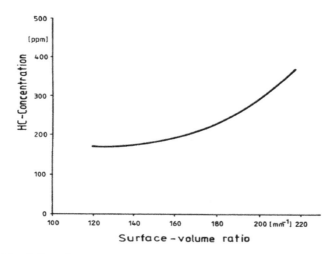

FIGURE 4.2 Effect of HC emissions on surface-to-volume ratio.

Obviously, lower CRs reduce engine efficiency and the consequence is greater fuel utilization and increased output. Moreover, it can be observed from Figure 4.2 that the variation of emissions is simultaneously influenced by speed, with increase in speed turbulence increase and thereby better combustion that helps in lowering HC emissions. With increase in turbulence, there would be increased heat losses and hence NOx emissions will reduce.

Figure 4.3 depicts the effect of increasing the CR on specific fuel consumption for two CRs. The lean-burn operation provides considerable fuel savings, while extending the operating range with regard to the air/fuel ratio, in turn, reduces CO_2 emissions. However, the exhaust temperature reduction achieved as the CR is raised may have a detrimental impact on the light-off performance of the catalytic converter in the warming-up phase. With retarded ignition timing, the drawback of catalytic converter performance can be overcome.

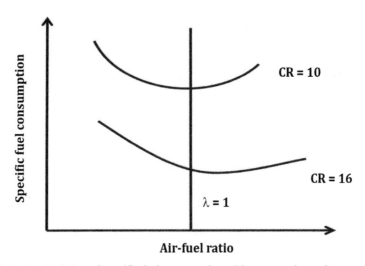

FIGURE 4.3   Variation of specific fuel consumption with compression ratio.

At higher CRs, with high peak in-cylinder temperatures, nitrogen oxide (NOx) emissions are increased, however, while with large gaps in the combustion chamber regions, the unburned hydrocarbon emissions are increased (with too large crevice regions). On the other hand, such designs respond better to lean-burn settings, or even maintaining $\lambda = 1$ control. The EGR rate can be raised at higher CRs as the misfire limit is moved further towards the "leaner" region these happen in systems with $\lambda = 1$ control. In addition, reduced NOx emissions and improved specific fuel consumption are obtained. The lean-burn operation can successfully be obtained with increased the CR with consequent reduction in fuel consumption and reduced CO_2 emissions.

4.4.1.2   BORE-STROKE RELATION AND CYLINDER VOLUME

Any variations in bore-to-stroke ratio and cylinder volume are greatly affects the SI engine emissions. By keeping longer stroke relative to the engine bore, lower HC emissions and part-load fuel consumption are achieved. However, with changes bore to stroke ratio, the surface/volume ratio gets impaired and with smaller surface to volume ratio, detonation sets-in.

Long-stroke engines are less favorable as such engine type engine yield high NOx emissions with consequent high in-cylinder temperatures. Larger swept volumes (displacements), in particular, also emit higher NOx concentrations in the exhaust. On the other hand, unburned hydrocarbons show opposite trends. It is well established through research studies that a displacement of about 500 c.c. is an optimum engine capacity in terms of fuel economy.

4.4.1.3   COMBUSTION CHAMBER SHAPE

Surface area of the combustion chamber is a factor of importance in controlling HC exhaust emissions. For a given volume, a spherical shape has a smallest surface area to volume ratio. The use of a hemispherical head and recessed bowl in the piston is a good approximation to achieve the objective. To mitigate detonation, hemispherical heads have commonly been used, without sacrificing the performance. However, from a practical point of view, a shallow angle pent-roof type is a good approximation and allows in addition good valve positioning and size.

As a yardstick of selection, compact chambers are preferable to extended chambers. It should be noted that these are also beneficial in reducing end-gas auto-ignition (detonation), also help in allowing faster bum rate and lower heat transfer losses thereby allowing higher CRs. The additional feature of such heads is that higher volumetric efficiencies and tumble within the cylinder could be achieved. Moreover, these chambers help in lessening quench zone thickness besides improving the burning rate with higher thermal efficiency, and, thereby mitigating HC emission levels. Also, twin overhead camshafts can best be designed with such chambers.

4.4.1.4   CYLINDER SIZE

In-cylinder pressures, temperatures, and wall temperatures are the indicators of quench zones of an engine. The larger cylinders have a smaller surface area to volume ratio. This will tend to reduce HC emissions. That is, for the same overall engine capacity, with all other parameters identical, an engine designed with fewer, larger cylinders should provide a benefit in relation to unburned HC emissions. The larger the cylinder, the greater the proportion of burning that takes place in the high-temperature gases away from the walls and that will mean that more NOx is formed. In the literature, very few studies on experimental data are available on the scaling effect of engine size on emissions.

4.4.1.5   ENGINE SPEED

The volumetric efficiency (VE) of an engine is strong function engine speed, with an increase in speed to a certain level it increases and then dips with further increase speed due to fluid friction losses dominating at higher speeds. This is due to the reason that the amount of charge or air intake varies with engine speed. Due to high residual gas, dilution at higher speed further drops in VE takes place. Although heat transfer rates increase with increase in engine speed as a result of higher turbulence, but total amount of heat transfer is lower due to shorter cycle time. Hence, cylinder gas temperatures increase at higher speeds. However, at high speeds, a shorter time is available for NO formation kinetics. The net result is a moderate effect of speed on NO, although this is specific to the engine design and operating conditions. An increase in exhaust gas temperatures at higher speeds enhances post flame oxidation of unburned hydrocarbons. About a reduction of 20 to 50% in HC emissions has been observed with an increase in speed from 1000 to 2000 rpm.

> *At high speeds due to shorter residence time, NOx formation is less. However, at higher speeds, exhaust oxidation with high temperatures results in lower HC emissions.*

It is revealed through numerous experimental results at constant load, the variation of HC emissions is significant with speed whereas the variation in NOx emission with speed is marginal. However, the load has a pronouncing effect on NO emissions and less effect on HC emissions (Heywood, 1988).

4.4.1.6   SPARK PLUG POSITION

The spark plug position, number of valves and valve gear layout and variable valve timing influence the exhaust emissions. Four or five-valve engines allow the spark plug to be positioned centrally. A compact combustion chamber with a central spark plug position that allows combustion time to be kept as short as possible thanks to short combustion paths and increased charge movement has a positive effect on HC emissions (Figure 4.4).

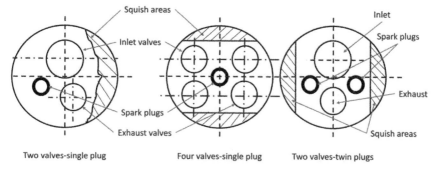

Two valves-single plug   Four valves-single plug   Two valves-twin plugs

FIGURE 4.4   Effect of spark plug position on squish zones.

Combustion noise constraints, however, tend to limit the speed of flame propagation in the mixture. Rapid flame propagation can be achieved by optimum sizing and location of squish areas (e.g., 10 to 15% of the piston area) that intensify this charge movement. By using two spark plugs per combustion chamber in two-valve engines, flame propagation can be accelerated. In modern-day, SI engine combustion chambers, spark plugs are being located in the cylinder head. Pistons of spark-ignition engines either have a slight recess or are flat-topped. Pent-roof combustion chambers are preferred and mostly used on multi-valve engines.

Figure 4.4 illustrates a typical design of spark plug position, squish areas, and valve layout for 2-valve and 4r-valve engines. Generally, in multi-valve engines, one inlet port is provided to improve mixture intake under low load condition and the other port is used as a charging port.

4.4.1.7 SPARK TIMING

The amount of power developed by the engine can be altered by varying start of spark-ignition timing and it affects NO and HC emissions. With advanced spark timing, more heat is released before and around the top-dead-center. Thus, higher peak in-cylinder pressures temperatures result with advanced ignition timings. These conditions favor for formation of high levels NOx emissions. Therefore, with advanced ignition timing, higher NO emissions are formed. However, with retarded ignition timing, the burning of mixture takes place in expansion stroke resulting in lower peak combustion pressures and a lower of mass of charge is pushed into crevice volume. Since the temperature and pressures are lowered with retarded timings, the engine efficiency suffers and hence there would be great loss of heat through exhaust gases. This condition helps in higher oxidation of HC and CO in the exhaust system. Therefore, with retarded ignition timing dual benefit is obtained, however, engine power, efficiency fuel economy suffers.

Therefore, spark-ignition timing and position of the spark plug have a vital effect on exhaust emissions. But post-reactions of HC and CO takes place with retarded ignition timing due to high exhaust temperatures, thus creating favorable boundary conditions for burning process in the exhaust system.

With retarding of ignition timing, fuel consumption increases and causes CO_2 emissions to increase. On the other hand, retarded ignition settings are an effective means of shortening the warming-up phase and ensuring early light-off of the catalytic converter. Increasing the ignition energy has no significant impact on exhaust characteristics, although it allows the operating limits to be shifted towards lean mixtures, resulting in the known benefits on NOx emissions. The exhaust recirculation rate can in this way be increased on engines with $\lambda = 1$ control.

To ensure optimum combustion of the air-fuel mixture, a large volume between spark plug and piston crown is required. A greater distance between spark plug electrode and piston crown therefore offers advantages. A central spark plug position giving short flame paths, as found on engines with four and five valves per cylinder, results in compact combustion chambers and high conversion speeds and therefore in low HC emissions and reduced fuel consumption. An opposite effect on HC emissions is observed at the higher CRs achievable in multi-valve engines.

If two spark plugs are used in two-valve engines, the flame paths can be shortened (as in multi-valve engines), although at the cost of higher NOx emission levels. Advancing the ignition timing produces increased NOx emissions across the entire range. In the stoichiometric range, in particular, advancing the ignition timing results in an excessive increase of NOx emissions.

As the exhaust, temperatures decrease when ignition timing is advanced, post-reactions in the exhaust system are reduced, and HC emissions increase. Based on the fact that the measures indicated improve lean operation characteristics and that, at the same time, the constant-volume process giving optimum efficiency is approached; fuel consumption is appreciably lower in the lean range. Knock characteristics prevent the optimum ignition timing from being matched to the effective efficiency, especially at full throttle and at high CRs. As a compromise, retarding the ignition timing is often essential.

> *HC emissions reduces with the spark retard because of increase in the exhaust temperature, which promotes HC oxidation. S[ark retard has little effect on CO levels except at very retarded timing where lack of time to complete CO oxidation leads to increased CO emissions. This increase is offset to some extent by the higher exhaust temeratures, resulting in improved CO clean-up in the exhaust system. By retarding the spark, the combustion temperatures decreases which then decreases the formation of NOx.*

4.4.1.8   COMBUSTION TIMING

Without redesigning the engine, only the initial spark timing can be controlled. Slower flame propagation rate takes place with greater ignition delay and the initial spark timing must be optimized. For typical gasoline engines, the optimal spark advance is 20 to 40° crankshaft rotation bTDC. Also, the optimal spark advance is also a function of engine speed-at higher speeds, a larger angular advance is required, and since the ignition, delay time remains roughly constant.

It is more common to retard the ignition timing as a means of reducing NOx emissions and is a trend adopted in emission-controlled engines. Excessively retarded ignition timing can increase hydrocarbon emissions, reduce power output, and increase fuel consumption. With lean mixtures, the ignition delay becomes longer and the flame speed becomes slower, so optimal spark timing is further advanced than for a stoichiometric mixture. The ignition/combustion timing is targeted for making smoother combustion in lean burn engines with $\lambda = 1$ control.

4.4.1.9   VALVE TIMING

The gas exchange phenomenon is affected by valve timing and, in turn, is a function of engine speed. Modern engines are incorporated with variable timing (especially of the inlet valves) have a flexible role to allow more charge in or more exhaust gases out depending on the engine speed.

With full load operation, lower levels of HC emissions are obtained with advanced inlet opening timing. At low loads, retarding the inlet opening is preferable as a means of minimizing HC emissions. Variable valve timing systems of this type may also help to stabilize the combustion process. Camshaft adjustment offers several benefits of reduced raw emissions, increased torque in the upper-speed range and good idle stability is achieved.

At idle and near the idle range, valve overlap should be small in order to ensure stable idling, contributing at the same time to reduced HC emissions by lowering the residual gas content. With increased valve overlap, charge dilution or loss of charge takes place thereby lower NOx emissions are obtained.

4.4.1.10   CREVICE REGIONS

Crevice regions provide a high proportion of the quench zones within an engine cylinder. Typical values are about 1% to 1.5% for the region above the upper piston ring, 0.25% for the sparkplug screw thread, and 1% to 1.5% for other regions (percent of the clearance volume). It is essential to maintain as less crevice regions as possible and should be taken into engine design consideration with respect to mechanical strength.

4.5   EFFECT OF ENGINE OPERATING PARAMETERS

4.5.1   AIR-FUEL RATIO (AF)

The numerical value of the stoichiometric AF for gasoline is 14.7:1~ 15:1, corresponding to an excess air ratio- λ of 1.00. This is a more prominent parameter and has an important effect on engine power, efficiency, and emissions. Engines using lean mixtures (λ >1.0) are more efficient than those using stoichiometric mixtures. Adoption of lean mixtures help in reduction of heat loss, higher CRs (lean mixtures knock less readily), lower throttling losses at part load, and favorable thermodynamic properties in burned gases. The generalized variation of emissions with AF for a spark-ignition engine is shown in Figure 4.5.

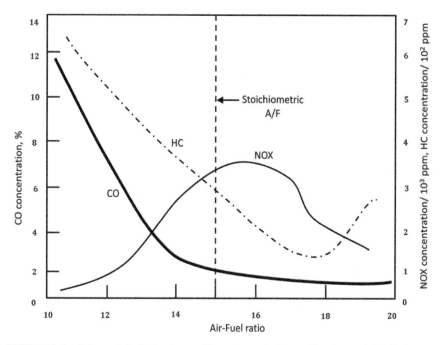

FIGURE 4.5   Effect of air-fuel ration on SI engine emissions with a lean misfire limit.

For controlling the SI engine emissions, the fuel/air equivalence ratio plays a vital role. Among the critical factors that affect emissions are the temperature of burned gases and oxygen concentration, as shown below. At slightly rich of stoichiometric air/fuel ratio, Ø ~ 1.1 maximum burned gas temperatures occur.

With reference to Figure 4.5, it can be seen that the major part of variation spread over lean mixtures. HC emissions are high for rich mixtures which can due to lack of oxygen for after burning any hydrocarbons that escape the main combustion process both in the cylinder and exhaust system. However, further leaning the mixture with increase oxygen concentration, the gas temperatures decrease and there is sharp increase in HC emissions with deteriorated combustion process and occurrence of misfires set in. The misfires happen even at the cost of high fuel consumption. The behavior of NOx emissions is different as too lean mixtures results in lower NOx levels with reduced burned gas temperatures with reduced probability of its formation.

The equivalence ratio is one of the most important variables in determining the emissions from SI engines. Leaner mixtures give lower emissions until the combustion quality becomes poor and eventually misfire occurs. Under this condition, HC emissions rise sharply and the engine operation becomes erratic. During warm-up, when the engine is cold, the rich mixture is supplied. With fuel rich mixture, there is not enough oxygen to react with all the carbon, resulting in higher emissions of CO and HC in the exhaust. At part load conditions, lean mixtures could be used which would produce lower CO and HC emissions and moderate NO emissions. The maximum value of NOx for a given engine load and speed invariably occurs at an equivalence ratio less than 1.0. The maximum cycle temperature with this lean mixture is lower than that with a richer mixture but the available oxygen concentration is much higher.

4.5.2   COOLING SYSTEM

Higher component temperatures may be obtained, e.g., if cylinder head and cylinder block are cooled separately, or by adopting hot or evaporation cooling which is a particularly efficient method. With more hot and hotter components, engine starts knocking and hence optimum cooling is essential.

Wall quenching is one of the main causes for formation of HC emissions and is formed with cylinder walls quenched at the cold cylinder wall. Therefore, coolant temperature and combustion chamber temperature have a significant role to play in HC emission formation. However, such condition of effect of low combustion wall temperatures can be compensated by retarding ignition timing. Nowadays techniques of faster heating of the coolant are being used to overcome issues with emission formation.

Also, engine temperatures both within the cylinders and in the induction system are important parameters in emission control.

4.5.3   EXHAUST GAS RECIRCULATION (EGR)

Mixing the fresh charge with burned gases is a well known technique for controlling NOx emissions. With increase charge dilution, the fresh charge heat capacity will be reduced thereby the in-cylinder peak temperatures will significantly fall thereby not providing favorable conditions for formation of NOx emissions. However, with this, the HC and CO emissions will rise in carbureted gasoline engine and there would increase PM emission in gasoline injected engines. The process of mixing homogeneous the fresh charge with burnt or spent gases is called EGR. Exhaust gas has a greater effect on flame speed and NOx emissions than the same quantity of excess air (Heywood, 1988). This is caused by the greater heat capacity of the carbon dioxide and water contained in the exhaust and the reduced oxygen content of the charge. As with excess combustion air, too much exhaust leads to unacceptable variation in combustion and increased hydrocarbon emissions. The degree of exhaust gas dilution that can be tolerated depends on the ignition system, combustion chamber design, and engine speed. Ignition systems and combustion chamber designs that improve performance with lean mixtures (lean burn—fast burn combustion systems) also improve performance with high levels of EGR. Many experiments have been tried out with variation of 0–30% EGR levels for observing drop in NOx levels. It is sometimes feasible to use supercharging in combination with EGR for restoring the loss of power and in such case; a cooled EGR is normally employed. A schematic arrangement of EGR system is shown in Figure 4.6.

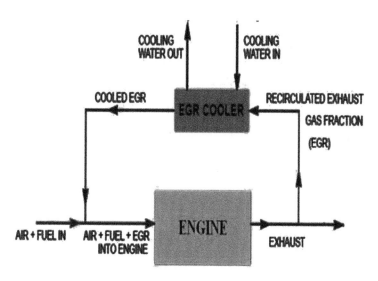

FIGURE 4.6 Illustration of exhaust recirculation system.

The effect of AF on engine emissions has already been seen in emission formation mechanism. Carbon monoxide results due to deficiency of oxygen during combustion and is reduced as the mixture weakens. CO emissions are reduced to very low values as the mixture is leaned to $\emptyset \sim 0.90$–0.95. Further, leaning of mixture shows a very little additional reduction in the CO emissions. With increase in AF, the initial concentration of hydrocarbons in the mixture is reduced and more oxygen is available for oxidation.

Hydrocarbon emissions, therefore, decrease with increase in AF until mixture becomes too lean when partial or complete engine misfire results which cause a sharp increase in HC emissions. For Ø< 0.8 engine would misfire more frequently thereby increasing HC emissions sharply. The highest burned gas temperatures are obtained for mixtures that are slightly (5 to 10%) richer than stoichiometric. On the other hand, there is little excess oxygen available under rich mixture conditions. As the mixture becomes lean, concentration of free oxygen increases but combustion temperature start decreasing. The interaction between these two parameters results in peak NO being obtained at about 0.9–0.95.

Burned residual gases left from the previous cycle or part of the exhaust gas recirculated back to engine act as charge diluents. The charge dilution by recirculation of part of the exhaust gas back to the engine is called EGR. The combustion temperatures decrease due to charge dilution caused by the residual burned gases or EGR, the decrease in combustion temperatures is nearly proportional to the heat capacity of the diluents as discusses earlier. The lower combustion temperatures resulting from the residual gas dilution/ EGR reduces NO formation and emissions (Figure 4.7).

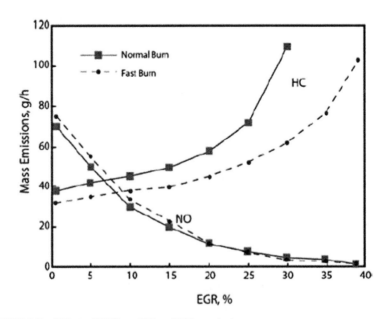

FIGURE 4.7   Effect of EGR on HC and NOx emissions.

As the EGR is increased the combustion reaction rates become more and more slow, and combustion becomes unstable. With increase in EGR cycle-to-cycle combustion variations increase and, more, and more engine cycles having only partial combustion are observed. Even though studies have been done with 30% EGR levels, an amount of charge dilution or EGR is limited to below 15% due to its adverse impact on engine performance causing power loss, high specific fuel consumption, and high unburned HC emissions.

Thus, a continuous but variable supply of exhaust gas to the inlet is required. Also, the substitution of exhaust gas has other effects. First, it replaces air in the inlet system and reduces the amount of fuel which can be burned. Hence, the fuel flow rate to the engine will need to be varied from that which would be required without EGR.

Second, the burning rate of the fuel-air mixture in the cylinder is lowered by the diluent, which may tend to reduce the thermodynamic efficiency of the engine and will certainly require a modification in the ignition timing. In order to obtain full engine power, it is customary to temporarily inactivate the EGR. This is possible as most vehicle engines do not require full power at any point in a typical driving schedule for emission testing, and thus does not affect vehicle emission rating, although it will mean that for brief periods in practice, higher NOx levels will exist. Generally, inactivation of EGR at idle conditions is allowed because with the low manifold pressures and existence of substantial residuals, combustion irregularity becomes evident with too much EGR.

In modern vehicles, with controlled inlet manifold pressure, the micro-processor enabled engine management may couple with other variables such as engine speed for EGR to be taken directly from an exhaust tapping via a control valve to the inlet manifolds.

4.6 EVAPORATIVE EMISSIONS

The essential characteristics of a good spark ignited engine fuel are high volatility, low viscosity, high heating, and high cetane number. As far as evaporative emissions from SI engines are concerned, they would be mostly of fuel leaking past certain narrow zones of the engine cylinder and fuel tank. These evaporative emissions are mostly hydrocarbons and also components that have been modified to increase octane number, alcohols may also be present. In general, the evaporative vapor comes from four sources:

1. Fuel tank venting system;
2. Permeation through the walls of plastics tanks;
3. The carburetor venting system; and
4. Through the crankcase breather.

4.6.1   *CRANKCASE EMISSION CONTROL*

In addition to tail pipe emissions, SI engine is also a source of evaporative emissions and about 25% and the fuel tank and 20% account from carburetor evaporation makes up the volatile organic fractions of the emissions. The evaporative emissions strongly depend on the ambient temperature and are high during summer season. Approximately 4–10% of the total pollutants amount to the unburnt hydrocarbons from these two sources. The evaporative emissions are regulated through positive ventilation system wherein the crankcase fumes are drawn into the induction manifold by a closed-circuit. This particular system eliminates the pollution from crankcase fumes and at an economical cost. In positive crank case ventilation system, one pipe is drawn from the interior of the air filter to the rocker cover, and another from the crankcase to the induction manifold. Thus, air that has passed through a filter is drawn past the rocker gear into the crankcase and thence to the manifold, whence it is delivered into the cylinders, where any hydrocarbon fumes picked up from the crankcase are burnt.

There are three requirements for such a system: first, the flow must be restricted, to avoid upsetting the slow running condition; secondly, there must be some safeguard to prevent blow-back in the event of a backfire and, thirdly, the suction in the crankcase has to be limited.

It is well established and wide adopted technique for all types of SI engines. In piston engines, a small quantity of the charge in the combustion space leaks past the piston rings to the crankcase due to the high pressure above the piston during the latter stages of compression and particularly during combustion.

Ventilation must, therefore, be achieved by removing the gases to where they can be oxidized to CO_2 and H_2O. This is achieved by sucking off the gases at a pressure lower than atmospheric pressure within the inlet manifold and returning them via the intake flow to the combustion chamber where they participate in the normal combustion process. The term positive crankcase ventilation (PCV) comes from this provision of a pressure drop to assist the flow.

Once the crankcase gases are drawn into the manifold, they mix with the incoming charge and are then burned in the cylinders. A one-way valve (the PCV valve) is required to ensure that flow can only take place from crankcase to manifold and that the flow is not excessive under high manifold vacuum conditions. In most modem systems, clean air is also ducted to the crankcase so that a continual purge exists. The emission advantages of a PCV system are that a significant source of UHC and CO is eliminated; the general advantages are in the slightly better ventilation and pressure reduction in the crankcase and the small saving of any wasted.

4.6.2 BLOW-BY CONTROL

The basic principle of crankcase blow-by-control system is recirculation of vapors back to the inlet manifold. Figure 4.8 shows a positive crankcase ventilation (PCV) system.

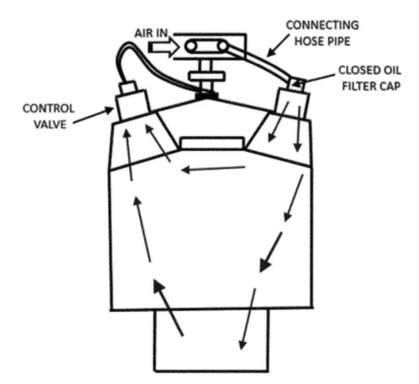

FIGURE 4.8 Positive crankcase ventilation (PCV) system.

The filtered air is drawn from the air cleaner and passes onto the crankcase. The air and the blow by gases passes through the flow calibrated PCV valve before drawn into the intake manifold in order to restrict the air flow while idling when the vacuum is high at the manifold and the blow by is less. PCV valve is spring loaded. At wide open throttle, the air flow gets unrestricted but flow rate is metered by the valve opening.

In view of the added blow by added blow by flow, the carburetor has to be calibrated. Since blow-by gas and air by-pass the carburetor in entering the manifold, the failure of PCV will give improper fuel ratios. Due to improper seating of the PCV valve at high vacuum, large quantity of air and also lubricating oil mist will enter the manifold. This will lead to lean mixtures. Also, the valve getting stuck to due to contaminations, will lead to rich mixtures. In case of valve failure, in this system, blow by is unable to go to atmosphere.

4.7 ELECTRONIC CONTROL SYSTEMS

To extract good catalytic conversion efficiency, it is required keep AF near to unity i.e., stoichiometric value as AF is predominant parameter in regulation of SI engine emissions. In addition to mixture strength, vacuum switches and other devices in earlier emission control systems are being controlled electronically. The engines have been provided with electronic controls incorporated with 3-way catalytic converters. The electronically controlled systems precisely monitor and regulate the mixture strength and always it maintains stoichiometric value. These systems also closely monitor with proper sensors for spark timing, idle speed, air injection systems, EGR, etc. But the precision and flexibility of the electronic control system can reduce emissions even in the absence of a catalytic converter. These electronically controlled systems are provided with on-board diagnostic tools which would self-diagnose engine and control system issues.

Engine electronic control systems become more complex with self-diagnostic capabilities are gaining popularity and increasingly sophisticated and important. Many modern cars in advanced countries are provided with on-board diagnostics (OBDs) is increasingly used in modern cars. It is not easy to tampering and maladjustment with computer-controlled engine systems and are resistant than mechanical controls. Also, the parameters that influence emission can be precisely controlled and monitored in computer-controlled vehicles.

4.8   SI ENGINE FUEL METERING SYSTEMS

AF plays a crucial role in formation of exhaust emissions due to its dependence on formation of emissions; hence, the role of fuel metering systems is crucial in the emission control systems of spark ignited engines. For SI engines, carburetors or fuel injection systems with which the engine fuel metering can be done. In old type and in single cylinder engines, carburetion is more prevalent and due to its difficulties, majority of manufacturers have now switched over to either multi-point or gasoline direct injection (GDI) systems.

Direct control of fuel quantity through electronic fuel injection (EFI) as a means of AF control systems for light-duty gasoline vehicles has been started in 1980s and has been evolved from mechanical systems such as carburetors. The EFI systems provide rapid and precise control of the AF. There are two basic types of fuel injection systems: central (throttle-body) injection systems, with one or two centrally-located fuel injectors; and multi-port fuel injection systems, with one fuel injector located at the inlet to each cylinder. The multi-port systems reduce cylinder-to-cylinder variations in AF and simplify intake manifold design, and ice formation was totally eliminated in intake manifold. Because they have fewer parts, central fuel injection systems are cheaper, while multi-port systems have better emissions and performance.

The electronic control unit (ECU) regulates the quantity of fuel injected by varying the length of time the valve remains open during each revolution of the crankshaft. Electronic multi-point fuel injection systems either activate at once all the fuel injectors or each injector is activated in series at the optimal time during engine rotation. The systems which activate at once are simpler and cheaper, while sequential fuel injection gives the most precise and flexible control over the injection process. Sequential systems allow better air-fuel mixing and thus better performance and emissions. As vehicle emission standards are becoming more stringent in developed countries, more vehicles are being equipped with multi-point, sequential fuel injection systems or even direct injection (DI) systems, popularly called as GDI systems.

The electronically controlled systems are provided with many sensors and actuators for precise control of quantities and parameters at strategic locations:

The sensors that are widely used include:

1. Airflow sensor;
2. Air inlet temperature sensor;
3. Camshaft position sensor;
4. Exhaust gas or oxygen sensor;
5. Engine temperature sensor;
6. Knock sensor;
7. Manifold air pressure sensor (MAP sensor);
8. NOx sensor;
9. Throttle position sensor.

4.9   FAST-BURN TECHNIQUES

The erratic behavior of SI engine is due to abnormal combustion that sets-in when the pressure, temperature, and density of the end charge attains critical values. This behavior can be understood when p-theta diagrams are traced at various speeds of engine operation. Also, one of the reasons for abnormal combustion in SI engines is slower combustion due to large stroke length associated with slower burn rate. The conditions for better and quick combustion can be obtained through inducing turbulence thereby improve efficiency, and to lessen the time required for combustion. These all issues point towards the better design of the combustion chamber to enhance burning velocity and burning rate or reduce the distance the flame has to travel, or both. In the conventional chamber, the flame spreads at a relatively slower rate, giving longer combustion duration. To overcome late or slower burn rate late in the expansion stroke, it is necessary to advance the timing so that a significant part of the fuel burns before the piston reaches top-dead-center. The hot gases produced during combustion, instead of expanding, must be further compressed. This subtracts from the network output of the engine, and also increases NOx emissions.

Due to increase in turbulence associated with high flame speeds, the burning will takes at a faster rates and also flame spreads uniformly throughout the combustion chamber with reduced combustion duration, this may happen in case even with delayed combustion. In delayed combustion but with fast burn rates, the burnt gases are not recompressed and hence there would be improved efficiency and thereby NOx are lowered. Since the energy release is faster and complete, there would be smaller mass of fuel taking part in combustion, and hence efficiency increases. Moreover, the remaining part of unburned mixture burns completely with better pre-flame reaction and hence chances of knock would also reduce. Also, in case fast-burn chambers with fast flame spread and reduced combustion duration, there would not adequate time for splitting and mixing of oxygen and nitrogen molecules to form obnoxious levels of NOx emissions and hence NOx formation reduces. Depending on the shape of the combustion chamber, the he flame area varies and is not same for all shapes of combustion chambers. Moreover, high rates of EGR and low cycle-to-cycle variability without combustion deterioration are the advantages associated with fast burn chambers (Heywood, 1988).

4.10   AIR INJECTION

The SI engine exhaust contains large amounts of unburned hydrocarbons and carbon monoxide. Since the temperature of exhaust, system is sufficiently high so that if additional is oxygen supplied, the probability of oxidation of HCs and CO will increases and thereby the chances of emission of such pollutant emissions will reduce. In the exhaust system, a minimum temperature range of 600°C–700°C for HCs and CO, is required to carry out oxidation. Hence, external supply of oxygen is needed and hence air in rich or stoichiometric condition, air is thus injected into the exhaust manifold. This air is provided either by a separate air pump or by a system of check valves that use the normal pressure pulsations in the exhaust manifold to draw in air from outside. The latter system, pulse air injection, is cheaper but provides a smaller quantity of air.

Air injection was first used as an emissions control technique in itself. It is still used for this purpose in heavy-duty gasoline engines and four-stroke motorcycles (both engine types operate under rich conditions at full load). Air injection is also used with oxidizing catalytic converters to ensure the mixture entering the catalyst has as an AF greater than stoichiometric. In vehicles with three-way catalytic converters, air injection before the catalyst must be avoided for it to control NOx emissions. So in three-way catalyst vehicles, air injection is used primarily during cold starts, and is shut- off during normal operation.

4.11   DILUENTS OR ADDITIVES

Availability of lean mixtures and high temperature are conducive condition for formation of NOx emissions. Oxides of nitrogen (NOx) emissions are essentially due to O_2 and N_2 combination that occurs at high temperatures. As noted in earlier sections NOx values peak at slightly lean equivalence ratios around i.e., Ø~ 0.9, which is the region where CO has become low and HC is also minimum.

One common method is to provide diluents to reduce the peak combustion temperature. The two major diluents that have been considered are:

- Water addition, either in the form of water injection or in an emulsified form with the fuel.
- Exhaust gas, in small proportions, recirculated back into the cylinder.

Extensive studies have been done by Quader (2017) with diluents such as CO_2, water (as steam), exhaust gas, nitrogen, helium, and argon. He targeted for reduction NO emissions.

With the addition of diluents, he observed that the specific heat of the diluent was the dominant factor and among the listed diluents, CO_2 being the best and argon the worst. However, that order was reversed for fuel consumption effects, with the CO_2 and H_2O increasing specific fuel consumption, exhaust gas and N_2 holding it roughly constant, while helium and argon reduced it.

Of these diluents, only water and exhaust gas are easily made available to an engine. The use of the former has some proponents, not only because of its ability to reduce NOx emissions but for a variety of combustion-related reasons and has been used in a number of situations in the past. The latter, because of its continuous availability in any engine, is by far the most commonly accepted diluent.

4.12   WATER ADDITION

Water was used as a coolant for many years in high duty engines, but of late, the water injection in few quantities could be used for lower in-cylinder temperatures. This technique is used to lower the NOx emissions by lowering the in-cylinder temperatures. The water injection also served as anti-detonant in SI engines and allowed engines to work with higher CRs. Spraying water into the cylinder or incoming fuel-air mixture to cool the combustion chamber of the engine, allowing for greater CRs and largely eliminating the problem of engine knocking. This effectively reduces the air intake temperature in the combustion chamber. The reduction of the air intake temperature allows for more aggressive ignition timing to be employed, which increases the power output of the engine. Depending on the engine, improvement in power and fuel efficiency can also be obtained solely by injecting water. Water injection may also reduce NOx or carbon monoxide emissions. The studies resulted in lower fuel consumption, lower CO levels, and lower exhaust gas temperatures (Shafee and Srinivas, 2017). The varying amounts of water mass (0–25%) was tried out to investigate the GDI engine performance and emission characteristics under light load conditions. The study revealed increase in the indicated mean effective pressure and efficiency with 15% water. Besides obtaining good performance, a decrease in the NOx emissions as well as soot emissions was also observed (Wei Mingrui et al., 2016).

In 1970s, rigorous work was done on the use of water addition and its effect on engine combustion and emissions and results were related to NOx reduction using either direct water injection into manifolds and ports or water-fuel emulsions. Many researchers concluded that with the help of typical city driving cycle tests, NOx emissions reduced significantly with CO remained unaltered and a penalty on unburnt hydrocarbons.

Water is added in small proportions to fuel, in homogenizer or emulsifier, a well-mixed homogeneous mixture (called an emulsion) is prepared and subsequently made use of the same in engines. However, since the viscosities and densities are different, water, and fuel use to separate out each other. Over longer intervals of its storage, emulsion stability has to be obtained even with small amounts. Moreover, with water addition the fuel also undergoes near-complete combustion with marginal reduction in HC levels in exhaust. With water addition, mainly the in-cylinder temperatures reduce and consequently reduction in NOx emissions takes place. The engine/vehicle should be equipped with addition infrastructure to carry and meter the emulsion. These days' surfactants are being used to prepare stable emulsions and its effect on emission needs an attention.

4.13   SI ENGINE EMISSION CONTROL BY FUEL VARIATION

The ability of a fuel to burn in mixtures leaner than stoichiometric ratio is a rough indication of its potential emission-reducing characteristics and reduced fuel consumption. If gasoline is replaced with propane as engine fuel, CO emission can significantly be reduced with additional reduction of HC and NOx emission moderately. Moreover, if gasoline fuel replaced with methane (its variants CNG/LNG), the CO, as well as HC emission levels can be substantially reduced.

While incorporating the fuel variants, proper care must be taken to accommodate lean mixture formation using gaseous fuel (such as propane or methane). Nowadays majority of vehicle in India being sold on two fuel variants fitted with retrofits which will easily accommodate gaseous fuels. However, while handling such fuels proper cooling and lubrication is essential, due to little higher heating value of the gaseous fuels compared to liquid hydrocarbon fuels. The issues with regards to fuel variation come under the purview of alternative fuels (AFs).

4.14   SUMMARY

It can be concluded that among the three major pollutants, CO, HC, and NOx emissions, the conditions of formation of CO, HC emissions are opposite to the formation of NOx emissions.

It can be summarized that NOx emissions in the exhaust of SI engines primarily due to availability of lean mixtures, high temperatures and secondarily due to time. Thus, the factors that lower these conditions, would reduce NOx emissions:

I. By decreasing the oxygen available in the flame region:

    i.   Use of rich mixtures

    ii. Reducing the homogeneity of the air-fuel mixture

    iii. Use of stratified charge engines

    iv. Use of exhaust gas recirculation and diluents.

II. By decreasing peak in-cylinder temperatures:

    i.   Use of retarded spark timings

    ii.   Adoption of lower compression ratios

    iii. Avoidance of knock

    iv.   Decreasing the charge temperatures

    v.   Decreasing the speed

    vi.   Decreasing the inlet-charge pressures

    vii. Use of exhaust gas recirculation and diluents.

    viii. Increasing air humidity and use of water injection.

    ix.   Use of very-lean or very-rich air-fuel ratios.

Major problem for formation of HC emissions is wall quenching and SI engine HC emissions can be reduced by:

I. *Higher temperature in the exhaust gas*

    i. *Use of retarded spark timings*

    ii. *Use of lower compression ratios*

    iii. *Increasing the charge and coolant temperatures*

    iv. *Use of high speed operation*

    v. *Increased charge pressure*

    vi. *Insulating the exhaust manifold*

II. *Use of more oxygen in the exhaust*

    i. *Use of lean mixture*

    ii. *Injection of air*

III. *Use of reduced quench zones*

    i. *Use of lower surface to volume ratios*

    ii. *Increased turbulence during combustion.*

    iii. *Increased charge and coolant temperatures*

    iv. *Use of compression ratios*

KEYWORDS

- **carbon monoxide**
- **electronic control unit**
- **exhaust gas recirculation**
- **highest useful compression ratio**
- **hydrocarbons**
- **positive crankcase ventilation**

REFERENCES

Heywood, J. B. (1988). *Internal Combustion Engines Fundamentals,* McGraw Hill.

Quader, M. A., & Ahmed, S. (2017). *Bioenergy With Carbon Capture and Storage (BECCS): Future Prospects of Carbon-Negative Technologies* Clean Energy for Sustainable Development, 91–140.

Shafee, S. M., & Srinivas, M. (2017) Improvement of fuel efficiency in a petrol engine by using water injection, *Int J Chem Sci. 15*(2), 123.

Wei, M., Nguyen, T. S., Turkson, R. F., Guo, G., & Liu, J. (2016). The effect of water injection on the control of in-cylinder pressure and enhanced power output in a four-stroke spark-ignition engine. *Sustainability 8*(10), 993, MDPI https://doi.org/10.3390/su8100993.

CHAPTER 5

SI-Aftertreatment Measures

5.1 INTRODUCTION: EXHAUST GAS AFTERTREATMENT OR POST-COMBUSTION TREATMENT TECHNIQUES

Chapter 4 dealt with the SI-in-cylinder measures to be considered for control of combustion generated pollutant emissions or tail pipe emissions from gasoline spark ignited fuelled engine. The methods described involved both engine design/geometrical and operational parameters. However, it can be inferred that these modifications sometimes poses manufacturing and operational difficulties and at high speed operation, hence these techniques will not serve effectively in controlling SI engine emissions. Due to these reasons and complying with stringent emission norms has become difficult and therefore many researchers and manufacturers switching over to adoption of exhaust gas aftertreatment methods. Therefore, effective control could be achieved by adopting post-combustion or exhaust gas aftertreatment techniques. In this method, the engine is allowed to undergo combustion and the burnt gases before being left to the atmosphere will be allowed to pass through a device called converter where in the exhaust gases are treated. The hydrocarbon gases that couldn't be oxidized completely will be allowed to undergo near-complete oxidation and oxides of nitrogen will be reduced to nitrogen in a device called a thermal or catalytic converter. A converter is typically a retrofit or an add-on arranged in the line before exhaust gases can exit from engine. A thermal converter adopts either hot exhaust gases or electric heater for further heating the gases before exiting through tail pipe of an engine.

Mostly, the converters were developed for complete oxidation of oxides of carbon, hydrocarbons, and reduction of oxides of nitrogen. Therefore, these converters are usually employed for pollutant emissions such as CO, HC, and NOx emissions.

In the world, the European Union emission norms are more stringent than in any other part of the world. The European Commission, in 2009, promulgated that car fleet CO_2 emission should be drastically brought down from 130 g/km to 95 g/km in 2015 and 2020, respectively. The commission is also hardly chalking out plans for bringing few unregulated pollutants hitherto into regulated emissions. In this lime, the emissions which pose a health risk and hence the levels of nitrogen dioxide (NO_2), N_2O, ammonia as well as the GHGs fall under this category. Therefore, the engine development and manufacturing units are trying hard to build engines that emit low levels of nitrogen oxides (NOx), carbon monoxide (CO), hydrocarbons (HC), particulate matter (PM) as well as greenhouse gases (GHGs) such as carbon dioxide (CO_2) and nitrous oxide (N_2O).

For the purpose of allowing complete oxidation, the gases are allowed to pass through a thermal converter, before letting of exhaust gases into the atmosphere. The converters resemble a heating source, permits further oxidation of exhaust gases, i.e., from more harmful products to less harmful exhaust species. In such a reactor or a converter, wherein secondary reactions occur much more rapidly and completely with prevailing high temperatures in the chamber. Hence, the engines are equipped with thermal converters as a means of lowering emissions. These are high-temperature chambers called thermal converters through which the exhaust gas flows. They promote oxidation of the CO and HCs present in the exhaust. A simple reaction for oxidation of CO into CO_2 is as under;

$$CO + 1/2\ O_2 \rightarrow CO_2$$

For this reaction to occur at a useful and fast rate, the temperature must be held around 700°C. Basically, the converters work on the principle of thermo-chemical reactions or catalytic reactions. The aftertreatment can be possible by further heating the exhaust gases through thermal converters or by employing suitable catalysts. The important function of a large enough thermal converters are therefore is to operate at a high temperature with adequate dwell time to promote the occurrence of secondary reactions. To allow such a design, most thermal converters are essentially accommodated in an enlarged exhaust manifold connected to the engine immediately outside the exhaust ports. For near-complete conversion or reduction of hydrocarbons into CO_2 and H_2O, a temperature above 600°C is required for at least 50 ms. A thermal converter should not dissipate heat for its effective functioning while keeping the exhaust gases from cooling to non-reacting temperatures.

To fit a thermal converter in an engine compartment, a designer of automobile should consider two aspects; firstly, it demands sufficient space, in modern, low-profile, aerodynamic automobiles and hence pose space constraint and is impractical. Secondly, because the converter must operate above 700°C to be efficient, even if it is insulated the heat losses create a serious temperature problem in the engine compartment.

It is essential sometimes to provide additional oxygen to react with the CO and HC to thermal converter systems and hence the thermal conversion process arrangement increases complexity, cost, and size of the system. Electronically controlled engine management system (EMS) is being provided for varying many parameters involved in thermal conversion. It is essential to provide additional air during rich operating conditions such as start-up.

Air addition is especially necessary as exhaust from engines is often at a lower temperature than sufficient for efficient operation of a thermal converter, it is necessary to sustain the high temperatures by the reactions within the system. Moreover, NOx emissions cannot be reduced with a thermal converter alone. Figure 5.1 depicts the simple structure of an aftertreatment device.

Untreated exhaust gases from engine

with pollutant emissions

(CO,NOx,HCs,PM ..)

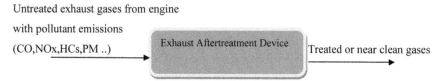

Exhaust Aftertreatment Device

Treated or near clean gases

FIGURE 5.1   Basic structure of exhaust gas aftertreatment device.

5.2   EXHAUST GAS AFTERTREATMENT APPROACHES

5.2.1   *THERMAL REACTORS*

The major exhaust emissions from SI engine are CO, HCs, and NOx. Among these pollutant emissions, CO, and HCs, before letting off them into atmosphere, can be allowed to under further oxidation to their respective fully oxidized form – CO_2 and H_2O by maintaining conducive condition in the exhaust line.

One of the primitive methods is to allow them to undergo secondary reactions in Thermal Reactor or a Chamber (or afterburner) in the exhaust system. This can be achieved by maintaining the reactor at a certain minimum temperature and there must be sufficient oxygen available to complete the oxidation process.

The question of maintaining higher temperatures arises to allow the reactor to function satisfactorily it to utilize the hot exhaust gases which are exiting with the additional energy released by the exothermic reaction. Normally, in a SI engine with rich mixture operation. A spark timing retarded from the optimal (MBT) setting will increase exhaust temperatures due to the later burning and lower power extraction during expansion, in spite of lower power, but will decrease the exhaust temperature as more heat is then transferred to the cylinder walls. In order to obtain the required reactor temperatures, it is usually necessary to run the engine rich so that the extra fuel can be burned in the reactor. Under certain circumstances, an additional spark retard would be used to provide high temperatures in the gas entering the reactor. With retarded spark timing, the gases which were expanding in the later part of expansion stroke could provide the additional thermal energy and hence can be made use a source for secondary reactions of oxidation of CO and HCs. Depending on the temperature range between from 500–800°C the oxidation percentages of CO and HCs would vary (Eran sher).

Usually, in a SI engine, combustion product gases leaving the exhaust port would lie between 350°C and 700°C. However, the temperature may reach as high as 900°C under a wide open throttle (WOT) condition and high-speed operation due to the greater mass flow rate at the higher inlet manifold pressures. This is due to the shorter time available at higher piston speeds for heat transfer to take place to the cylinder walls.

On the lower side of range given, the HCs would oxidize however; the CO will oxidize from 700°to 750°C. For achieving the high temperatures and permitting the reactor to work at high temperature, the red heat would pose difficulties in design and the chamber material frequently undergoes thermal oxidation and raises the problems in design, construction, durability, operation, and safety.

However, by allowing lower residence time for thermal oxidation for exhaust species, the use high temperatures, i.e., reactor temperature near the maximum limits can be necessitated and thereby achieving good conversion values. A time range between 50–100 ms is required to allow thermal reactions near completion stage.

Since the highest exhaust temperatures are achieved under stoichiometric conditions, the thermal reactor therefore must be placed close to the manifold outlet. The exhaust temperatures vary throughout an engine cycle by about ±200°C and that the exhaust cools rapidly in a normal exhaust pipe as it moves downstream from the exhaust port, dropping by around 50°C to 100°C/meter.

As observed, the rich mixtures are required to fulfill thermal reaction and there would be exhaust gases participating in thermal oxidation that would be deficient of oxygen and hence additional air is then required. The additional requirement may vary from 10% to 20% of the normal engine air supply. This condition necessitates use of an air pump and air nozzle to introduce the air into the Thermal Reactors. The additional air may need heating and the flow rate needs to be controlled for different conditions. Therefore, use of thermal converter for lessening partially oxidized components of exhaust gases into fully oxidized form adds to the complexity, cost, and weight. Also, as said earlier, the retarded timing were required to allow secondary reaction, it would be associated with loss of brake power. Even if all such additional conditions are maintained, it is not guaranteed that the exhaust gases would undergo high levels of oxidation. However, as far as CO and HCs treatment with Thermal Reactor is concerned, it is suitable for stationary engines.

5.3 CATALYTIC EXHAUST GAS AFTERTREATMENT

5.3.1 PURPOSE AND FUNCTION OF A CATALYTIC CONVERTER

It is understood from the previous section that CO and HCs oxidation (i.e., thermal aftertreatment) is possible only when high temperatures are provided and could best be possible through electrical heating. Another, drawback with the use of thermal reactors is that they cannot be used for treating oxides of nitrogen. The control of NOx with thermal conversion was remained a question. In the days of use of thermal reactors, the oxides of nitrogen didn't pose emission difficulties and late on it was understood that NOx was mainly responsible for photochemical smog. Moreover, the maintenance all additional items are adding to the complexity of an automobile and hence the manufactures have abandoned the use of thermal reactors for exhaust gas treatment. The researchers looked into control of all the three exhaust gases-CO, HCs, and NOx in a single unit. This has led to the development of catalytic converters.

The main issue with thermal reactors is to use higher temperatures in the downstream of exhaust line. The use of thermal reactors was totally eliminated. By employing certain oxidizing catalysts, complete oxidation of CO and HCs was made possible even at lower temperatures. Also, with the use of oxidizing catalysts, the rate of oxidation or reduction reactions can be brought down heavily.

A catalyst is a chemical that modifies or accelerates a reaction without itself being modified or consumed. There lies a skill in matching a suitable catalyst and the exhaust species with available exhaust temperatures in the exhaust gases depending on the operating condition profile such as equivalence ratio, speed, and load for effective pollutant exhaust abatement systems. The use of catalysts should be aimed at without sacrificing the performance of engine and also without unregulated emission species. Therefore, a device that is used for converting into more harmful emissions into less harmful levels is done outside the engine, is an aftertreatment device called catalytic converter. The catalyst is an important component in catalytic converter that helps and completes the chemical reaction without being consumed itself. The catalyst would enhance the rate of reactions to convert the pollutant emissions to a less harmful level or to non-harmful levels.

Catalytic converter systems will best be at its working with proper catalyst and its location in an automobile plays a vital role. It should usually be positioned beneath the passenger compartment and installed in the exhaust system between the exhaust manifold and the muffler. The location of the converter is important since. Also, it should be as nearer to the exhaust line of the engine as possible and is important as much of the exhaust heat as possible must be retained for effective operation.

The efficacy of catalyst is measured in terms of conversion efficiency. The temperature also plays a role in improving conversion efficiency.

The catalytic converter conversion efficiency is defined as the ratio of mass removal in the catalyst of a pollutant to the mass flow rate of that pollutant into the catalyst.

$$\text{Efficiency of a catalytic converter, } \eta_{\text{catalytic converter}} = \frac{m_{Pollutant,in} - m_{Pollutant,out}}{m_{Pollutant,in}}$$

Generally, it is mentioned as a percentage and normally varies from 80–95% with better functioning of a catalytic converter. The minimum operating temperature is typically between 250°C and 300°C.

The use of catalyst makes the conversion hydrocarbon based fuel emissions into carbon dioxide (CO_2) and water vapor (H_2O). The catalytic action takes at relatively lower temperatures when compared to thermal converter. With the strict emission norms being announced and followed, the use of catalytic converter has become an integral part of automobile. In India even in two-wheelers, the catalytic converters are being used extensively. To achieve best performance the vehicle should be kept running at least for one hour, in few cases of very frequent start and stop operation of vehicle will weaken the function of catalytic converter.

5.3.2   CONSTRUCTION DETAILS OF CATALYTIC CONVERTER

The main harmful pollutant emissions from spark ignited engines are CO, HC, and NOx. Among these three, the first two require oxidizing agent to oxidize the partially combusted products into fully oxidized form i.e., CO_2 and H_2O. For performing the catalyst action, a two-way catalytic converter is generally used. Also, NOx is required to be reduced to N_2 with the help of a reducing agent and when in combination, the oxidizing and reducing agents are employed in a single converter, is called a three-way catalytic converter. In the early days of catalytic converters, the NOx was not seen as harmful pollutant hence only two-way converters were used. However, over the years, NOx has become a threat and has been started gaining interest for its abatement. Hence, nowadays only a three-way catalytic converter is prevailing in the market is popularly called as TWC.

Whether it is a 2-way or a 3-way converter, normally a chromium stainless steel is used for housing the catalyst for carrying out catalytic action. The important components of a typical catalytic converter are a substrate, a washcoat, and finally a catalyst.

For the oxidation or reduction, reactions to take place inside a catalytic converter mainly depend on the available residence time. Moreover, the surface area of exposure to exhaust and the amount of catalyst requirement are important while designing a catalytic converter. For this, typically the design of a catalytic converter, therefore a honeycomb monolithic structure that exposes maximum surface area of catalyst to the exhaust stream while minimizing the amount of catalyst needed.

Two configurations namely ceramic honeycomb structure and spherical pelletized designs are used for making of catalytic converters. Figure 5.1 represents the monolithic honeycomb structure and schematic of commonly used catalytic converters. For the exhaust gases to flow through this structure it is made with number of channels for the gases to flow and hence the honeycomb structure is packed within insulating and protective masks. Aluminum oxide (alumina) is applied on the inner surface of the channels as a washcoat which acts a protective mask. The thickness of highly porous alumina washcoat is of 20 µm and normally the monolyth has inside dimensions of about 1 mm square cross section passages, the weight the washcoat varies between 5 and 15% of total weight of monolith. There have wide spread developments in the choice of catalyst and reader is referred to the review of the same (Jan Kaspar et al., 2003). The development of catalysts and its use in Automobile Catalytic Converters and its effects on emissions have been well studied (Kathleen, 1984).

The catalysts are embedded or suspended in the washcoat prior to applying to the core or substrate which acts as a carrier for the catalysts over a large surface area. The important part of catalytic converter (cat-con) is a catalyst and for carrying out oxidation and reduction reactions (i.e., redox reactions), the noble metals such as platinum, palladium, and rhodium are wide commonly preferred. The noble metals exhibit high intrinsic activity and more preferred catalyst materials. While platinum serves as both a reduction and oxidation catalyst, palladium, and rhodium are exclusively preferred as oxidation catalyst and reduction catalyst respectively. Since these catalyst materials are precious and expensive and a care should be taken to minimize the amount of the noble metals. In place of noble metals, base metal oxides can also be use but would not effective in conversion process. Of late cerium oxide (ceria), a non-expensive catalyst is also being added with the alumina in the washcoat. It acts as a buffer, absorbing or releasing oxygen, depending on the partial pressure of oxygen around it.

Iron, manganese, nickel, and copper can also be used. Due to reactive nature of these metals with the exhaust gases, few nations have stopped the use of some of these materials.

An exhaust pipe is connected to the manifold or header to carry gases through a catalytic converter and then to the muffler or silencer. V-type engines can use dual converters or route the exhaust into one catalytic converter by using a Y-exhaust pipe (Figure 5.2).

FIGURE 5.2   Photograph of catalytic converter.

5.3.3   OPERATION OF CATALYTIC CONVERTER

As described above, a catalyst is an element that starts a chemical reaction without becoming a part of, or being consumed in the process. The converter substrate contains small amounts catalysts of rhodium, palladium, and platinum. In a three-way (catalytic) converter (TWC), all three exhaust emissions (NOx, HC, and CO) are converted to carbon dioxide (CO_2) and water (H_2O). As the exhaust gas passes through the catalyst, oxides of nitrogen (NOx) are chemically reduced (that is, nitrogen, and oxygen are separated) in the first section of the catalytic converter. In the second section of the catalytic converter, most of the hydrocarbons and carbon monoxide remaining in the exhaust gases are oxidized to form harmless carbon dioxide (CO_2) and water vapor (H_2O).

5.3.3.1   CONVERSION REACTIONS

1. Conversion reactions in two way catalytic converter:

   $CxH_4x + 2xO_2 \rightarrow xCO_2 + 2xH_2O$ (conversion of hydrocarbons)
   $2xCO + O_2 \rightarrow 2xCO_2$ (conversion of carbon mono-oxides)

2. Conversion reactions in three-way catalytic converter:

   $CxH_4x + 2xO_2 \rightarrow xCO_2 + 2xH_2O$ (conversion of hydrocarbons)
   $2xCO + O_2 \rightarrow 2xCO_2$ (conversion of carbon mono-oxides)
   $2NOx \rightarrow N_2 + xO_2 [O_2 + 2H_2 \rightarrow 2H_2O]$ (Decomposition of oxides of nitrogen)

Many converters contain cerium, an element that can store oxygen. Its purpose is to provide oxygen to the oxidation bed of the converter when the exhaust is rich (oxygen deficient) and lacks enough oxygen for proper oxidation. When the exhaust is lean, the cerium absorbs the extra oxygen. For the most efficient operation, the converter should have a 14.7:1 air-fuel ratio (AF) but can use a mixture that varies slightly.

Excess oxygen enabled exhaust is required to oxidize HC and CO (combining oxygen with HC and CO to form H_2O and CO_2) whereas deficient oxygen exhaust is required for reduction-stripping the oxygen (O_2) from the nitrogen in NOx. Tuning of AF is essential to extract good catalytic converter performance and thereby the catalytic converter efficiency increases.

5.3.4   CONVERTER LIGHT-OFF TEMPERATURE

Besides the use of catalyst, a minimum temperature of exhaust is to be maintained for catalyzing the conversion and in this way, temperature of close to

300°C is required to attain for obtaining a conversion efficiency of 50%. The temperature at which minimum of 50% conversion efficiency is obtained is called as light-off temperature (Figure 5.3).

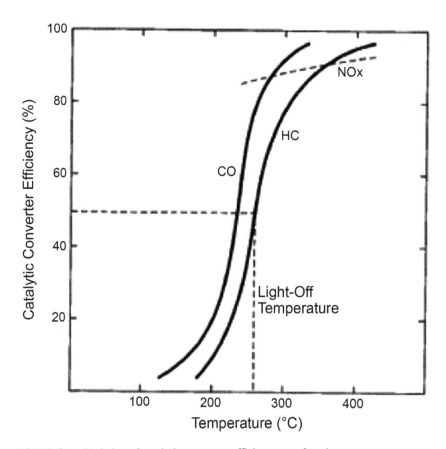

FIGURE 5.3   Variation of catalytic converter efficiency as a function temperature.

$$CO+H_2O \quad CO_2+H_2$$

From Figure 5.4 it can be inferred that hen a converter is working at temperatures above 400°C, about 98–99% of CO, 95% of HC, and more than 95% of NO_x from exhaust gas flow are obtained through oxidation and reduction reactions.

It is essential to maintain proper equivalence ratio for effective conversion of exhaust species such as CO, HC, and NOx in to their oxidized and reduced forms which possibly occurs through stoichiometric or lean mixture operation of engine.

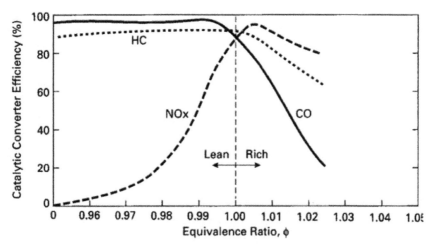

FIGURE 5.4 Catalytic converter conversion efficiency as a function of fuel equivalence ratio.

It is known that highest engine efficiency occurs when engines operate near stoichiometric conditions. Lean mixtures are not good for effective reduction of NOx emissions. Such conditions are more critical of stratified charge SI engines generally operate very lean overall. Generally, the cyclic variability of SI engines is prone for ill functioning of catalytic converter. It has been found that this cyclic variation lowers the peak efficiency of a catalytic converter but spreads the width of the equivalence ratio envelope of operation in which there is acceptable emissions reduction.

It is essential for a catalytic converter be operated hot to be efficient. Frequent stop-conditions of engine malfunctions can cause poor efficiency. Too hot running of engine would also lead to overheating of converters and thereby degrades the conversion efficiency. Moreover, a catalytic converter results in poor efficiency when a turbocharger lowers the exhaust temperature by removing energy.

A poorly tuned engine can have misfires and periods of too lean and/or too rich conditions. Such affects the converter to be inefficient at the very time emissions are very high and maximum converter efficiency is needed.

It is desirable that catalytic converters have an effective lifetime equal to that of the automobile or at least 200,000 km. Converters lose their efficiency with age due to thermal degradation and poisoning of the active catalyst material.

A temperature to an extent range of 800–1000°C is possible under certain conditions of operation, and then the catalytic converter is fully effective. In spite of the intense heat, however, catalytic reactions do not generate a flame associated with a simple burning reaction. Most vehicles use a series of heat shields to protect the passenger compartment and other parts of the chassis from excessive heat. This is most likely to occur if the heat shields have been removed from the converter. Converter usage for the effective functioning of a catalytic converter is required to be located as close as possible to the exhaust manifold. The farther back the converter is positioned in the exhaust system, the more the exhaust gases cool before they reach the converter. Since positioning in the exhaust system affects the oxidation process, vehicle manufacturers that use only an oxidation converter generally locate it underneath the front of the passenger compartment. The oxygen released during this action help provide extra oxygen to help oxidize the HC and CO into harmless water (H_2O) and carbon dioxide (CO_2).

5.3.5   CONVERTER-DAMAGING CONDITIONS

Catalytic converters do not have any moving parts in its structure and doesn't require periodic service. However, it is poisoned with excess engine oil, antifreeze agents, sulfur (from poor fuel quality), and other chemical substances that may exist in the exhaust system of an engine on the upstream of converter.

1. **Excessive Temperatures:** The presence of excessive lean mixtures or unburned fuel raises exhaust temperatures during long idling periods or high speed operation. The existence of excessive temperature cause damage to the internal parts by clogging resulting in the restricted exhaust flow severely and thus eventually reduces engine power.
2. **Improper Air-Fuel Mixtures:** Too rich or too lean mixtures cause that runs the manifolds with high temperatures causes damage to the catalytic converter. The conditions that enhance HC/CO emissions such as presence of excess fuel increase the operating temperature of the catalyst. Moreover, an air injection into exhaust stream would quickly destroy a new converter. In addition, engine ignition misfire will cause damage to the converter.

As far as stoichiometric AF is concerned, with a mixture near to 15:1 are better suited for effective functioning of converter.

3. **Sulfur Free Fuels:** Majority of automotive catalysts are sulfur sensitive. Adoption of fuel with minimal sulfur as sulfur forms sulfate, a particulates exit from engine and would block the honeycomb structure. Additional sulfate particulates originate from lubricating oil and stop the catalytic action; also, presence of sulfur may also retard or even stop the engine by increasing the back. This is one of reasons for introducing ultra-low sulfur fuel in EURO VI norms (sulfur`10 ppm).

A good transient response of the catalyst is also essential. Catalysts may lose performance in the very early stages of usage. A process of "de-greening" is usually applied in production to settle the emission conversion at its durable efficiency.

For the effective working of a catalytic converter, the fuel should be free from heavy distillates otherwise; it will form particulates and may block the catalytic converter (Figure 5.3). This will prevent free flow of exhaust gases after catalytic oxidation and exert back pressure and hence the performance of an engine gets deteriorated with fuel penalty.

The following points should be taken as precautionary for better working and long life of the converter:

1. Avoidance of fuel additives or cleaners.
2. Cranking of engine beyond 40 seconds when it is flooded or misfiring.
3. Ignition switch should not be turned off when the vehicle is in motion.
4. Place the converter as near as possible to exhaust of engine.
5. Proper loading of converter with enough catalyst.
6. Avoidance of sulfur-based fuel.

5.3.6   *LEAN NITROGEN-OXIDE CATALYSTS*

These days' lean-burn engines are an attractive technology because of their superior fuel efficiency and low carbon monoxide emissions. In advanced lean-burn engines in passenger vehicles, the conventional three-way catalysts are ineffective for reducing NOx under lean conditions. Thus, the catalytic converters have become in effective. However, zeolite catalytic materials that reduce NOx emissions, using unburned hydrocarbons in the exhaust as the reductant. Although the lean nitrogen-oxide catalyst is typically about 50% effective—considerably less than a three-way catalyst under stoichiometric conditions—the benefit is still significant. Japan has introduced lean nitrogen-oxide catalysts in few of automobile fleet.

5.4   CRANKCASE EMISSIONS AND CONTROL

The blow-by of compressed gases past the piston rings consists mostly of unburned or partly-burned hydrocarbons. In uncontrolled vehicles, the blow-by gases were vented to the atmosphere. Crankcase emission controls involve closing the crankcase vent port and venting the crankcase to the air intake system via a check valve. Control of these emissions is no longer considered a significant technical issue.

5.5   EVAPORATIVE EMISSIONS AND CONTROL

Gasoline is a relatively volatile fuel compared to other liquid fuel used in IC engines. Even at normal temperatures, significant gasoline evaporation occurs if gasoline is stored in a vented tank. The four primary sources of evaporative emissions from vehicles are diurnal (daily) emissions, hot-soak emissions, resting losses, and running losses. Evaporative emissions are controlled by venting the fuel tank (and, in carbureted vehicles, the carburetor bowl) to the atmosphere through a canister of activated charcoal. Hydrocarbon vapors are absorbed by the charcoal, so little vapor escapes to the air. The charcoal canister is regenerated or "purged" by drawing air through it into the intake manifold when the engine is running. Adsorbed hydrocarbons are stripped from the charcoal and burned in the engine.

5.6   SI ENGINE PARTICULATE EMISSIONS

The earlier SI engines were developed with carburetors and later found that due to the problems of mal-distribution of charge, later carburetors were discarded and switched over electronic fuel injection (EFI) systems. However, of later similar to diesel injection systems, the SI engines are being equipped with gasoline direction systems (GDI). Due to this, particulate matter emissions started emitting from such engines. The particulates are complex mixtures of volatile and non-volatile materials containing soot, organic carbon, and hydrocarbons.

As far as particle size, the particles are composed mainly of nucleation mode (<50 nm) and accumulation mode (50–200 nm). Nucleation mode particles have contains solid particles and accumulation mode particles comprise carbonaceous soot particles with an elemental carbon (EC) structure and adsorbed volatile material.

The mode of mixture preparation is a reason for particulate formation in GDI engines. A homogeneous, fully vaporized fuel-air mixture will give low levels of PN emissions. Any fuel that remains in a liquid state at ignition, particularly on combustion chamber surfaces, will burn as a diffusion flame leading to high levels of PN emissions known as a pool fire. Such liquid impingement may be caused by injection strategy, fuel composition (particularly a high enthalpy of vaporization), or cold conditions in-cylinder (notably during start-up).

Therefore, careful design of the combustion system and optimization of the injection and ignition strategy is needed to avoid high levels of engine out PN.

In modern times, the measure of particulate matter has become a serious concern with regards to its size and instead of denoting as soot density, it is being given in terms of particulate number (PN).

Fundamentally, it is understood that a good mixture preparation (i.e., fully vaporized fuel, well mixed with the air) would reduce PN emission levels. The parameters such as early injection, high fuel pressure and warm inlet air/EGR are observed to influence particulate emissions. Also, the AF is a key parameter for PN emissions.

Post-flame oxidation reactions increase exhaust temperature (such as late ignition) plays a role in reduction in PN.

Another factor that influences engine-out PM emissions is engine's fuel composition. The high levels of aromatic components present in fuel have been found to increase PN emissions, an aromatic ring being an early stage of the fundamental particulate formation process.

In the event of reducing emissions and improving octane rating, gasoline is being blended with ethanol, high levels of oxygenate (ethanol) have been shown to reduce PN emissions, with fuels such as E85 capable of giving extremely low levels of particulate emissions.

For reducing PN emission levels, a trade-off between the measures such as high inlet air temperatures, high exhaust temperatures, or low aromatic (and hence low RON) fuels), and also factors that are known for efficiency gains are observed to lower PN levels and enhance the use of GDI engines. Also, besides this, the particulate filters may also be adopted for reduction particulate emissions from gasoline fuelled SI engine.

A large number of studies have been performed on understanding the impact of fuel injection pressure (FIP) on formation particulate emissions from SI engines (Raza et al., 2018). The summary from these studies is that with higher injection pressure the particle number emissions are decreased. As a result of increasing injection pressure from low to high, the significant declined in PN concentration wasobserved. The impacts of FIP s on particle number (PN) emissions are presented in Figure 5.5.

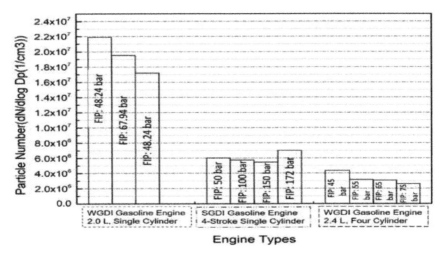

FIGURE 5.5 Effect of fuel injection pressure (FIP) variation on particulate number emissions.
Source: Raza et al. (2018). Open access.

KEYWORDS

- **carbon dioxide**
- **engine management system**
- **gasoline direction systems**
- **greenhouse gases**
- **hydrocarbons**
- **particulate matter**

REFERENCES

Kaspar, J., Fornasiero, P., & Hickey, N. (2003). Automotive catalytic converters: current status and some perspectives. *Catalysis Today 77,* 419–449.

Raza, M., & Chen, L. (2018). A review of particulate number (PN) emissions from gasoline direct injection (GDI) engines and their control techniques. *Energies 6*(1417).

Taylor, K. C. (1984). *Automobile Catalytic Converters,* Springer-Verlag Berlin Heidelberg.

CHAPTER 6

Diesel-in-Cylinder Measures

6.1 INTRODUCTION: DIESEL ENGINE COMBUSTION AND EMISSIONS

In previous chapters, the emission control methods being adopted for SI engine emissions are dealt with in detail. Normally, In India, the SI engine finds application in light-duty application such as passenger vehicles and in small load carriers. In the rest of the world especially in developed countries, the SI engines are extensively used both passenger and goods transport. Its counterpart-diesel fuelled CI engine basically makes use of high compression ratios (CR) and it's sturdy and bulky in design. Since its invention, it has undergone multitude of developments and is finds its use in light, medium, and heavy-duty application. It is also used in some power generating units. It is most commonly found engine as a prime mover in automobiles especially in India, where the fuel price is competitive. Though, the diesel engine emits pollutant emissions such as CO, NOx, and HCs similar to gasoline engine but another species-Particulate Matter or soot or smoke is very exclusive of diesel tail pipe emissions. In addition, however small the sulfur present in fuel, leads to formation of sulphur compounds (SOx) are formed and may lead to corrosion related problems. The sulfur also promotes PM formation. Since the diesel engine runs with excess air, the CO emission levels are relatively far lower compared to gasoline engines. Moreover, the concern about PM is that it doesn't have fixed structure and the measurement and its treatment poses entirely a different issue. Another difference of complexity with diesel emission control is that except PM all other emissions are gaseous phase species. It interferes with visibility issues. Tackling diesel fuel emissions is little complex than petrol engine emissions. As far as diesel-fuelled vehicles are concerned, particulate matter, oxides of nitrogen and hydrocarbons are the most significant emissions which require attention for their control. The levels of particulate matter emissions from uncontrolled diesel engines are six to ten times more than those from gasoline engines. All these can be attributed to the difference in mode of combustion; the mixture is heterogeneous in diesel fuelled engines with somewhat difficult to control emissions.

Typically, the phenomenon working of diesel engine is also different from SI engines. Normally a diesel engine is supplied with fuel from a fuel injection nozzle at a pressure of around 50 MPa in case of indirect injection (IDI) engines, and 100 MPa or higher for direct injection (DI) engines (as CRDI engine are of late introduced). Such high fuel delivery pressures are essential to allow intimate mixing of fuel with highly compressed air in the engine cylinders at the end of the compression stroke, and to break up the liquid fuel, since the diesel fuel less volatile and more viscous relatively compared to petrol/gasoline.

Through proper modifications in fuel injection systems, turbocharging, inlet charge air cooling, fuel quality, a reduction of about 50% in emissions of NOx and hydrocarbons and more than 75% in particulate matter emissions can be achieved in contrast to its uncontrolled levels. These changes would also improve fuel economy (diesel fuel-efficiency has improved by 30% since the 1970s) but with increased engine costs. Particulate matter and hydrocarbon emissions can be further reduced through the use of low-sulfur fuel and an oxidation catalytic converter. The lowest particulate matter emissions (a 95% reduction from uncontrolled levels) are possible with the use of reliable and durable particulate trap-oxidizers (Brijesh and Sreedhara, 2013).

The conditions for formation of NOx and PM emissions are opposite to each other and since the PM is not pure gas and can be filtered and it interferes with visibility and public nuisance issues and moreover thus it has to be dealt entirely differently. For control of diesel fuelled engine emissions both in-cylinder measures and exhaust gas aftertreatment methods can be used. To meet ever stringent emissions which are periodically promulgated by different nations in tune with climate change obligations, there is great need for emission control strategies.

This chapter presents the in-cylinder measures adopted to control diesel engine emissions.

As discussed in basic review of internal combustion engine (ICE) fundamentals, the following are well known factors that are responsible for formation harmful pollutant emissions especially from combustion generated pollution species:

1. Heterogeneous and non-stoichiometric air-fuel ratio (AF) mixture in participating in combustion.
2. High speed operation of engines giving minimal time for near complete combustion of mixture.

3. Fuel rich mixtures taking part in combustion under high load operation leading to heterogeneities present in mixture.
4. In complete combustion due to working of non-equilibrium conditions.
5. Improper Driving habits of vehicle operating personnel.
6. Improper road infrastructure and frequent stop and start of engine/vehicle.

Ever since the limits on emission levels recommended by EURO norms, there is a continuous drop in emission levels, there observed to be significant reduction from EURO-I to EURO-IV, as shown in Figure 6.1, the periodic reduction in exhaust emissions per test kilometer.

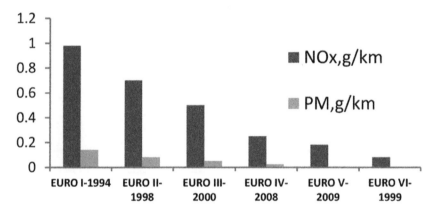

FIGURE 6.1 Status of European emission standards for light commercial diesel vehicles. *Source:* Brijesh and Sreedhara (2013).

Since the factors are different in formation of NOx and PM emissions, many efforts are put in to arrive a trade-off and has become an urgent issue for simultaneous reduction of NOx and PM in order to meet the requirements of future emission standards with these technologies.

A map plot is drawn between the equivalence ratio (Φ)-temperature (T) for soot and NO concentrations using the SENKIN code with a surrogate diesel fuel, a mixture of n-heptane and toluene (Alriksson and Denbratt, 2006).

Homogeneous mixtures of various Φ and T values are given as the inputs for the code. The results showed that the local combustion temperature should be kept below 2200 K to avoid high NO concentrations for low equivalence ratios. At high equivalence ratios, it becomes necessary to further decrease the maximum allowable temperature to avoid soot formation. Adopting these conditions cal NOx and soot production during combustion process in diesel engines is shown on Φ-T plane by Kitamura et al. (2002) (Figure 6.2).

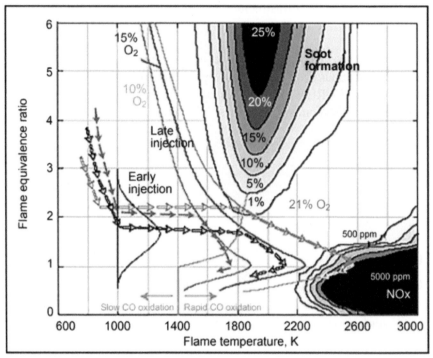

FIGURE 6.2 Variation of equivalence ratio and flame temperature with regimes of emissions.
Source: Reprinted from Johnson, 2008. Open access.

To comply with upcoming stringent emission standards, the low-temperature combustion (LTC) suitable with a uniform mixture ratio $\phi < 1$ (Potter and Durrett, 2006). From Figure 6.2, it can be inferred that by keeping the combustion temperature very low (~1650 K) and equivalence ratio less than 2, the NO and soot formation areas can completely be avoided. The conventional diesel combustion process is also shown on Φ-T plane in Figure 6.2. It can also be understood from Figure 6.2 that the scope for simultaneous reduction of NOx and soot emissions with LTC. Low Temperature Combustion can be achieved and NOx emissions would be kept low by retarding the injection timing, decreasing CR and increasing amount of charge dilution. However, soot emissions are increased because of low temperature. Soot formation may be reduced by having a homogeneous mixture of fuel and air with $\Phi \leq 1$.

As far as diesel emission control is concerned, fuel injection parameters such as fuel injection pressure (FIP), injection timing, injection duration, etc. are important parameters to achieve better performance and low emissions. The variation of these parameters has been studied extensively on diesel engine performance and exhaust emissions (Borman, 1998; Heywood, 1988).

6.2 DIESEL ENGINE IN-CYLINDER MEASURES TO CONTROL EMISSIONS

Targeting in-cylinder measures as a means for controlling emissions involves both engine geometrical parameters and engine operating parameters. Figure 6.3 briefs various engine design/geometrical and operating parameters that would help in controlling diesel engine tail-pipe emissions.

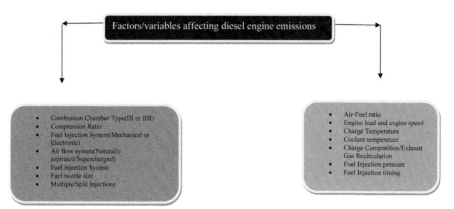

FIGURE 6.3 Schematic of diesel engine design/geometrical and operating parameters to control emissions.

6.2.1 ENGINE GEOMETRICAL PARAMETERS

6.2.1.1 COMPRESSION RATIO (CR)

CR is an important parameter that dictates engine's performance and emissions. Diesel engine's thermodynamic efficiency increases with increase in CR and thus it is an important geometrical or design parameter that affect both performance and emissions as diesel fuel take part in combustion by virtue of air attaining high temperature during compression sufficient enough to ignite the fuel and hence CR is a significant parameter to affect engine's performance. With variations in CR, the engine fuel economy, cold starting, and maximum cylinder pressures are also affected.

Ignition delay is one issue that affects quality of diesel engine combustion on which engine performance and emissions heavily depends. Higher temperatures and higher pressures are realized with increase in CR and thus make the charge to be at higher in-cylinder temperature. The higher the temperatures during the start of injection of fuel and thus in a shorter ignition delay and higher flame temperature. The higher the in-cylinder temperatures, a shorter ignition delay, the quicker the fuel to ignite and thereby combustion continues. On the other hand, if a diesel fuel experiences higher ignition delay, the more fuel is accumulated during the delay period and once the delay period ceases, there would be sudden release of energy with consequent sharp rise in NOx emissions. This is due to the reason of excess air present in initial stages combustion and high temperatures favoring NOx emission formation quickly. Therefore, reduced CR will help in reduction of NOx emissions. However, the higher the CRs for a given fuel with proper mixture ratio, the prevalence higher temperatures reduce HC emissions significantly.

Generally, smaller cylinders high speed engines require higher CRs for adequate cold starting. It is essential to have higher CR—in the range of 15–20 or more as against CRs of 12–15 to allow engine's starting ability under cold conditions, and hence most diesel engines are designed within this range of CRs.

The performance of diesel engine can be improved with supercharging or turbocharging. However, with consequent higher in-cylinder temperatures with turbocharging, the CRs are voluntarily reduced to keep NOx emissions low. Further reduction in the CR prevents due to cold starting requirements. Use of higher CR results in a shorter ignition delay period. With shorter ignition delay 'over-mixing' of fuel and air can be avoided and thus lower HC emissions be achieved. Also, increased oxidation of unburned HC emissions can be obtained at higher combustion temperatures owing to higher CRs. However, with a low CR, a fuel with low quality experiences a longer ignition delay, consequently more fuel being accumulated and results in spontaneous burning of large amount of fuel resulting higher peak pressures and temperatures and thereby high NOx emission formation takes place. Therefore, advanced fuel injection timing, higher CRs with higher combustion temperatures all leading to formation of high NOx emissions.

On the other retarded injection timings with lower combustion pressure and temperature due to combustion, occurring in the beginning of expansion would be favorable for lower NOx emissions.

The NOx emissions can be lowered with a long ignition delay leads but results in poor fuel efficiency. This can be achieved through a low CR which results in too long a delay during engine warm up under cold conditions. But this changes leads to appearance of "white smoke" associated with high emissions of unburned fuel. However, a high CR leading to high combustion temperatures would increase soot formation while on the other hand, it increases soot oxidation. Therefore, simultaneous reduction of PM-NOx emissions is a solution by an optimum CR. The CR of passenger vehicle diesel engines and swirl-chamber or pre-chamber engines is normally high in a range 20 to 23; normally IDI engines have slightly higher CRs than DI diesel engines.

> *Therefore, increasing the compression ratio, for a give a fuel and given fuel injection parameters, increases NOx emissions but lowering the compression ratio has an adverse effect on HC emissions. The best possibility could be drawing a trade-off for obtaining simultaneous reduction of NOx and PM emissions. Increasing the compression ratio no doubt improve HC and particulate emissions but increases friction work and, hence, fuel consumption and NOx emissions. The limits are imposed by soot emissions at full throttle.*

6.2.1.2 COMBUSTION CHAMBER SURFACE TO VOLUME RATIO

The diesel engines are broadly classified as DI and indirect injection (IDI) Engines. The combustion in IDI engines takes place in two stages; a rich mixture burns in the pre-chamber where all the fuel is injected and the partially burned rich fuel-air mixture from the pre-chamber is passed onto the main chamber where due to presence of excess air combustion is nearly completed. The jet of high pressure partially burned gases from the pre-chamber enters the main chamber generating high turbulence that causes rapid mixing and most of the fuel burns as lean mixture. Thus, the IDI engines are preferred for lowering emissions of HC, CO, and smoke. Since near complete combustion is obtained with IDI engines and work with high NOx emissions. Earlier, mostly in high speed small engine applications, the IDI type diesel engines were employed. Now almost, the However, IDI engines are phased out due to their poor fuel economy with new advancements in high speed direct injection (HSDI) engine technology. As a comparison, the levels of NOx emissions in IDI engines are far lower than equivalent DI type diesel engines.

Since there are two chambers in an IDI engine, the surface to volume ratio of combustion chamber is higher than in DI engine and hence is a reason for greater fuel consumption.

In an IDI engine, major portion of NOx would form in the pre-chamber. However, extra NOx formation takes place at higher loads in the main-chamber. During and at the end of pre-mixed mode of combustion (also called uncontrolled combustion phase) of DI type diesels, NOx is formed in near stoichiometric mixtures due to high peak pressures and temperatures compared to IDI engines. Also, slightly lower NOx emissions are formed in mixing controlled phase due to post-combustion gases. Thus, IDI engines emit lower NOx. In fuel rich zones of DI type diesel engines, higher CO is formed due to low turbulence and insufficient oxygen than required for complete combustion even though the temperatures are high. Therefore, CO emissions are higher in DI type than in IDI type diesels.

6.2.1.3   DESIGN OF COMBUSTION CHAMBER-CREVICE REGIONS

The crevice volume includes the clearance between the top of the piston and the cylinder head, and the 'top land'-the space between the side of the piston and the cylinder wall above the top compression ring. The air in these spaces contributes little to the combustion process. The smaller the crevice volume, the larger the combustion chamber volume can be for a given CR. Thus, reducing the crevice volume increases the amount of air available for combustion. To reduce these dead volumes, engine makers have reduced the clearance between the piston and cylinder head through tighter production tolerances, and have moved the top compression ring toward the top of the piston. However, this results in higher temperature of top piston ring. To withstand high piston temperature, demands higher quality engine oil to reduce formation of gummy deposits.

In a diesel engine, there are regions such as piston bowl, top land crevice; piston-cylinder head clearance, valve recess and head gasket clearance are the regions wherein air is contained in several different volumes.

In DI diesel engines, the piston bowl accounts for slightly more than 50% of total clearance volume at TDC. The other narrow regions wherein air is trapped are poor regions of air utilization during combustion. Hence, the degree of utilization air is a measure of completeness of combustion.

Also, the air contained in the volume between piston crown and cylinder head at TDC is poorly utilized. Therefore, a good combustion chamber should aim at a minimizing the regions where there would be retention of air and finally results in better utilization of supplied/available air.

6.2.1.4   MULTI-VALVES AND AIR MOTION

In addition to use of supercharging for increasing volumetric efficiency (VE) of the engine, use of multiple valves (3 or 4) per cylinder increases flow area has become a practice. It has become common to find turbocharged as well as multi-valve DI diesel engines in modern diesels and valves are being located in centralized location of injector and combustion bowl in the piston. The injector can be placed more centrally and vertically. Generally, with two valves designs the injector is always offset and inclined (injector inclination ~ 20 and 10° from vertical) and results in high PM emissions due to poor fuel distribution in the cylinder compared to a vertically and centrally located injector. The multi-valve engines necessitates use of higher number of spray holes, improved swirl, improved mixture formation and increased injection pressures and retarded injection timing, combination of all these finally improve the particulate-NOx trade-off significantly (Diesel Engine Reference Book, 1999).

Also, depending on the swirl provision, the engines are classified as induction induced swirl, compression induced swirl, or combustion gener-ated swirl type. Therefore, a Provision should be made to incorporate swirl by properly designed inlet ports and intake manifold, masking of inlet valves. These are essential for a successful air-dispersion combustion process. By achieving proper swirl, the fuel- air mixture gets thoroughly mixed and results in near complete combustion. Hence the more the swirl the high is the combustion temperature. However, beyond certain value of swirl, it is prohibitive to increase swirl in order to keep NOx emissions low, and also, a sharp increase of the smoke number has to be reckoned with and hence swirl must be reduced or optimized. Moreover, too high swirl is not good in a sense that it erodes the air film that acts as insulation on the inner walls of cylinder. The effect of swirl on fuel consumption and HC emissions is not very well established. In addition, engine power gets affected with increase in swirl as the cylinder charge gets reduced with higher swirl thus the output gets lowered (Figure 6.4).

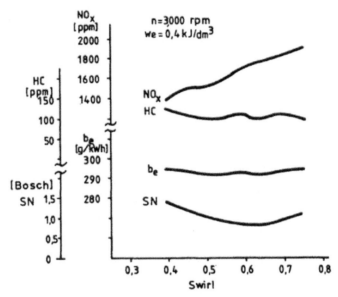

FIGURE 6.4   Effect of swirl on exhaust emissions and fuel consumption.

Lower specific fuel consumption and lower NOx emission levels can be obtained with centrally placed combustion bowl and injector in multi-valve engines due to lower swirl requirements and aided by more equal fuel distribution and availability of equal air to each spray for mixing. Moreover, with four-valve engines compared to conventional two-valve engines, more fuel sprays with better distribution of air and fuel point towards lowering both HC and smoke emissions.

6.2.1.5 SHAPE OF PISTON BOWL

Combustion in diesel-fuelled CI engines largely depends on the intimate mixing of fuel and air. These two are allowed to mix inside the engine cylinder. Many trials are made to provide proper swirl, squish, tumble, and turbulence. The swirl and tumble are organized motion provide in line and perpendicular to cylinder axis respectively. The swirl could be low, medium, and high depends on how vigorous the movement is with respect to engine speed. In majority of DI diesels, the fuel is injected onto the top of the piston and the air should have acquired high swirl, high pressure, and temperature by the time the fuel is injected and then allowed to mix. The swirl imparted results in thorough mixture formation. The swirl could be induction induced, compression induced and combustion induced one. The induction induced swirl can be impacted by properly orienting the intake port and intake manifold. The top of piston could be a single bowl, toroidal bowl or reentrant bowl (Heywood, 1988). The shape decides the way the fuel impinges and takes swirling action. The shapes of piston bowls are illustrated in Figure 6.5.

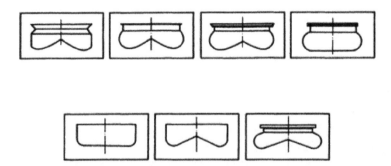

FIGURE 6.5   Types of diesel engine piston bowls.

In contrast with modern spark-ignition engine designs, the bowl is located in the piston. Figure 6.5 shows some combustion chamber bowl configurations. Bowl characteristics are particularly favorable if the bowl is rotationally symmetrical to the piston axis since this allows swirl disturbances to be avoided. For reasons of geometry, however, this is only possible on multi-valve engines as this design requires the injection nozzle to be located centrally. A bowl restricted along its edges (Figure 6.5) offers benefits with regard to NOx emissions. A raised section in the center also promotes favorable emission characteristics. Other design-related effects on NOx and HC emissions are dictated by the position of the uppermost piston ring, i.e., the height of the top piston land and the resulting dead volume.

The thorough and rigorous mixing of fuel enables the combustion to be near complete and hence a large amount of heat will be released with consequent rise in temperatures. The higher the heat release the more is the temperature and better is quality of combustion. The diesel engines normally work with lean mixtures. For a given fuel and fuel injection parameters, the high temperatures are favorable for lowering CO, HC, and PM emissions but high NOx emission will result. The lean mixtures and high temperatures are conducive conditions for NOx emissions to be high.

Moreover, fuel injection parameters do affect a lot for formation of diesel exhaust emissions. To allow better mixing, high injections pressures with small nozzle orifice diameter are adopted. These condition result in high

NOx emissions due to high heat release whereas low injection pressures and relatively low injection pressures would not provide good environment for release of heat and hence results in poor combustion and hence high CO, HC, and PM emissions will result but low NOx emissions. Instead, one can employ retarded injection timings with high injection pressures, small nozzle holes for lowering NOx emissions, without compromising on high swirl levels.

6.2.1.6   *TURBOCHARGING AND INTERCOOLING*

For improving the power output of a diesel engine, supercharging is employed with some degree of boost pressure. The energy of exhaust gases is utilized to allow supercharging-called turbocharging. By including, boosting the in-cylinder pressures and temperatures will rise with subsequent increase in peak pressures and increased NOx levels due to further compressed charge. However, to lower the temperature of compressed charge after the boosting, an inter cooler is generally be adopted to reduce temperature levels and hence modestly reduced NOx emissions and low HC emissions will take place.

By cooling the compressed air in an intercooler before it enters the cylinder, the adverse thermal effects can be reduced. This also increases the density of the air, allowing an even greater mass of air to be confined within the cylinder, and thus further increasing the maximum power potential. Turbocharging and inter cooling offer an inexpensive means to simultaneously improve power-weight ratios, fuel economy, and control of NOx and PM emissions.

Increasing the air mass in the cylinder and reducing its temperature can reduce both NOx and PM emissions as well as increase fuel economy and power from a given engine displacement. Most heavy-duty diesel engines are equipped with turbochargers, and most of these have intercoolers.

Because of the inherent mismatch between engine response characteristics and those of a fixed geometry turbocharger, a number of engine manufacturers are considering the use of variable geometry turbines instead (Wallace et al., 1986). In these systems, the turbine nozzles can be adjusted to vary the turbine pressure drop and power level in order to match the engine's boost pressure requirements. Thus, high boost pressures can be achieved at low engine speeds without wasteful over boosting at high speed. The result is a substantial improvement in low speed torque, transient response, and fuel economy, and a reduction in smoke, NOx, and PM emissions.

6.2.2 EFFECT OF ENGINE OPERATING PARAMETERS

6.2.2.1 AIR-FUEL RATIO (AF)

The increase in fuel-air ratio (FA) makes mixture richer and increase fuel/air ratio and diffusion burning fraction, thus, increasing particulates, but improving particulate oxidation rates.

In diesel engines, the fact that fuel and air must mix before burning means that a substantial amount of excess air is needed to ensure complete combustion of the fuel within the limited time allowed by the power stroke. Equivalence ratio is a key parameter that affects engine load, speed, and emissions as well. The quality of combustion heavily depends on equivalence ration (Figure 6.1). Diesel engines, therefore, operate with overall air-fuel ratios that are considerably lean of stoichiometric ($Ø \ll 1$). The minimum AF for complete combustion is about $Ø = 1.5$. This ratio is known as the smoke limit, since smoke increases significantly at AFs lower than this limit. For a given amount of air in the cylinder, the smoke limit establishes the maximum amount of fuel that can be burned per stroke, and thus the maximum power output of the engine. The brake power also exhibits a relation with safe smoke limit (Figure 6.6).

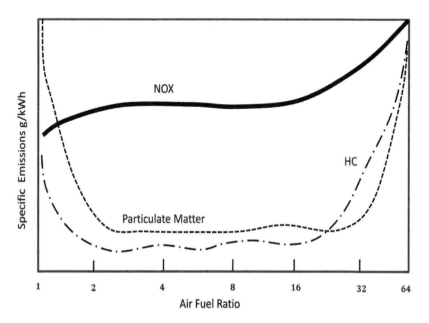

FIGURE 6.6   Effect of air-fuel ratio on emissions of a typical diesel engine.

At very high AFs (corresponding to very light load), the temperature in the cylinder after combustion is too low to burn out residual hydrocarbons, so emissions of gaseous hydrocarbons and particulate SOF are high. At lower AFs, less oxygen is available for soot oxidation, so soot emissions increase.

In conventional turbocharged engines, an increase in the amount of fuel injected per stroke increases the energy in the exhaust gas, causing the turbocharger to spin more rapidly and pump more air into the combustion chamber. For this reason, power output from turbocharged engines is not usually smoke-limited. Instead, it is limited by design limits such as the allowable turbocharger speed and mechanical and thermal loading of the engine components.

The air supply during the first few seconds of full-power acceleration is less than would be supplied by the turbocharger in steady-state operation. Unless the fuel is similarly limited during this period, the AF will be below the smoke limit, resulting in a black 'puff' of dense smoke. The resulting excess particulate emissions can be significant under urban driving conditions.

6.2.2.2   AIR-FUEL MIXING

The rate of mixing between the compressed charge in the cylinder and the injected fuel is among the most important factors in determining diesel performance and emissions. The mixing rate during the ignition delay period determines how much fuel is burned in the premixed burning phase. The higher the mixing rate, the greater the amount of fuel burning in the premixed mode, and higher the noise and NOx emissions. In the subsequent mixing-controlled combustion phase, the rate of combustion is limited by the mixing rate. The more rapid and complete this mixing, the greater the amount of fuel that burns near the top-dead-center, the higher the efficiency, and the lower the PM emissions.

In engine design practice, it is necessary to strike a balance between the rapid and complete mixing required for low soot emissions and best fuel economy, as the too-rapid mixing results in high NOx emissions.

The primary factors affecting the mixing rate are the FIP, the number, and size of injection orifices, any swirling motion imparted to the air as it enters the cylinder during the intake stroke, and air motions generated by combustion chamber geometry during compression. Much of the progress in in-cylinder emission control over the last decade has come through improved understanding of the interactions between these different variables and emissions, leading to improved design of fuel injection systems and combustion chambers.

6.2.2.3 FUEL INJECTION PRESSURE (FIP) AND INJECTION TIMING

The diesel engines are quality governed engines wherein only the amount of fuel will be varied keeping the amount of air unaltered. Hence, engine power can be regulated by varying the amount of fuel alone. Hence, the fuel injection parameters are the more crucial one for extracting good amount of energy from a diesel engine. The parameters include FIP, fuel injector nozzle orifice size and injection timing and injection profile. The time at which fuel is injected will greatly affect the amount of heat released by the combustion and thus it is the injection timing a critical determinant of the combustion events and the resulting emission products. The relation of injecting timing with other parameters is listed below:

- Start of injection timing influences the ignition delay period.
- The residence time for the fuel to air in contact inside the combustion chamber.
- Too advanced timing with high FIP and smaller holes results in release large amount of heat. Advanced timings results in more pre-mixed mode of combustion. The retarded timing yield low power and hence more diffusive mode of combustion with high particulates but with low NOx emissions.
- Injection timing influences greatly the fuel spray and air interaction, FIP, fuel injection rate and duration.
- Injection timing can be spilt so as to increase or decrease the pre-mixed mode or diffusive mode of combustion.

Provision of these components in a diesel engine increases the level of complexity and expenditure needed to reduce both these is high.

Along with variation of fuel injection parameters, it has become customary to include a small dose of exhaust gas recirculation (EGR) to achieve trade-off between PM and NOx emissions. The higher the dosages of EGR the lower are NOx levels in combination with very high injection pressures (~200 MPa) for soot reduction. There is a misunderstanding between the terms -soot and particulate matter. A distinction is made here between soot and particulate emissions. Soot consists of pure carbon compounds, whereas particulates also include fuel or lube-oil droplets, ash, rubbed off metal, corrosion products, and sulfate compounds. Of late, multiple or split injection systems have been employed which can be better controlled by electronic fuel injection (EFI) controls (Borman).

The beginning of fuel injection has an important effect on diesel engine emissions and fuel economy. The essential feature of diesel fuel combustion is that diesel fuel experiences ignition delay (normally 5 to 15° of crankshaft rotation) and it is customary to allow combustion to begin a well before TDC, even prior to this happens; the fuel should overcome its delay period. This way the best fuel economy can be achieved before). Advanced injection timings are possible to give more time for better mixing of fuel with available air and hence the mixture undergoes compression heating. The longer ignition delay provides more time for air and fuel to mix, increasing the amount of fuel that burns in the premixed combustion phase. More fuel burning at or just before top-dead-center also increases the maximum temperature and pressure attained in the cylinder. Both of these effects tend to increase NOx emissions. On the other hand, earlier injection timing tends to reduce PM and light-load hydrocarbon emissions. Fuel-burning in premixed combustion forms little soot, while the soot formed in mixing-controlled combustion near top-dead-center experiences a relatively long period of high temperatures and intense mixing, and is thus mostly oxidized.

Compared to uncontrolled diesel engines, modern diesel engines with emission controls generally have moderately retarded injection timing to reduce NOx, in conjunction with high injection pressure to limit the effects of retarded timing on PM emissions and fuel economy. Great precision in injection timing thus becomes necessary—a change of 1° crank angle can have a significant impact on emissions. The optimal injection timing is a complex function of engine design, engine speed, and load, and the relative stringency of emission standards for different pollutants. The fuel injection manufacturers are facing a major challenge in attaining the required flexibility and precision of injection timing with FIPs and other allied parameters. Another important parameter is injection duration, for this purpose, pump, pipe, and injection nozzle should ensure that the fuel quantity is injected within a specified span of time. There is no single fixed setting of injection duration, at the same time, that permit minimizes all pollutant emission components as well as fuel consumption.

The in-cylinder characteristics are varied by varying fuel injection timing. The HC and NOx emissions are affected by its variation quite similar to that of spark timing in SI engines. The effect of injection timing on NOx, HC, smoke emissions, and fuel consumption is shown in Figure 6.7.

With retarded injection timing, the NOx emissions decrease sharply. On the other hand, an increase in smoke results with retarded injection timing. If the injection timing is retarded too much, HC emissions in naturally aspirated engines also may increase sharply. An increase in injection pressure results in higher NOx and HC, but yields lower smoke and PM emissions.

Indirect-injection engines use pintle nozzles with only a single injection jet. The nozzle injects the fuel into the pre-combustion or whirl chamber in such a way that the glow plug is protruding in the injection jet by just a fraction. The injection direction is matched precisely to the combustion chamber. Any deviations in injection direction result in poorer utilization of combustion air and, therefore, to an increase in soot and hydrocarbon emissions.

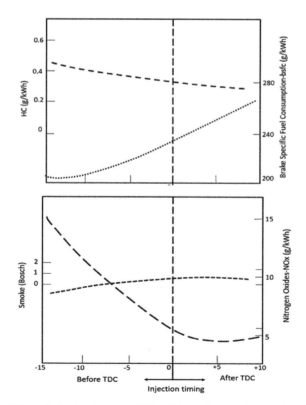

FIGURE 6.7   Effect of injection timing on NOx, HC, smoke emissions, and fuel consumption.

The hydraulics of fuel injection is another essential parameter having an impact on pollutant formation in the diesel engine. Direct-injection diesel engines, in particular, benefit from high injection pressures with regard to fuel consumption and pollutant emissions. Different injection systems such as common rail injection systems, unit injectors, pump-line-nozzle systems, etc. are available for this purpose.

The injection nozzle geometry also is an essential feature in conjunction with injection hydraulics. Direct-injection engines use either single-hole or multi-hole nozzles. HC emissions can be improved by reducing the pollution volume (sac hole volume) of the injection nozzle as this limits dripping of the nozzle.

The number of injection holes has a considerable influence on smoke formation and on NOx emissions. At an identical smoke number, NOx emissions are lower if, for example, a 4-hole nozzle is replaced by a 5 or 6-hole nozzle.

The nozzle sac is a small internal space in the tip of the injection nozzle. The nozzle orifices open into the sac so that fuel flowing past the needle valve first enters the sac and then sprays out of the orifices. The small amount of fuel remaining in the sac tends to burn or evaporate late in the combustion cycle, resulting in significant PM and HC emissions. The sac volume can be minimized or eliminated by redesigning the injector nozzle.

6.2.2.4 ENGINE LOAD AND SPEED

As described above, the diesel engine load or speed can be varied by varying the amount of fuel injected into the engine. This, in turn, depends on the fuel injection timing. For quality governed engines, for developing higher power, more fuel is required to be injected. For a good quality of fuel (high Cetane number) with shorter ignition delay, the higher is the peak pressure with high heat release with consequent high rise in temperatures. With proper swirl, the fuel will undergo thorough mixing and with increase in amounts of fuel, near-complete combustion, lower CO and PM (HC) emissions and higher NOx emissions are obtained. With these conditions, if fuel injection timing is retarded, the NOx emissions can best be reduced. However, with retarded timings higher exhaust temperatures are attained with soot oxidation resulting n lower HC emissions.

Normally under part-load operation, the diesel doesn't emit CO emissions but with increase in load, the increased FA leads to higher CO emissions start increasing again and rise sharply with more fuel is injected to increase engine power output. At maximum load, NOx, CO, and soot are also at their maximum level. The HC emissions reduce with increase in engine load as higher gas temperature lead to an increase in the oxidation rates.

Engine brake thermal efficiency increases with engine load because the ratio of friction to break power reduces. Interaction among these factors results in lowest value of brake specific fuel consumption (BSFC), and optimum value of brake specific nitrogen oxides (BSNOx) and particulate emissions at an intermediate load. The variable speed engines are designed to give lowest fuel consumption at about 2/3rd of maximum speed at which heavy duty engines are normally operated. In turbocharged engines, the boost pressure is reduced at low engine speeds resulting in higher FA. At high speeds pumping losses increase and cooling decreases. The coolant and residual gases are hotter as the speed increases. Both these factors increase NOx at high engine speeds. The HC and PM have an optimum at an intermediate speed because time available for oxidation decreases at the higher engine speeds.

The effect of engine speed and load is little complex due to complicated engine governor system and its fuelling system response. For example, in on-road vehicles engine torque is prevented from falling off with small decreases in speed resulting from increased load (due to a hill, for example); this reduces the need for downshifting. The speed-load operating factors are also affected by other system parameters such as turbocharger performance map.

At low engine speeds, in case of turbocharged engines, the turbocharger cannot provide adequate boost, results in higher fuel/air ratios. At high speeds the pumping work increases, as do other loss factors. The engine system is typically designed to give optimum fuel economy at the moderate speeds at which it is usually operated. Temperature levels increase with speed due to reduced cooling and increased hot residuals; this increases NOx.

A higher engine speed means greater friction losses in the engine and a higher power input by the ancillary assemblies (e.g., water pump). Engine efficiency, therefore, drops as engine speed increases. If a specific performance is produced at high engine speed, it requires a greater fuel quantity than if the same performance is produced at low engine speed. It also produces more pollutant emissions.

Nitrogen oxides (NOx) as the time available to form NOx in the combustion chamber is shorter at higher engine speeds, NOx emissions decrease as engine speed increases. In addition, the residual-gas content in the combustion chamber must be considered since it causes lower peak temperatures.

As the residual-gas content normally drops with rising engine speed, this effect runs counter to the interdependence described above.

Hydrocarbons (HC) and carbon monoxide (CO) as engine speed rises, HC, and CO emissions rise as the time for mixture preparation and combustion shortens. As piston speed rises, combustion-chamber pressure drops faster in the expansion phase. This results in poorer combustion conditions, especially at low loads, and combustion efficiency suffers. On the other hand, charge movement and turbulence increase the rate of combustion as engine speed rises. Combustion time becomes shorter, and this compensates at least partly for the poorer marginal conditions.

With increased speeds, there exist more turbulence generated inside an engine, better mixing of fuel and air is achieved. This tends to better combustion and hence lower HC, CO emissions but higher NOx emissions occur.

Soot normally, soot becomes less as engine speed increases, since charge movement is more intense, thus resulting in better mixture formation. Soot occurs due to localized oxygen deficiency due to thermal cracking of the hydrocarbon molecules at local temperatures above approx. 1,500 K. Consequently, enhanced VE leads to the formation of less soot, or it allows the injection of larger quantities of fuel, and thus increased performance for the same soot factors.

6.2.2.5   QUALITY OF FUEL

For improving the heterogeneous nature combustion in diesel engines, a high cetane number fuel is preferred. This results in a shorter delay. The diesel has poor volatility and the primary effect of fuel volatility is to increase the rate of premixed burning and thus increases NOx, and to create more lean regions which do not burn, thus increasing hydrocarbons. Longer ignition delay gives to combustion of accumulated fuel sharply and suddenly with high combustion noise. With highest possible cetane number, i.e., optimized ignition quality and short ignition delay, and has a favorable impact on combustion noise. The fuel should also possess good lubricity, and a low water and impurity content to ensure proper functioning of the fuel-injection system throughout its service life.

Generally, the petroleum based liquid fuels contain sulfur and with increased amounts of sulfur more oxides of sulfur results resulting in corrosion of components on one hand and rise in soot levels on the other. Therefore, the new emission norms are aiming at lower sulfur levels. For example, sulfur-dioxide emissions have dropped to negligible values in road traffic since the introduction of low-sulfur or sulfur-free fuel.

The demands placed on fuel quality have also risen due to constantly rising engine performance. Various additives increase the cetane number, enhance fuel lubricity and flow ability, and protect the fuel system from corrosion.

For petroleum fuels, many of the properties such as hydrocarbon composition, natural cetane number, volatility, viscosity, and density are interdependent. On the other hand, with higher fuel volatility a larger lean flame out 'over mixing' region may result and due to faster fuel evaporation the fraction of fuel burned during premixed combustion is also higher. Therefore, an increase in NOx as well as HC may be observed with more volatile diesel fuels.

It is also researched the extent of aromatic compounds and H/C ratio influencing the emissions. An addition of aromatics to diesel fuel observed to lower the peak flame temperatures and hence low NOx emissions can be obtained. However, due to lower temperatures the soot/particulate emissions either remain unchanged or increase modestly. The higher the hydrogen content or high H/C of a fuel tends to have higher heating value thereby more heat release and hence lower soot emissions. But a little higher heating value may not prompt formation of NOx. It is observed that addition of oxygenated components-such as dimethyl ether (DME) and methanol point towards formation lower NOx emissions.

6.2.2.6   CHARGE TEMPERATURE

The process of compressing the intake air in turbocharged engines increases its temperature. Reducing the temperature of the compressed air charge going into the cylinder has benefits for both PM and NOx emissions. Reducing the charge temperature directly reduces the flame temperature during combustion, and thus helps to reduce NOx emissions.

In addition, the relatively colder air is denser, so that (at the same pressure) a greater mass of air can be contained in the same fixed cylinder volume. This increases the AF in the cylinder and thus helps to reduce soot emissions. By increasing the air available while decreasing piston temperatures, charge-air cooling can also make possible a significant increase in power output while remaining within the engine thermal limits. For this reason, many high-powered turbocharged engines incorporate charge-air coolers even in the absence of emission controls.

6.2.2.7   COOLANT TEMPERATURE

Increased coolant temperature reduces heat loss and increases temperature levels, especially during expansion. The federal test procedure requires both a cold (room temperature) and a hot test. The test cycle is the same for both. At the cold start the engine is warming up with time; thus friction decreases and fuel vaporizes better as time goes on, but the temperature levels, causing increased NOx but decreased particulates.

6.2.2.8   CHARGE COMPOSITION

It is understood from the formation mechanism of pollutant emissions, oxides of nitrogen emissions are heavily dependent on flame temperature. By altering the composition of the air charge to increase its specific heat and the concentration of inert gases, it is possible to decrease the flame temperature significantly. The most common way of accomplishing this is through EGR.

At moderate loads, EGR has been shown to be capable of reducing NOx emissions by a factor of two or more with little effect on PM emissions. Although soot emissions are increased by the reduced oxygen concentration, the soluble organic portion of the PM and gaseous HC emissions are reduced due to the higher in-cylinder temperature caused by the hot exhaust gas. EGR has been used for some time in light-duty diesel engines in order to reduce NOx.

While increasing the FIPs, reducing the nozzle orifice size and increasing the number of orifices all point towards high NOx emissions. Keeping these parameters intact, NOx, and PM emissions can be reduced with small dosages of exhaust gases obtained through EGR.

6.2.2.9   AMBIENT TEMPERATURE

In most turbocharged engines, there would be an air-to-air intercooler so that increased ambient temperature does increase the intake temperature, thus decreasing the inlet air density. This enables reduction in trapped mass as well as increasing compression temperature. Use of turbocharging besides increasing ambient temperature to engine intake and decreases the ignition delay, however, increase in temperatures also raises possibility of higher products temperature and thus increases NOx emissions.

6.2.2.10   LUBRICATING OIL CONTROL

A significant fraction of diesel particulate matter consists of oil-derived hydrocarbons and related solid matter, ranging from 10 to 50%.

Further reductions in oil consumption are possible through careful attention to cylinder bore roundness and surface finish, optimization of piston ring tension and shape, and attention to valve stem seals, turbocharger oil seals, and other possible sources of oil loss. Some oil consumption in the cylinder is required with present technology, however, for the oil to perform its lubricating and corrosion retarding functions. The reduction in diesel fuel sulfur content has reduced the need for corrosion protection, and has opened the way for still greater reductions in lubricating oil consumption. Changes in oil formulation can also help to reduce PM emissions by 10 to 20% (Dowling, 1992).

6.2.2.11   START OF INJECTION AND DELIVERY

The point at which injection of fuel into the combustion chamber starts has a decisive effect on the point at which combustion of the air/fuel mixture starts and, therefore, on emission levels, fuel consumption, and combustion noise. For this reason, start of injection plays a major role in optimizing engine performance characteristics.

Start of injection specifies the position stated in degrees of crankshaft rotation relative to crankshaft top-dead center (TDC) at which the injection nozzle opens, and fuel is injected into the engine combustion chamber. The position of the piston relative to top-dead center at that moment influences the flow of air inside of the combustion chamber, as well as air density and temperature. Accordingly, the degree of mixing of air and fuel is also dependent on start of injection. Thus, start of injection affects emissions such as soot, NOx, unburned hydrocarbons (HCs), and carbon monoxide (CO).

The start-of-injection set points vary according to engine load, speed, and temperature. Optimized values are determined for each engine, taking into consideration the impacts on fuel consumption, pollutant emission, and noise. These values are then stored in a start of-injection program map of electronic diesel management system.

A decisive factor to obtain an ideal torque curve with low-smoke operation (i.e., with low particulate emission) is a relatively high injection pressure adapted to the combustion process at low full-load engine speeds. Since the air density in the cylinder is relatively low at low engine speeds, injection pressure must be limited to avoid depositing fuel on the cylinder wall. Above about 2,000 rpm, the maximum charge-air pressure becomes available, and injection pressure can rise to maximum.

To obtain ideal engine efficiency, fuel must be injected within a specific, engine-speed-dependent angle window on either side of TDC. At high engine, speeds (rated output); therefore, high injection pressures are required to shorten the injection duration.

6.2.2.12   EXHAUST GAS RECIRCULATION (EGR)

EGR is a highly effective internal engine measure to lower NOx emissions on diesel engines. Internal EGR, which is determined by valve timing and residual gas. External EGR, which is routed to the combustion chamber through additional lines and a control valve (Figure 6.8).

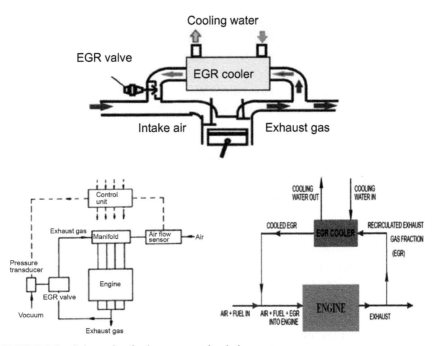

FIGURE 6.8   Schematic of exhaust gas recirculation system.

By adopting EGR levels, the peak cycle temperatures can be significantly reduced to the level where NOx may not be finding conducive environment to form. By increasing EGR concentration, drop in the rate of combustion,

and thus local peak temperatures due to an increase in the inert-gas component in the combustion chamber are obtained. This can be attributed to
reduction in partial oxygen pressure or local excess-air factor. The specific
heat capacity of dead gases present in EGR may not allow releasing effective
amount of energy and hence there could be some drop in power is noticed.
While adopting EGR levels, it is essential to employ turbocharging.

Also, in place of EGR, the engine can run with retarded injection timing
wherein the fuel would not release its full energy and thus NOx levels can
be substantially reduced. There are two ways of using EGR; one is with hot
and second one is to use cooled EGR.

Reducing the reactive components in the combustion chamber also leads
to a rise in black smoke, which limits the quantity of recirculated exhaust
gas. The quantity of recirculated exhaust gas also affects the period of ignition lag. If EGR rates are sufficiently long in the lower part load range,
ignition lag is so great that the diffusive combustion component, that is so
typical of diesel engines, is strongly diminished, and combustion only starts
after a large percentage of the air and fuel has been mixed.

In order to enhance the effects of EGR, the recirculated exhaust gas
quantity is cooled in a heat exchanger cooled by engine coolant. This raises
gas density in the intake manifold and causes a lower final compression
temperature. In general, the effects of higher localized excess-air factors
cancel each other out as a result of increased charge density and reduced
peak temperatures due to the lower final compression temperature. At the
same time, however, EGR compatibility rises to produce possibly higher
EGR rates at much lower NOx emissions.

Since diesel-engine exhaust gas already has a low temperature at very
low load points anyway, cooling the recirculated exhaust gas at the high
EGR rates required to reduce NOx emissions leads to unstable combustion.
This then results in a significant rise in HC and CO emissions. A switchable
EGR cooler is very effective to increase combustion-chamber temperature,
stabilize combustion, reduce untreated HC and CO emissions, and raise
exhaust gas temperature. In particular, this occurs in the cold start phase of
the car emission test, during which the oxidation-type catalytic converter
has not reached its light-off temperature. It also helps the oxidation-type
catalytic converter to reach its operating temperature much faster (Bosch
Hand Book).

Direct-injection diesel engines achieve recirculation rates of more
than 60%, and pre-chamber engines reach figures of up to 30%. Control
should focus on NOx as they can be reduced to a particularly great extent

in the part-load range. To a lesser degree, this also applies to HC emissions. Depending on the recirculation rate in the higher-load range, particulate emissions, on the other hand, may increase due to the reduced oxygen supply available during combustion (Figure 6.9).

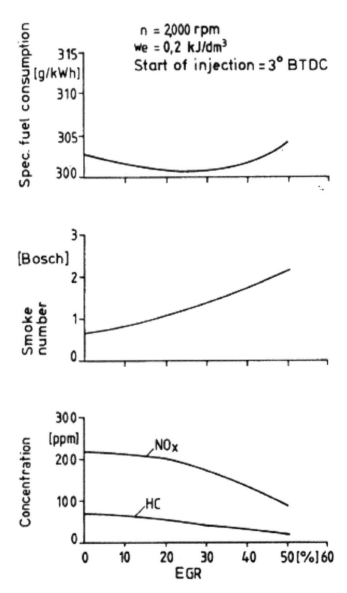

FIGURE 6.9 Effect of EGR on exhaust emissions and fuel consumption.

6.2.2.13   DE-RATING OF ENGINES

Since diesel engines work with lean mixtures and the load/speed is regulated by, regulating the amount of fuel injected into the engine. The air is sufficient enough for allowing combustion at part loads where as at high loads, the engine combustion will experience fuel rich zones due to which there would be soot formation. Hence, the engine may be prevented from exiting high soot levels by allowing running at loads less than those are problematic. Therefore, engine with applied partial load operation can be regarded as De rated engine.

6.2.2.14   USE OF SMOKE SUPPRESSANT ADDITIVES

The use of barium based compounds as additives to the diesel fuel would reduce the tendency to run with lower smoke levels. However, barium based compounds would leave deposits in filter system leading malfunctioning fuel filter frequently.

6.2.2.15   FUMIGATION

Fumigation involves injection of smaller amounts of fuel into the intake manifold. This will reduce the chemical delay of fuel while combusting and hence the ignition delay reduces thereby modest reduction in soot levels can be achieved.

Also, fumigation refers to induction of small amount of alcohol fuels such as ethanol or methanol. Due to the high latent heat of vaporization of these alcohols, there would a drop in temperature and hence VE increases and combustion will be complete. Studies have been performed for optimizing amounts of ethanol fumigation resulted in increase in brake thermal efficiency with significant reductions in CO and HC emission and soot mass (Abu-Qudais et al., 2000).

6.3   ELECTRONIC MANAGEMENT SYSTEM FOR DIESEL ENGINES-USE OF ELECTRONIC CONTROLS

6.3.1   USE OF ADVANCED FUEL INJECTION SYSTEMS

Conventional diesel engines have served many years with mechanical fuel injection systems and for reducing emissions and improving the power from diesel engines, common rail direct injection (CRDI) systems have been introduced. These systems were introduced in few vehicles but have gained a lot of attraction with the advent of electronic controls. The CRDI system was extensively start appearing with the use of very high FIPs to the tune of ~ 2000 bar. The CRDI enabled engine undoubtedly reduce PM emissions but trials have also been made to simultaneously reduce NOx emissions, the CRDI systems have been integrated with turbocharging and EGR levels.

Since 1990s, the following key technologies are driving the dramatic progress in diesel engine performance, fuel economy, and emissions:

- Electronic controls and on-board diagnostics (OBDs) (engine management systems (EMS) and OBD);
- Advanced electronic fuel injection systems (EFI);
- Exhaust gas recirculation (EGR);
- Variable geometry turbo-charging (VGT);
- Advanced exhaust after-treatment techniques.

6.3.2   ENGINE MANAGEMENT WITH ELECTRONIC CONTROLS

Development of EFI has marked significant improvements in diesel engine's performance and emissions. The electronic controls has improved the working of common rail injection.

The following are the essential differences between conventional and electronic diesel fuel injection systems:

1. Use of very high injection pressure (up to 2000 bar) – with a short injection period and acceptable fuel atomization and mixing.
2. Precise and accurate injection timing for controlling peak combustion pressure and emissions.
3. Uniform fuel metering for all cylinders for obtaining uniform power developed.
4. Handle varying degree in viscosity, lubricity, and density diesel fuel.
5. Quick delivery of fuel injection and initial injection rate for controlling noise and emissions.
6. Delivery of required fuel for cold start.

By introducing EFI, the following things can be precisely controlled:

1. Injection timing;

2. Fuel injection quantity;
3. Injection rate during various stages of injection;
4. Injection pressure during injection;
5. Nozzle opening speed;
6. Pilot injection timing and its quantity.

The above parameters have crucial in achieving simultaneous reduction PM and NOx emissions of a typical diesel engines and it has become mandatory to provide ECU in all modern multi-cylinder diesel engine. In the name of multiple injections or split injection, the use of EFI enabled diesels has become wide spread. By providing EFI, the following things can be easily achieved. Also, the fuel injection system with piezo-electric sensors has become integral part of modern diesels with CRDI systems:

1. Very high injection pressure;
2. Sharp start and stop of injection;
3. Cylinder cut-off;
4. OBD capabilities;
5. Turbocharger control.

EFI systems or electronically controlled diesel fuel injection systems use following parameters as inputs:

1. Ambient air temperature;
2. Engine speed;
3. Intake air temperature;
4. Intake air mass flow rate;
5. Lubricating oil temperature;
6. Turbocharger boosts pressure;
7. Crankshaft position;
8. Accelerator pedal position.

6.4 USE OF REFORMULATED DIESEL FUEL

Use of reformed fuels rich in hydrogen is gaining lot of attention by the researchers to contain emissions from IC engines. There are different ways of producing hydrogen including electrolysis of water, steam reforming (SR), etc. wherein high-temperature steam separates hydrogen from carbon atoms. Even though this process is potential but is energy intensive and it involves endothermic reactions.

Studies have been performed with exhaust gas assisted fuel reforming in diesel engines. The process involves hydrogen generation by direct catalytic interaction of diesel fuel with engine exhaust gas. About 16% hydrogen in the reactor product gas was obtained at a reactor inlet temperature of 290°C with a laboratory scale mini-reactor. The studies revealed that sufficient hydrogen proportion can be obtained by appropriate control of the reaction parameters at low load diesel exhaust temperatures without the aid of external heat source or air and steam supply. Also, reduction of NOx and smoke emissions was achieved with the use of simulated reformed fuel of engine exhaust emissions.

Effective heat transfer from burner surface to the reactor has to be provided. The reaction for a typical diesel fuel is:

$$C_{12.31}H_{22.17} + 12.31H_2O \rightarrow 12.31CO + 23.4H_2$$

Process parameters and its effect were investigated to achieve production of hydrogen at low temperatures viz.; diesel engine exhaust gas temperatures at part load engine operation. The engine emissions were sampled with various percentages of exhaust gases containing hydrogen by supplying simulated reformed fuel into the engine intake. The composition of the simulated reformed fuel was determined on the basis of the composition of the reforming reactor product gas (Tsolakis et al., 2004).

KEYWORDS

- **brake specific fuel consumption**
- **electronic fuel injection**
- **exhaust gas recirculation**
- **low temperature combustion**
- **nitrogen oxides**
- **variable geometry turbo-charging**

REFERENCES

Abu-Qudais M., & Osamah Haddad, M. (2000). Qudaisat Effect of alcohol fumigation on diesel engine performance and emissions. *Energy Conversion and Management 41*(4), 389–399.

Alriksson, M., & Denbratt, I. (2006). Low temperature combustion in a heavy duty diesel engine using high levels of EGR. SAE Paper No. 2006–01–0075.(Cross ref in Brijesh and Sreedhara).

Borman, G. L., & Kenneth, W. (1998). *Ragland Combustion Engineering,* McGraw-Hill International Edit.

Brijesh P., & Sreedhara S. (2013). Exhaust emissions and its control methods in compression ignition engines: a review. *International Journal of Automotive Technology, 14*(2), 195–206.

Challen, B., & Baranescu, R. (1999). *Diesel Engine Reference Book,* Second Edition; Butterworth-Heinemann.

Dowling, M. (1992). *The Impact of Oil Formulation on Emissions from Diesel Engines*, SAE Paper 922198, SAE International, Warrendale, Pennsylvania.

Heywood, J. B. (1988). *Internal Combustion Engines Fundamentals*, McGraw Hill.

Johnson, T., (2008). Engine emissions and their control: An overview. *Platinum Metals Rev., 52*(1); https://www.technology.matthey.com/pdf/23-37-pmr-jan08.pdf

Kitamura, T., Ito, T., Senda, J., & Fujimoto, H. (2002). Mechanism of smokeless diesel combustion with oxygenated fuels based on the dependency of the equivalence ratio and temperature on soot particle formation. *Int. J. Engine Research,* 3(4), 223–247. (Cross ref in Brijesh and Sreedhara).

Potter, M., & Durrett, R. (2006). Design for compression ignition high-efficiency clean combustion engines. *2006 Directions in Engine-Efficiency and Emissions Research (DEER) Conf.* Detroit, Michigan, USA. (Cross ref in Brijesh and Sreedhara).

Wallace, F. J. (1986). Variable geometry turbocharging – optimization and control under steady-state conditions. In: *Proceedings of the 3rd IMechE International Conference Turbocharging Turbochargers.*

CHAPTER 7

Diesel-Aftertreatment Measures

7.1 INTRODUCTION: DIESEL EXHAUST GAS AFTERTREATMENT

The diesel engine is serving the purpose of a prime mover in light, medium, and heavy duty applications. Ever since the invention of diesel engine, it has undergone multitude of design changes and developments and probably the gasoline engines could not have undergone to the same scale. Major issue with diesel engines is that of heterogeneous and diffusion mode of combustion wherein both fuel and air do allow mixing within the engine cylinder. Even though diesel work with lean mixtures, the high load operation with fuel-rich mixture combustion lead to the particulate formation. This can be attributed to the governing of diesel-fuelled engines.

Diesel engines yield high fuel conversion efficiencies with better control over fuel satisfying in both passenger and goods transport applications. Exhaust gases from diesel engines include species similar to gasoline engine exhaust species (CO, HCs, and NOx); in addition, diesel exhaust consists of particulate matter (or soot) emissions which comprise of solid carbon particle. The common man also distinguishes easily with the kind of exhaust being emitted by the two engines. The serious problem associated with the diesel engine is containing particulate matter as it is not gaseous pollutant and is a concern for visibility issues. That's why the control of diesel engine emissions poses altogether different issue. In this context control, techniques include both in-cylinder measures as well as aftertreatment techniques. Similar to SI engine's emissions control, it is mandatory to use aftertreatment devices also with diesel engine emission control modules. Moreover, the simultaneous control of highly pollutant gases like NOx emissions and PM emissions is a major cause of concern in the light of stringent emission control legislation. It is both manufacturers and researchers striving hard to arrive at a trade-off between NOx and PM emissions. The vehicles with diesel-powered engines used for passenger applications are called light-duty diesels (LDD). Over a period of time, the diesel engine emission regulations have undergone severe changes both regarding PM and NOx emissions, so also the vehicles are named as low emission vehicles, ultra-low emission vehicles or super ultra-low emission vehicles.

Number of control strategies are employed to make clean diesel engines. These approaches include advanced common rail fuel injection, electronic engine controls, combustion chamber modifications, air boosting, improved air/fuel mixing, and reduced oil consumption, all aimed at achieving ultra-low exhaust emission targets. Engine manufacturers are focusing on ways to control engine operation to reduce engine-out emissions as low as possible and reduce the burden on the exhaust emission control systems.

Nowadays gasoline fuelled engines (gasoline direct injection (GDI) Engines) are also producing PM emissions but lower than diesel fuel power engines. Particulate matter emissions from diesel engines are considerably higher than from direct injection or GDI spark-ignition engine. Diesel engine emissions with black smoke, a major source of high ambient concentrations of particulate matter in most large cities of the developing world. Sulfur in diesel fuel further adds to formation of Particulate Matter. In addition to combustion generated pollution, uncontrolled diesel engine is also a source of noise pollution resulting from the combustion process. The experience of heavy diesel exhaust odor and smog has become a serious concern in Countries like India especially in cold weathers as people are heavily dependent on diesel fuel since it is available at a cheaper cost than gasoline fuel.

Especially diesel particulates cause many health-related issues such as blockage of nostrils, respiratory disorder and hence WHO has also raised concerns about PM emissions and therefore regulations are stricter for the PM emissions. As per EURO norms, the PM emissions for diesel-fuelled passenger cars has reduced significantly from EURO I to EURO III by 65% and 96% by EURO VI. From EURO IV norms onwards, cap has been put on emission of particulate number (PN) also. The size of particulates emitted from diesel engines are smaller than 2.5 microns in diameter, generally expressed as PM_{10}, $PM_{2.5}$, etc. These particulates are very complex in structure containing carbonaceous matter at the core adsorbed hydrocarbons from engine oil and diesel fuel, sulfates, water, and inorganic materials produced due to engine wear.

The fuel economy exhibited by diesel-powered is about 40% greater than that of gasoline engine and hence there would be about 20% reduction in CO_2 emissions.

The diesel engine emissions can be controlled by adopting modifications in both engine geometrical, operating parameters as well as using exhaust gas aftertreatment methods. Of these, the exhaust gas aftertreatment methods (add-ons) significantly reduce the diesel engine emissions. In diesel engines, the exhaust gas mass flow also largely depends on engine operating conditions, particularly engine speed, the supercharging rate, and the exhaust gas recirculation (EGR) rate.

Majority of diesel engine's aftertreatment devices works either through chemical processes or physical separation or removal with filters or traps. The aftertreatment systems include catalytic converters, sensors, particulate filters/particulate traps along with auxiliary items such as regeneration systems. Due to low temperatures prevailing in exhaust process of diesels, the catalytic converters used in diesel engines may be ineffective therefore the chemical processes are limited kinetically, however, maybe the quality of converters.

The engineering and material selection include corrosion protection as engine exhaust gas contains water vapor, which condenses while the exhaust gas system cools after the engine has been stopped and also exhaust condensate also contains corrosive constituents formed in reactions with the nitrogen oxides (NOx) and sulfur dioxide contained in the exhaust gas.

Thus, the exhaust aftertreatment comprises any form of exhaust gas processing aimed at reducing the emission of one or more exhaust components. Aftertreatment, devices for diesel engines include diesel catalysts and soot filters-sometimes known as particulate traps.

7.2   DIESEL ENGINE EXHAUST COMPOSITION

As stated above, the diesel-fuelled engines emit CO, HCs, NOx, and PM emissions. Among these, PM doesn't have fixed chemical formula. This is the main drawback to predict and measure the PM emissions. The diesel engine exhaust is a complex mixture of organic and inorganic compounds in solid, liquid, and gaseous phases. The fuel and lubricating oil are the sources of formation of organic compounds originate during combustion with few of which reach the tailpipe without modification, others are modified by combustion or by reactions in the catalyst if present. The H:C ratio of fuel plays a vital role in the formation of pollutant diesel emissions. The emissions are typically hydrocarbons ranging from lightweight fractions with high hydrogen: carbon (H:C) ratios, through heavier fractions with very low H:C ratios. The hydrocarbons usually contain aldehydes, alkanes, alkenes, aromatics, etc.

Also, the inorganic compounds are largely produced from the fuel and lubricating oil, although other sources include the intake air and wear products from the engine hardware. Depending on the fuel formulation, sulfur in the fuel can be a significant source of exhaust particulate. This can take the form of gaseous sulfur dioxide (SO_2), sulfates (SO_3, SO_4), and sulfuric acid (H_2SO_4). Fuel and lube oil provides a source of carbon which can form soot particles in the combustion process through gas- or liquid-phase reactions. Typical major constituents of diesel engine exhaust emissions are carbon monoxide (CO), oxides of nitrogen (NOx), particulate matter (PM) and hydrocarbons (HCs).

For developing catalysts or filters for diesel emission control, care should be taken to consider the composition of diesel particulate matter. In diesel exhaust, soot, soluble organic fraction (SOF), and sulfates are the three vital parts that comprises diesel particulates and thereby make the emission control process more complex. The SOF consists of the hydrocarbons adsorbed on to the soot particles, and liquid hydrocarbons. SOF is so-called because it may be measured by extraction using an organic solvent.

During the diesel combustion process, in the high-temperature flame regions with conducive air-fuel atmosphere the oxides of nitrogen form from nitric oxide (NO) and nitrogen dioxide (NO_2), collectively these two gases are known as NOx. NO is the predominant form since the reactions from which it is produced proceed at higher rates than the reactions for NO_2 at the conditions prevailing in the diesel combustion process. The diesel engines are normally operated with lean mixtures and the levels CO emissions is not a constraint. However, the major cause of concern is HC, NOx, and PM emissions.

7.3   DIESEL ENGINE AFTERTREATMENT TECHNOLOGIES

The viscosity of diesel fuel is relatively higher than gasoline and its volatility is also poor and hence the evaporative emissions are little less than gasoline fuelled engines, i.e., volatile organic compounds at ambient conditions are less due to evaporation. Chapter 6 dealt with in-cylinder techniques for mitigating diesel engine tail-pipe emissions. The fact that if control of emissions could not be possible with in-cylinder measures, the use of add-ons such as catalytic converters and particulate filters or traps is imminent with modern diesels and these devices have proven beneficial in lowering harmful pollutant emissions, especially PM emissions. After the passage of exhaust gases through the aftertreatment devices, the levels of pollutant emissions will be significantly reduced once they come out of these devices. These add-ons or aftertreatment devices have become integral part of modern diesel engine vehicles. In India, the traps enabled vehicles haven't yet entered the market. However, they may certainly appear for complying EURO VI emission norms. The exhaust gases before leaving into the atmosphere are treated and then let-off into the atmosphere.

The emission species far above the permissible limits, such as exhaust gases like CO, HC, and NOx are chemically treated with oxidization and reduction processes (redox reactions), outside the engine before letting it into the atmosphere.

The concern for developing clean diesel engine started with the emergence of smog-related issues, however, it was not steady at its pace of growth, by employing a number of strategies. These approaches include advanced common rail fuel injection, electronic engine controls, combustion chamber modifications, air boosting, improved air/fuel mixing, and reduced oil consumption. Achieving ultra-low exhaust emission targets requires a systems approach.

The technological developments in diesel engine research enabled viable aftertreatment technologies, to lessen diesel exhaust emissions from both new vehicular engines and, as well as in-use engines and such devices are existing in the form of add-ons or the use of retrofit kits. The PM-NOx paradox is a major concern of diesel engine emission control approaches. The following are major technologies with regards to control of PM and NOx control. The major problem lies with the control of HC, NOx, and PM emissions from diesel engines. However, to some extent, HCs would be seen as PM emissions. Therefore, the onus lies in effective control of PM-NOx emissions only. The techniques employed for these emissions control are depicted in Figure 7.1.

7.4 DIESEL OXIDATION CATALYSTS (DOC)

In the earlier days, in-cylinder measures were only the means of controlling diesel-engine emissions. For doing this, number of trial and error techniques used to be adopted to arrive at compromised solution with respect to emissions and performance. However, in-cylinders require multiple parameters to taken into account for optimization and hence, the effective mitigation of diesel emissions be achieved with aftertreatment methods.

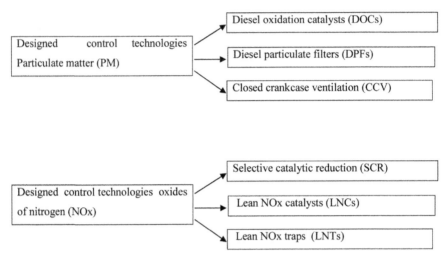

FIGURE 7.1   The techniques employed for HC, NOx, and PM emissions control.

It is the presence of oxygen that matters for the mitigation of exhaust species-CO, HC, PM, and NOx emissions. One way the higher proportion of oxygen in mixture is good for oxidation whereas it is detrimental for reduction reactions. Over the years, for effective reduction of NOx emissions, NOx storage catalyst, or an SCR catalytic converter (selective catalytic reduction) have been developed.

In diesels, Soot or PM emissions form majorly due to presence of fuel rich zones during combustions which occur mostly under high load condition. Diesel particulate filters or traps (DPF or DPT) are being used to limit the flow of PM into atmosphere. The DPFs also contain regeneration system. In European countries, the SCR and DPF work in conjunction with each other.

7.4.1   THREE-WAY CATALYTIC CONVERTERS

Similar to SI engines, catalytic converters can also be employed for diesel fuelled CI engine for both oxidation and reduction reactions (redox reactions). These reactions are carried out in a three-way catalytic converter to oxidize HC and CO emissions and reduce NOx emissions. It is observed from the literature that good conversion is possible under stoichiometric mixture condition. The modern vehicles are being incorporated with electronic sensors and actuators for effectively sensing the excess oxygen in both the upstream and downstream of exhaust systems. For this oxygen sensor or lambda control (λ ~1 is preferred) device is extensively used (Allan Bonnick). The converter consists of a ceramic matrix or corrugated metal, bundled with a washcoat to provide a large surface area for chemical reactions. The washcoat is composed of active catalyst materials mostly of noble metals such as platinum (Pt), rhodium (Rh) and palladium (Pd). Pt and Rh are used as reduction catalyst, while Pt and Pd are used as oxidization catalyst (Twigg, 2011). Platinum works both as oxidation and reduction catalyst. Normally in the temperature range of 400°C to 800°C, a three-way catalytic converter achieves good conversion (~ 80%) and also lasts longer. At lower temperatures, they are ineffective. The converters also have shown degraded performance when vehicles run for shorter durations and frequent start and stop conditions. Particulates in diesel exhaust poison the catalytic converters. The benefit of NOx reduction by catalytic converter is poor as diesel engines always run on overall lean mixtures.

With poorly maintained catalytic converters, only 40–50% is possible, however, with computer-controlled systems it is conveniently possible of conversion efficiencies about 95%. Three-way closed-loop catalytic converters currently are the most efficient emission control systems available for internal-combustion engines. The components of 3-way catalytic converters are similar to that explained in Chapter 6.

7.4.2 DETAILS OF OXIDATION CATALYSTS

The presence of a proper catalyst can significantly increase the rate at which the oxidation reaction proceeds and hence catalysts are inherently employed in exhaust aftertreatment of diesel exhaust. Active sites are developed with the action of the catalyst when adsorbed one or more of the reactants onto the surface of the catalyst. Oxidation catalysts are used for oxidizing large portions of HC (including aldehydes) and CO emissions and few of the soluble organic portions of PM emissions and mutagenic emissions.

Catalytic converter cannot be enough to oxidize solid particulate matter (however small the size they may be) and also as particulates in the exhaust are mostly adsorbed with sulfates from sulfur dioxide. The use of Oxidation catalytic converters in both light-duty vehicles (LDV) and heavy-duty applications are proven to be successful. With the help of oxidation converters 80% and even higher conversion, efficiencies are possible with volatile organic compound and carbon monoxide emissions but they are of insignificant use for reducing NOx emissions. For improved performance of converters, it is mandatory to use as low-sulfur diesel as possible. To comply with future emission norms, a diesel fuel should be made with 10 ppm so as to meet ultra-low sulfur diesel standards (ULSD).

Fuel with less than 0.05% sulfur by weight is required for diesel catalysts to perform well. Due to high 350°C temperatures, the sulfur can easily be converted to SO_2 (gaseous) to SO_3 (particulate matter) in vehicles equipped with oxidation catalysts but the oxides of sulfur may be detrimental in view of life of certain engine components.) This established through static dynamometer tests. Lower rate of sulfate particle emissions is obtained on actual on-the-road operating conditions even with lower catalyst temperature.

Usually, chemical adsorption will happen with some kind of chemical breakdown of the molecule at the catalyst surface. For example, an oxygen molecule might be adsorbed onto the surface of a catalyst and in the process dissociate into two separate atoms of oxygen. The actual behavior at the surface of a diesel oxidation catalyst is complex and is believed to support several different adsorption processes.

Diesel catalysts in volume production for automotive applications primarily control emission of the soluble organic fraction of particulate (SOF), gaseous hydrocarbons (HC), and carbon monoxide (CO). Some catalysts also reduce NOx emissions in relatively small amounts.

A diesel oxidation catalyst (DOC) is a passive device, installed in the engine exhaust system, through which all of the exhaust gas flows. It consists of a metal outer casing that contains a core or substrate, usually of ceramic. The substrate forms a matrix of flow passages parallel to the direction of flow, having as many as 60 cells/cm². The object of the matrix design is primarily to provide a large surface area exposed to the exhaust gas. The substrate is coated with a washcoat consisting of base metal oxides, for example, silica, and alumina, and small quantities of precious metals-typically a combination of platinum, palladium, and rhodium which, together with the base metal oxide, act as the catalyst. Figure 7.2 represents a simple model of catalytic action in a diesel oxidation catalyst.

The challenge in any catalyst application is to maximize the efficiency of desired reactions while minimizing the efficiency of any undesired reactions. The problem of formulating catalysts that promote desired reactions and inhibit undesired reactions, known as selectivity, is one of the most challenging areas of catalyst development.

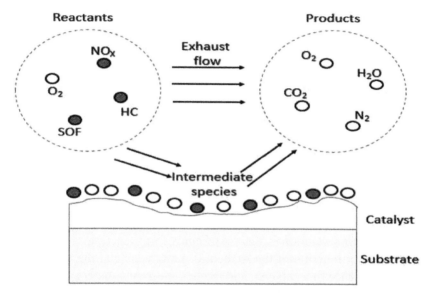

FIGURE 7.2 Simplified model of catalytic action in a diesel oxidation catalyst.

In addition to the selection of catalyst materials, the geometry of the catalyst plays a very important role in the overall performance. The most important geometric factors are:

1. Catalyst volume;
2. Cell density and active surface area;
3. Net flow cross-sectional area;
4. Inlet diffuser design.

Space velocity is defined as volume rate of gas flow divided by the volume of the catalyst. A key design requirement in all vehicle applications where space is at a premium for good performance with a high space velocity enables the catalyst volume to be reduced. For both heavy-duty and light-duty applications, it is important that the catalyst does not unduly restrict the gas flow, since this will apply back-pressure to the engine and increase fuel consumption. A uniform flow through the catalyst core is essential for good performance and durability and can be achieved by proper inlet diffuser design.

Diesel oxidation catalysts (DOCs) installed on a vehicle's exhaust system can reduce total PM. Diesel engine CO and HC emissions can best be reduced to more than 90% and more than 70% of the toxic hydrocarbon emissions in diesel exhaust.

7.4.3   FLOW-THROUGH THE DIESEL OXIDATION CATALYST

The diesel oxidation catalyst (DOC) is one of the oldest forms of controlling diesel engine exhaust pollutants. The reactants before entering the catalyst zone would be partially oxidized hydrocarbons along with soot or PM and the particulates will be engulfed with soluble organic fractions (SOF), the SOF originate from unburned hydrocarbons out of fuel and lube oil that have condensed on the solid carbon particles.

DOCs are most often based on a flow-through honeycomb substrate (either metallic or ceramic), coated with an oxidizing catalyst such as platinum and/or palladium. The typical functions of DOCs in exhaust gas treatment are significant reduction in CO and HC emissions, particulate mass and oxidation of NO to form NO_2.

The oxidation reactions follow Arrhenius form reaction rates, the rate of reaction is a function of the Arrhenius parameters and the temperature and in the absence of a catalyst, and the Arrhenius parameters are constant for a given reaction. The temperature is exponentially varied in the Arrhenius equation, the Arrhenius equation also relates to the activation energy of the overall reaction such that the reaction proceeds at a high rate even at low temperatures. Mostly, catalysts increase the rate of a given reaction by providing alternative paths with lower activation energies. The conversion of pollutant emissions to oxidized or reduced emissions is depicted in Figure 7.3.

Input:
CO
HCs
PAHs*
SO_2
NOx
PM/EC/SOF

Output:
CO_2
H_2O
SO_2/SO_3
NOx/NO_2
PM/EC**

* Polyaromatic hydrocarbons or other toxic hydrocarbon species
** Elemental carbon

FIGURE 7.3   Diesel oxidation catalytic conversion—a representation of three channels of a straight through, flow path honeycomb.

The oxidation catalysts work effectively if an exhaust system is maintained hot enough to enhance the rate of reactions and majority of vehicle manufacturers allowed thermally insulated the exhaust manifold exhaust pipes. To provide insulation, few of the designs include double-walled, stainless steel exhaust pipe containing an air gap within the tube walls.

The engine out exhaust gases (reactants) enter the channels from the left and as they pass over the catalyst coating and finally oxidized to the oxidized products on the right. The level of total particulate reduction is influenced in part by the percentage of SOF in the particulate (Figure 7.4).

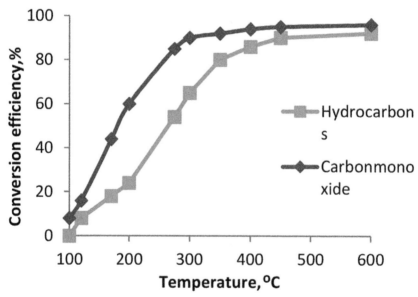

FIGURE 7.4 Catalytic conversion efficiencies of HC and CO as a function of temperature.

Figure 7.4, is a typical conversion curve for CO oxidation, which is obvious with the use of DOC. A very steep increase in conversion is clearly seen. At around 200°C of the exhaust gas temperature at least a conversion of 50% is possible is referred as light-off temperature it depends on the catalyst composition, flow velocity and exhaust gas composition. Very temperatures are required for oxidized methane if present in exhaust line however, short-chain alkenes can be oxidized at low temperatures.

The DOC at times resembles a catalytic burner. In few cases, fuel may either required to be post-injected (normally takes place in a multiple injection system) or introduced by an engine by a device downstream from the engine. Depending on the temperature requirement for oxidation, the fuel quantity is required to be calculated. The introduction prior to and post-combustion are precisely controlled/monitored in CRDI systems equipped with ECUs.

It must be ensured that when fuel is introduced downstream of the engine, the required quantity should disperse throughout the flow cross-section as homogenously as possible and reach the catalytic converter fully vaporized. The fuel can be introduced in the form of fine diesel spray with a metering valve, vaporized diesel fuel, or introduction of gases with low light-off temperatures.

7.5   DIESEL PARTICULATE FILTERS OR TRAPS (DPFS OR DPTS)

The serious concern of diesel engine is emission of high levels of PM and NOx emissions. Over a period of time, the diesel emission regulations have undergone strict norms. Figure 7.5 depicts periodic revision of diesel engine exhaust PM and NOx emission standards.

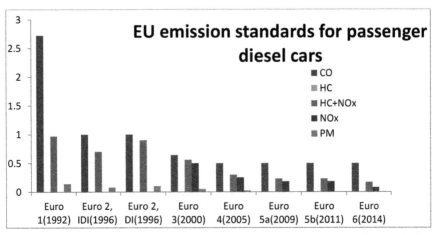

FIGURE 7.5   Periodic revision of diesel engine exhaust PM and NOx emission standards.

From Figure 7.5, it is obvious that in early 1990s, precisely from 1992 onwards, the EURO emission norms came into force. As far as diesel engine emission norms are concerned, till EURO 3, emphasis is given for CO, HC + NOx and PM emissions. From EURO 3 onwards, specific control of NOx started entering into emission control due to its obvious hazardous nature to environment and life on the planet. However, the diesel particulates caused lot of concern from researchers and therefore particle number count started emerging from Ibrahim (2014) onwards. Figure 7.5 also clearly shows the reduction in PM emission as well as NOx emission with continuously getting stringent emission norms through the time span.

The diesel oxidation catalyst may not be suitable for total removal of visible pollutant i.e., soot of diesel engines, it is PM which is not a gaseous emission makes difference in diesel emission control approaches compared to gasoline engine control. The PM emission (soot) is critical of diesels especially under full load operation. For this purpose, many of the advanced countries are planning to use filters for collection of particulates. The function of DPF is to separate a very large fraction of particulates from the exhaust gas flow. As the name implies DPF collects/remove particulate matter from diesel exhaust by filtering exhaust from the engine. They are called as particulate traps as they arrest the flow of particulate matter without allowing to get released into atmosphere. They can be installed both on road vehicles or stationary diesel engines.

In the process of filtering the particulate matters from diesel exhaust, lot of particulates get accumulated into it and unless it is burned off, the filtering process would get slowed down. The method used for this purpose is to oxidize accumulated particulate matter and then to burn within the filter when exhaust temperatures are adequate. In such a case, the filter is cleared off the particulate mass (unwanted particulates) and the filter is cleaned or "regenerated." There are two types of filters in use- passive filters and active filters. When regeneration is carried on with available exhaust heat, such filters are regarded as "passively regenerated" filters. On the other hand in "actively regenerated" filters, an energy input, like injection of diesel fuel into an upstream of filter is done. It is customary to use DPFs that employ a combination of passive and active regeneration strategies to ensure that all vehicle operating conditions satisfy filter regeneration effectively. Active regeneration strategies employ various engine controls to achieve filter regeneration conditions on demand.

DPF can be combined with EGR NOx adsorber catalysts or SCR to achieve significant NOx and PM reductions.

Few engines are equipped with multiple injections wherein the fuel injection can be split into two or more phases to facilitate both reductions of NOx emissions initially and later to help in active regeneration of filters. Therefore, simultaneous reduction of NOx and PM emissions can be achieved (Sindhu et al., 2017).

7.5.1 FILTRATION MECHANISM

The diesel particulate filter resembles like any other filter to separate non-miscible components such as in Tea preparation from tea powder. The principle of diesel particulate filter involves separation and collection by deposition on a collecting surface of gas borne particles called particulates while the gas stream flows. A porous barrier of DPF retains the particulates. Based on the type of barrier DPF are divided into:

- deep-bed filters; and
- surface-type filters.

The wool and foam of Deep-bed filters collect particles throughout of their structures whereas particulates collect on its porous walls in Surface filters. The mean pore size of filter of deep-bed filters is larger than the mean diameter of collected particles. The particles are deposited through a combination of depth filtration mechanisms driven by various force fields. The force fields may be related to velocity or concentration gradients in the gas. In contrast, the pore diameter is less than the particle diameter of surface-type filters. The types of DPFs are depicted in Figure 7.6.

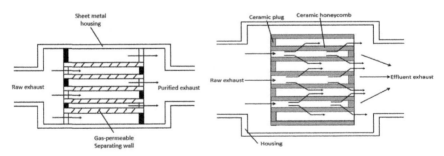

FIGURE 7.6   (a) Ceramic monolith filter, (b) ceramic particulate filter.

In most dust collection devices including majority of diesel filters, depth filtration is the primary filtration mechanism. Also, the diesel engine intake-air cleaners work primarily through the depth filtration. In the surface type filters, the layer of collected dust, called "filtration cake," of diesel particulates itself is the principal filter medium and the process is named "cake filtration." In liquid filtration, cake filtration is very common. Such a process of cake filtration is not observed in diesel filters and in most of the dust collecting filters. Depth and surface filtration are combination mechanisms in ceramic wall-flow monoliths.

7.5.2 FLOW-THROUGH OR PARTIAL DIESEL PARTICULATE FILTERS (DPF)

The flow-through filter technique is a new method for reducing diesel PM emissions. These types of filters use catalyzed metal wire mesh structures or tortuous flow, metal foil-based substrates with sintered metal sheets for filtering diesel PM emissions (Figure 7.7).

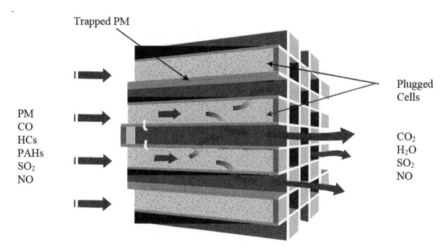

FIGURE 7.7   Wall-flow diesel particulate filter.
Source: http://www.meca.org/technology/technology.

In most of the heavy-duty and LDVs Partial Filters with no active regeneration or ash removal are commonly installed due to maintenance-free operation, necessary, and exhibit filtration efficiencies >50–84%, and may vary depending on the engine operating characteristics.

Catalyst materials can also be coated to assist in oxidizing the soot or used in conjunction with an upstream diesel oxidation catalyst to oxidize diesel soot as the exhaust flows through these more turbulent flow devices. Flow through filter, due to their relatively smaller package size, are advantageous in applications requiring special shapes or having space limitations. These filters do not accumulate inorganic ash constituents present in diesel exhaust. In many applications, even ash passes through such filters eliminating the need for filter cleaning. The wall-flow filters can be made from a variety of ceramic materials. It is most common with high-efficiency based on a porous wall, square cell, honeycomb design where every alternate channel is plugged on each end. The filters also are made of sintered metal fibers and the characteristic feature of this wall-flow filters do posses high strength and thermal durability.

Figure 7.7 depicts the operation of a wall-flow DPF. A particulate-laden diesel exhaust enters from the left side of filter. In a checkerboard like configuration, the porous cell wall channels are capped at opposite ends to make the exhaust flow through easily. The gas passes through the porous walls of filter cells (thus the wall-flow filter designation); the particulate matter is deposited on the upstream side of the cell wall. Thus, exhaust gas cleared off particulates exits the filter on the right. A back-pressure builds up over a period of working of filter as the soot deposited on the cell wall and is required to be cleared by a process known as "regeneration."

In olden days, the soot is measured in terms of Smoke Opacity or Soot or Smoke Density. However, soot density has lost its significance and Particulate size and number in addition to the total mass is gaining increasing interest among the research community. It has become mandatory while promulgating new emission regulations on the measurement and characterization of ultrafine particles and particle number. The Association for emissions control by catalysts (AECC) conducted a test program for particle size and number on light-duty and heavy-duty vehicles using the procedures outlined in the European particle measurement program (PMP). The heavy-duty testing was conducted using both the European transient cycle (ETC), as well as, the world harmonize test cycle (WHTC) and looked at the total number of particles within the range of 25 nm to 2,500 nm (2.5 µm). The wall-flow DPFs employ ceramic materials and most commonly used are cordierite and silicon carbide. The desirable characteristics of a good ceramic material are strength, low thermal expansion coefficient and low cost and cordierite, it posses high porosity has been commonly used as a substrate material in exhaust catalysts.

Most large heavy-duty diesel applications use cordierite filters whereas filters made from silicon carbide (SiC) are mostly used in light-duty diesel passenger cars as SiC offers much higher temperature tolerance than cordierite for use with filters where exhaust or regeneration temperatures are prevailed. The high strength of SiC makes it suitable in high thermal stress exhaust environments. These properties of SiC enables a higher soot loading limit as the higher thermal conductivity of SiC controls the peak temperature during regeneration. However, a segmented architecture within the filter element is required for SiC due to its higher thermal expansion coefficient than cordierite. The other candidate wall flow ceramic materials are aluminum titanate (Al_2TiO_5) and mullite (Al_2SiO_5) because of their high temperature capability and thermal shock properties. To extract maximum benefit from filters, the filter efficiency has to target by aiming at the following especially with regards to wall-flow filter designs:

1. Minimizing the back pressure for optimizing filter efficiency.
2. Radial flow of oxidation should be improved filter regeneration.
3. Mechanical strength of filter designs should be improved.
4. Ash storage capacity of the filter should be increased.

DPF design includes advancements in cell shape and cell wall porosity optimization in the wake of technological developments by minimizing engine back-pressure and extends the life of filter.

Advances such as higher pore volume, increased pore connectivity along with thinner web designs facilitate catalyst coating while maintaining longer times between soot regeneration events.

7.5.3 FILTER REGENERATION

The basic function of diesel particulate filter or trap is to collect the diesel engine exhaust particulates without obstructing of flow through its own structure. In the process of doing this a lot of soot gets accumulated and it should be cleared off without obstruction, i.e., to minimize the back-pressure.

For achieving the clean filtration, the accumulated soot should be burned off continuously or should be oxidized without exerting pressure on the free flow. The exhaust gas temperatures are also to be maintained high to allow burn off therein the filter gets regenerated.

This is referred to as passive regeneration and will be covered in more detail below. Filters that regenerate in this fashion cannot be used in all situations, primarily due to insufficient exhaust gas temperatures associated with the operation of some types of diesel engines, the level of PM generated by a specific engine, and/or application operating experience. To ensure proper operation, filter systems are designed for the particular engine/vehicle application and account for exhaust temperatures and duty cycles of the specific vehicle type. In low exhaust temperature, operation an active regeneration strategy may need to be implemented to raise the exhaust temperature sufficient for oxidizing the soot.

The performance of catalyst-based DPFs is significantly affected by Sulfur in diesel fuel as it interferes with the reliability, durability, and emissions. Sulfur affects filter performance by inhibiting the performance of catalytic materials upstream of or on the filter. Sulfur content in diesel fuel main reason for increasingly high levels of particulate matter through catalytic sulfate formation. It is preferable to have less than 15 ppm sulfur level in diesel when it is used in catalyst-based diesel particulate filter technology, in order to obtain better filtration.

An essential feature for proper functioning of any diesel particulate filtration (DPF) system relies on a prescribed regeneration to occasionally burn soot collected in the filter and reduce the back-pressure of the exhaust stream. Many exhaust control systems rely on a DOC or regeneration catalyst upstream of the DPF to assist with regeneration. This strategy can be applied to either coated or uncoated DPFs and essentially performs two functions. The first is to oxidize unburned HC and CO in exhaust and utilize the exothermic heat of combustion to raise the temperature of t exhaust gas entering the DPF to temperatures sufficient to combust the captured carbonaceous soot. This can be done by enriching the fuel/air ratio going to the cylinder or injecting a small amount of fuel into the exhaust ahead of the DOC. A second regeneration function is to oxidize some of the NOx in the exhaust to nitrogen dioxide (NO_2) which oxidizes carbon at a lower temperature than oxygen. The presence of higher concentrations of NO_2 thus facilitates filter regeneration at lower exhaust temperatures.

Besides use of the correct catalyst, the main factors governing the efficiency of exhaust gas treatment or catalytic conversion process depends on the correct operating conditions. They are adjustable by the engine management system (EMS) within a wide range.

If the operating temperatures are excessively high, sintering processes will occur, i.e., several small noble-metal particles will clump together to form a larger particle with a correspondingly smaller surface area, and thus reduced activity. The function of exhaust gas temperature management is, therefore, to enhance the service life of the catalytic converter by avoiding excessive temperatures.

7.5.4 PASSIVE REGENERATION

The type of passive regeneration is the simplest of filtration efficiency enhancer system as its design doesn't require driver or engine intervention to combust the soot on the filter. In this case, the ceramic or metal filter substrate is coated with a high surface area oxide and precious or base metal catalyst. The catalyst acts to reduce the ignition temperature of the accumulated particulate matter up to high degree. The reduction in burn temperature allows the DPF to regenerate passively on some applications, but there will be others for which the exhaust temperature is too low to regenerate the filter. As minimum exhaust temperatures are required, passive regeneration cannot be applicable to all types of engines and vehicles.

Besides collecting particulates, the catalyzed filters also help in reducing odor and soluble organic fraction of the particulate, but some catalysts may increase sulfate emissions. Companies utilizing these catalysts to provide regeneration for their filters have modified catalyst formulations to reduce sulfate emissions to acceptable levels.

Catalyst-based passive regeneration also relies on an upstream oxidation catalyst to facilitate oxidation of nitric oxide (NO) to nitrogen dioxide (NO_2). Nitrogen dioxide is a much stronger oxidizer than oxygen allowing filter regeneration at lower temperatures. The nitrogen dioxide oxidizes the collected particulate thus substantially reducing the temperature required to regenerate the filter.

7.5.5 ACTIVE REGENERATION

To a much larger range of applications, the actively regenerated, high-efficiency filter systems can be employed. The active regeneration systems are more expensive than passive filters. Some of the active technology options are Burners, injection of diesel fuel into the exhaust stream for oxidation across a DOC upstream of the DPF, or electrical heaters are few of the options to work as active regenerating units. In addition, the methods such as (i) by air throttling, exhaust temperature can be increased to facilitate filter regeneration; (ii) fuel injection after the piston crosses the TDC (Post TDC fuel injection) to further increase the exhaust gas temperatures; and lastly (iii) post injection of diesel fuel in the exhaust upstream of an oxidation catalyst and/or catalyzed particulate filter. The last method provides heat to combustz accumulated particulates by oxidizing fuel across a catalyst present on the filter or on an oxidation catalyst upstream of the filter.

The above techniques can be used in combination with a catalyzed or uncatalyzed DPF. In special applications, where sufficient exhaust temperatures cannot be reached using the above techniques it may be necessary to use external means such as on-board fuel burners or electrical resistive heaters to heat the filter element and oxidize the soot. These can be used with catalyzed or uncatalyzed filter elements. In some cases, regeneration can be accomplished while the vehicle is in operation, whereas in other cases the engine must be turned off for regeneration to proceed. In some situations, installation of a filter system on a vehicle may cause a very slight fuel economy penalty. This fuel penalty is due to the backpressure of the filter system.

Filter efficiency depends on pore size, cell thickness, and wall thickness. Ceramic monolith filters can be coated with catalytically active materials. The catalytic effect and the resultant exothermic processes offer benefits with regard to regeneration. The drawbacks of such a system are a tendency of fracturing due to temperature differences (heat tension) and increased exhaust backpressure that increases even further during the service life of the system (due to the effect of deposited non-oxidizing soot components).

Particulate filters, developed based on ceramic monoliths, ceramic fibers and metal substrate, are used to absorb particulates of very small size ($\sim$ 0.1 μm). Particulate deposits block and obstruct the particulate filter which affects engine output and increases fuel consumption due to increased backpressure.

With more and more accumulated soot in the filters, the back pressure increases and finally the engine efficiency is degraded. It is mandatory to carry out filter regeneration for removal of blocking. The high temperatures ($\sim$ 600°C) required to burn off soot can only be achieved under full load condition. Further development is required with alternative catalytic coatings, additive supported regeneration, electric heating and diesel combustor to achieve lower soot ignition temperatures.

A trap-oxidizer system has a particulate filter (the trap) in the engine exhaust stream and some means of burning (oxidizing) collected particulate matter from the filter. The manufacturing of traps is made simple in the recent times. The main problem of trap-oxidizer system development is how to remove the soot effectively and regenerate the filter. Diesel particulate matter consists of solid carbon coated with heavy hydrocarbons. This mixture ignites at 500 to 600°C, well above the normal range of diesel engine exhaust temperatures (150–400°C). Special means are therefore needed to ensure ignition. Once ignited, however, this material burns at temperatures that can melt or crack the particulate filter. Initiating and controlling regeneration without damaging the trap is the central problem of trap-oxidizer development.

A number of trapping media have been tested or proposed, including cellular ceramic monoliths, woven ceramic-fiber coils, ceramic foams, corrugated multi-fiber felts, and catalyst-coated, stainless-steel wire mesh. The most successful trap-oxidizer systems use either the ceramic monolith or the ceramic-fiber coil traps. Many techniques for regenerating particulate trap oxidizers have been proposed and invested. Regeneration techniques can be divided into passive and active approaches.

Passive trap systems attain the conditions required for regeneration as a result of normal vehicle operation. This requires a catalyst (as either a coating on the trap or a fuel additive) to reduce the ignition temperature of the collected particulate matter. Regeneration temperatures of 420 °C have been reported with catalytic coatings, and lower temperatures can be achieved with fuel additives.

Active trap systems monitor particulate matter in the trap and trigger specific actions to regenerate it when needed. A variety of approaches to trigger regeneration have been proposed, including diesel-fuel burners, electric heaters, and catalyst injection systems. Passive regeneration is difficult on heavy-duty vehicles.

Regeneration temperatures must be attained in normal operation, even under light-load conditions. At present, for heavy-duty applications, no purely passive regeneration system is available. Some manufacturers are working on quasi-passive systems, in which the system usually regenerates passively without intervention, but the active system remains as a backup. No catalytic coating has sufficiently reduced trap regeneration temperature to permit reliable passive regeneration in heavy-duty diesel service. But catalyst coatings have a number of advantages in active systems. The reduced ignition temperature and increased combustion rate resulting from the catalyst imply that less energy is needed from the regeneration system.

Regeneration will also occur spontaneously under most duty cycles, greatly reducing the number of times the regeneration system must operate. A trap catalyst may simplify a regeneration system. To date, trap-oxidizer systems have been used in only a few engine and vehicle models. Regeneration must be carried out approximately every 500 kilometers; Regeneration takes about 10 to 15 minutes. The additive system takes slightly less time. It is also dependent on engine operating conditions.

The filter is regenerated by burning off the soot that has collected in the filter. The particle carbon component can be oxidized (burned) using the oxygen constantly present in the exhaust gas above a temperature of approx. 600°C to form nontoxic CO_2. Such high temperatures occur only when the engine is operating at rated output. It is highly rare in normal vehicle operation. For this reason, measures must be taken to lower the soot burn off temperature and/or raise the exhaust gas temperature. Soot oxidizes at temperatures as low as 300.450°C using NO_2 as oxidizer. This method is used industrially in the continuously regenerating trap (CRT®) system.

7.5.6   REGENERATION WITH ELECTRICAL ENERGY

This particulate trap system consists of two parallel monoliths that are used in an alternating manner. If a preset charging level is reached in one filter (approx. 40 g of soot), the system switches to the other monolith. The charged system is regenerated electrically by adding secondary air. An electronic control system processes, among others, signals from pressure transducers, temperature sensors, and air flow meters and controls a bypass flap. The filter is regenerated every 4 to 5 hours; this process requires approx. 15 minutes. Fuel consumption increases by approx. 1 to 2%, this being caused primarily by added regeneration energy and to a lesser extent by increased exhaust backpressure. The system is used mainly for commercial vehicles (buses). Service life expectancy of all soot burn-off filters still remains to be improved.

7.5.7   REGENERATION WITH OXIDANTS

By adding certain special additives the ignition temperature of soot can be reduced down to 250°C. A number of constituents based on Cu, Fe (e.g., Ferrocen), calcium, manganese, and cerium are suitable for this purpose and are added to the fuel or are introduced directly ahead of the particulate trap. The required substances can be added, for instance, across a separate tank (at a rate of 6 liters per approx. 25,000 miles). However, the use of such type of trap regeneration is currently being tested for its environmental compatibility.

7.5.8   CATALYTIC REGENERATION

This type of regeneration is done by a catalytically active substance, e.g., copper oxide coated filter. The filter is activated using an additive (such as acetyl acetone) when a particulate /filter trap is loaded with a good amount of soot, with reduced about 250°C, and the regeneration occur through whole engine operating range. However, due to the need for complex control system it is not used widely. Also, with vanadium oxide coating soot burn temperature can be lowered. The passenger cars are not accommodative to Particulate Filters because of their large size of, even though lot many types of filters were researched.

To improve comprehensiveness and accuracy of DPF many simulation models with powerful simulation packages have been developed in the last three decades. However, simulation of diesel particulate filter is still not a reliable resort to fine-tuning of diesel particulate filter and much effort is still needed. While reviewing various models available for diesel filters. Their characteristics and application occasions are discussed. Some key sub-models are introduced, which are pressure drop model in the wall, filtration model, and soot oxidation model. Finally, some conclusions are made and further researches are recommended (Shichun et al., 2016; Creative Commons CC-BY). They review 1D, 2D, and 3D models depicting DPF, however, felt that the most popular model is still the 1D model, because it loses little accuracy compared with the 3D model, and it is computation efficient. They were of the opinion that in fundamental understanding the complex interaction of convection, diffusion, and reaction in DPF should be obtained. Intrinsic soot cake properties such as packing density and permeability should be more clearly characterized. This maybe realized by identification of soot deposition mechanisms and soot cake microstructures. The model will describe the actual physics taking place in the DPF by less number of assumptions and more accurate model parameters. High accuracy and standard measurement are required to calibrate the model and diminish the divergence of the data.

7.6   SELECTIVE CATALYTIC REDUCTION (SCR) OF NITROGEN OXIDES

As far as diesel engine exhaust, the two main emissions are PM and NOx. The best way to control of NOx emission has been understood to lower in-cylinder peak temperatures and use of stoichiometric mixtures. However, with advanced technology aftertreatment methods for control of NOx emissions has been widely researched. With suitable catalysts, the NOx emissions are lowered. In this context, urea as a reducing agent in SCR method is observed an effective control technology for reducing NOx emissions. With SCR technology, certain oxides of nitrogen are reduced effectively to its molecular form in the presence of oxygen using selected reducing agents. The SCR has exhibited conversions of up to 90% while simultaneously reducing HC emissions by 50 to 90 and PM emissions by 30 to 50%. The method is a tried and tested method of waste-gas denitrification. Instead of reacting with molecular oxygen present in exhaust due to use lean mixtures, the reducing agent oxidize selectively with the oxygen contained in the NOx, and hence the name " selective" has been referred to. Among chemicals, ammonia (NH_3) is proven to be a highly selective reducing agent for use SCR. this case. Even though the toxic nature of ammonia, it can be produced from nontoxic carrier substances, such as urea or ammonium carbamate. Urea [$(NH_2)_2CO$] is biologically compatible with groundwater and chemically stable for the environment, a good catalyst carrier, well known to works as a fertilizer, produced on an industrial scale. It is highly soluble in water be added to the exhaust gas as an easy-to-meter urea/ water solution. At a mass concentration of 32.5% urea in water, the freezing point has a localized minimum at $-11°C$, a eutectic solution forms, but does not separate when frozen.

The method of SCR has been in use over 20 years for control of NOx emissions from stationary sources. This is started emerging in automobile sector as it provides a way for simultaneous reductions of NOx, PM, and HC emissions. SCR offers a high level of NOx conversion with high durability. Open loop SCR systems can reduce NOx emissions from 75 to 90%. Closed loop systems on stationary engines have achieved NOx reductions of greater than 95%. For complying with EPA's 2010 heavy-duty highway emission standards, engine manufacturers have started adopting combined DPF+SCR system in vehicle designs. The SCR method is known to clean up exhaust line with significant reductions in NOx levels even with long hours of operation.

7.6.1 OPERATING CHARACTERISTICS AND CONTROL CAPABILITIES

An SCR system uses a metallic or ceramic wash-coated catalyzed substrate, or a homogeneously extruded catalyst, and a chemical reductant to convert NOx to molecular nitrogen and oxygen. In mobile source applications, an aqueous urea solution is a preferred reductant. In open loop systems, the reductant is added at a rate calculated by a NOx estimation algorithm that estimates the amount of NOx present in the exhaust stream. The algorithm relates NOx emissions to engine parameters such as engine revolutions per minute (rpm), exhaust temperature, backpressure, and load. As exhaust and reductant pass over the SCR catalyst, chemical reactions occur that reduce NOx emissions. In closed loop systems, a sensor determines the amount to be injected and helps in reduction of NOx concentrations in the exhaust.

SCR catalysts formulations based on vanadia-titania and base metal-containing zeolites have been commercialized for both stationary and mobile source applications. Urea first forms ammonia before the actual SCR reaction starts. The reaction is called as Hydrolysis happens in two-steps. This technology is generally used in stationary units, in which NO is converted to N_2 by reacting with ammonia (NH_3). The maximum NOx conversion window for SCR catalysts is a function of exhaust gas composition, in particular the NO_2 to NO ratio. The three common NOx reduction reactions are:

$$4NH_3 + 4NO \rightarrow O_2 + 4N_2 + 6H_2O \tag{1}$$

$$2NH_3 + NO \rightarrow NO + 2N_2 + 3H_2O \tag{2}$$

$$8NH_3 + 6NO_2 \rightarrow 7N_2 + 12H_2O \tag{3}$$

Low temperature SCR is promoted by NO_2. Of the three competing reactions over a vanadia catalyst, reaction 3 is the slowest and reaction 2 is the fastest. Titania supported vanadia catalysts have been used for many years and are effective at temperatures less than 500°C. Modern zeolite based SCR catalysts aged at 700°C for 50 hours show little deterioration whereas the vanadia catalyst degrades rapidly at these temperatures.

Ammonia is hazardous in gaseous form and poses safety constraints to use the technology in the transport vehicle. NH_3 may be extracted from urea, $(NH_2)_2CO$, as one of the options but has serious disadvantages related to handling, storage, and injection into vehicles. Shimizu and Satsuma (Shimizu and Satsuma, 2007) have specified AdBlue (32.5% aqueous urea solution) as the standard precursor of ammonia for vehicle applications.

In the process of new developments in SCR methods, base metal zeolite SCR catalysts are also emerging for reduction of NOx emissions. The studies related to control of the ratio of NO_2 to NO present at the inlet of the catalyst, and improving the urea decomposition process at low exhaust temperatures.

Figure 7.8 compares the conversion window for a vanadia SCR catalyst and two zeolite-based SCR catalysts.

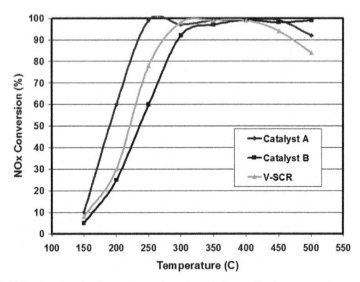

FIGURE 7.8   Catalyst A: Cu-zeolite and catalyst B: Fe-zeolite is compared to a vanadia-based SCR catalyst.
Source: Reprinted from Johnson, 2008. Open access.

Catalyst A represents a copper zeolite catalyst having the lowest temperature light-off characteristics of the three catalysts shown. The vanadia based catalyst shows better low temperature conversion that the iron zeolite system (Catalyst-B) however the conversion efficiency drops off above 400°C whereas the iron zeolite maintains peak efficiency above 500°C. Both zeolite-based catalysts show better high temperature conversion than the conventional vanadia catalyst.

SCR is managed by the engine ECM. The urea is contained in aqueous/water form (32%) in a replenishable container. AdBlue™ is a sole manufacturer of aqueous urea. It is injected upstream from the converter/DPF/muffler assembly sometimes assisted by on-chassis compressed air. The urea injection has to be precisely metered by the ECM. Too much urea can result in ammonia discharge through the exhaust system, while too little results in NOx emissions.

SCR catalysts based on vanadia exhibit a strong sensitivity of NOx conversion to the NO_2:NOx ratio of the exhaust gas. Optimum conversion is achieved at a ratio of 1:1 or a 50% NO_2 composition. Zeolite based catalysts have shown less sensitivity to NO_2 concentration as shown in Figure 7.9 shows a typical arrangement where the SCR is downstream of the DOC/DPF. The final catalyst in the exhaust system is an oxidation catalyst designed to remove any ammonia slip that might occur in the SCR.

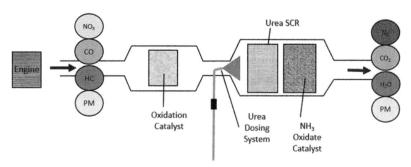

FIGURE 7.9 SCR (green) catalyst downstream of DOC (light gray)/DPF (pink) catalyst, urea dosing nozzle.

In addition to NOx, SCR systems reduce HC emissions up to 80% and PM emissions 20 to 30%. They also reduce the characteristic odor produced by a diesel engine and diesel smoke. Like all catalyst-based emission control technologies, SCR performance is enhanced by the use of low-sulfur fuel. Combinations of DPFs and SCR generally require the use of ultra-low sulfur diesel to achieve the highest combined reductions of both PM and NOx.

Significant advancements have been made not only to improve the catalyst performance and durability but also in the urea injection hardware to insure an accurate and well distributed supply of reductant. This insures that the entire catalyst volume is being utilized and the ammonia slip is minimized. Manufacturers are developing high precision injectors and mixer systems to disperse the reductant upstream of the catalyst. Urea injector suppliers are moving away from air driven injectors to airless designs to eliminate the need for air pumps specific to the urea supply. In Europe, a urea-based reductant, more than 100,000 new, SCR-equipped trucks are in use.

7.7 OTHER NOX REDUCTION TECHNIQUES

The diesel engine because of its lean mixtures exhibit high thermal efficiency and good fuel economy. The oxygen-rich combustion environment in combination with high combustion temperatures results in the formation of NOx in the combustion process. The techniques used for SI engine NOx control can also be used for diesel engine but to its lean mixture working nature, it is difficult to manage. These approaches are generally not employed on diesel engines in order to maintain the significant fuel economy and low CO_2 benefits of these engines.

Among known techniques, use of EGR and retarded injection timing showed promising as in-cylinder measures to reduce NOx emissions. About 50% reduction in NOx emissions is capable of achieving through adoption of EGR, however, due to charge dilution and reduced volumetric efficiency (VE) and an increase in engine-out PM emissions. The researchers employed NOx reduction catalysts similar to DOCs. NOx catalysts have demonstrated NOx reductions of 10 to 40%, whereas NOx adsorbing catalysts (also known as NOx traps) are capable of 70% or more NOx reduction. These NOx catalysts also provide oxidation capabilities that result in significant reductions in exhaust hydrocarbons, CO, and the soluble fraction of PM.

7.7.1   LEAN NOX CATALYSTS

It is difficult to chemically reduce NOx to molecular nitrogen in a diesel engine of oxygen-rich environment of diesel exhaust, even though direct NOx decomposition is thermodynamically attractive, but the activation energy is very high for this method and no catalysts have been developed for wide-spread use.

Catalysts have been developed that use a reductant like HC, CO, or H_2 to assist in the conversion of NOx to molecular nitrogen in the diesel engine exhaust stream. They are generally called "lean NOx catalysts." Because sufficient quantities of reductant are not present to facilitate NOx reduction in normal diesel exhaust, most lean NOx catalyst systems inject a small amount of diesel fuel, or other reductant, into the exhaust upstream of the catalyst. The added reductant allows for a significant conversion of NOx to N_2. This process is sometimes referred to as hydrocarbon selective catalytic reduction (HC-SCR). Currently, NOx conversion efficiencies using diesel fuel as the reductant are around 10 to 30% over transient test cycles. Other systems operate passively without any added reductant at reduced NOx conversion rates.

Lean NOx catalysts often include a porous material made of zeolite having a microporous, open framework structure providing trapping sites within the open cage network for hydrocarbon molecules along with either a precious metal or base metal catalyst. These microscopic sites facilitate reduction reactions between the trapped hydrocarbon molecules and NOx. Lean NOx catalyst systems have been demonstrated and verified for diesel retrofit application and thousands have been commercially applied. However, due to the relatively low NOx conversions (20–30% over transient cycles) and corresponding fuel economy penalties associated with the operation of these systems, lean NOx catalysts are generally not being considered for upcoming new vehicle regulations for either light-duty, or heavy-duty applications where NO_x conversions of at least 60% are expected to be required. Nonetheless, researchers are developing methods to improve the conversion efficiencies and hydrothermal durability of lean NOx catalysts to identify formulations that meet the needs of the industry, perhaps in combination with advanced engine technologies like HCCI. One such program has identified several promising catalyst formulations using combinatorial screening techniques with conversion efficiencies as high as 75%.

7.7.2 NOX ADSORBER CATALYSTS

Research in the light development of effective ways of reduction of NOx emissions has led to NOx adsorber catalysts, also referred to as lean NOx traps (LNT); provide another catalytic pathway for reducing NOx in an oxygen rich exhaust stream. NOx adsorber technology removes NOx in a lean (i.e., oxygen rich) exhaust environment for both diesel and also in gasoline lean-burn GDI engines. The mechanism is shown in Figures 7.10:

1. Catalytically oxidizing NO to NO_2 over a precious metal catalyst.
2. Storing NO_2 in an adjacent alkaline earth oxide trapping site as a nitrate.
3. In a two-step regeneration step, the stored NOx is periodically removed by temporarily inducing a rich exhaust condition followed by reduction to nitrogen by a conventional three-way catalyst reaction.

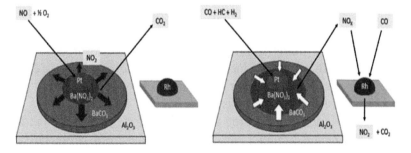

FIGURE 7.10 NOx trapping mechanisms under lean operating conditions.

7.7.3 NOX STORAGE AND NSC REGENERATION

During regeneration, the stored NOx are discharged from the storage components and converted into the harmless components of nitrogen and carbon dioxide. The processes of NOx discharge and conversion proceed separately.

If λ is regarded as excess-air ratio, oxygen deficiency (λ < 1, also called a rich exhaust gas condition) is required to be done in the exhaust gas. The components of carbon monoxide (CO) and hydrocarbons (HC) present in the exhaust gas serve as the reductant. Discharge—presented below in an example with CO as the reductant-proceeds so that CO reduces the nitrate (e.g., barium nitrate $Ba(NO_3)_2$) to NO and, together with barium, reforms the carbonate originally present. CO_2 and NO are produced in the process. In a method familiar from three-way catalytic converters, a rhodium coating subsequently uses CO to reduce the NOx to N_2 and CO_2: Less NOx is discharged as regeneration advances and thus less reductant is consumed.

A NSC is coated with chemical compounds with a high propensity to enter stable but chemically reversible bonds with NOx. Examples are oxides and carbonates of alkali and alkaline earth metals. Barium compounds are employed particularly frequently because of their thermal characteristics.

NO is oxidized to form nitrate. NO is initially oxidized to NO_2 in a catalytic coating. The NO_2 reacts with the storage compounds in the coating and afterward with oxygen (O_2) to become nitrate:

$$Ba\ CO_3 + 2NO_2 + 1/2\ O_2 \rightarrow Ba\ (NO_3)_2 + CO_2 \qquad (4)$$

$$Ba\ (NO_3)_2 + 3\ CO \rightarrow Ba\ CO_3 + 2NO + 1/2\ CO_2 \qquad (5)$$

$$2NO + 2CO \rightarrow N_2 + 2CO_2 \qquad (6)$$

Thus, a NOx storage catalyst stores the NOx emitted by an engine. Storage is only optimal in a material-dependent exhaust gas temperature interval between 250 and 450°C. The NO oxidizes to NO_2 very slowly. The nitrate formed is unstable above 450°C and NOx is discharged thermally. Under normal lean diesel engine operation, the NOx adsorber stores the NOx emissions. In order to reduce the trapped NOx to nitrogen, called the NOx regeneration cycle, the engine must be operated rich periodically for a short period of time (a few seconds).

Development and optimization of NOx adsorber systems is continuing for diesel engines. Adsorber systems have demonstrated NOx conversion efficiencies ranging from 50 to in excess of 90% depending on the operating temperatures and system responsiveness, as well as diesel fuel sulfur content. An important consideration in designing a NOx adsorber emission control system is the effect on fuel economy.

The method-LNT would lead to fuel economy penalty as a result of the fuel necessary to generate a rich exhaust environment during regeneration of the catalyst. However, there is a potential to overcome this associated penalty by utilizing system engineering and taking advantage of all components. An approach to minimize the fuel economy penalty associated with the NOx regeneration step may be to calibrate the engine for maximum fuel economy at points on the engine map where the NOx adsorber is performing at its peak conversion efficiency. Although such a calibration results in higher engine-out NOx emissions, with the NOx adsorber functioning at its peak conversion efficiency, NOx emissions could still be kept low. LNTs can achieve even higher NOx reduction (>90%) when regenerated with on-board generated hydrogen via a fuel reforming reaction over an appropriate catalyst.

Specifically, formulations, and on-vehicle configurations that improve low temperature performance and lower temperature sulfur removal. NOx adsorber technology offers tremendous potential for providing a high level of NOx reduction across a wide range of operating conditions (temperature and NOx concentration) which are consistent with the diversity in engine-out exhaust associated with both light- and heavy-duty diesel applications. Figure 7.11 shows the improvements that are achievable through advances in NOx storage compounds. Advanced storage components have resulted in lower light-off temperatures and wider operating windows for NOx conversion.

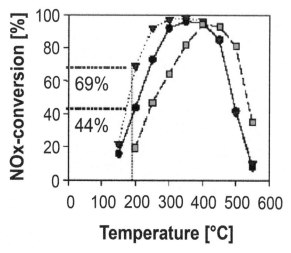

FIGURE 7.11 Conversion temperatures and broaden operating window for NOx adsorber catalysts.

Advanced NOx storage materials can lower conversion temperatures and broaden operating window for NOx adsorber catalysts (square-potassium-based, circle-barium-based, triangle – advanced barium-based technology) (SAE 2006-01-1369).

7.7.4 EFFECT OF FUEL SULFUR AND DURABILITY

NOx storage catalysts suffer with a problem of their sensitivity to sulfur. Sulfur compounds contained in fuel and lube oil oxidize to form sulfur dioxide (SO_2). The coatings used in the catalytic converter for forming nitrates ($BaCO_3$), however, have a very great affinity to sulfate, i.e., SO_2 is removed from the exhaust gas more effectively than NOx, and is bound in the storage material by forming sulfate. Sulfate bonding is not separate when the storage is regenerated normally. The quantity of the stored sulfate, therefore, rises continuously over service time. This reduces the number of storage places for the NOx, and NOx conversion decreases. For better and sufficient NOx storage capacity, the use desulfated (sulfur regeneration) catalytic converter must be ensured.

During the desulfating process, the catalytic converter is heated to a temperature of over 650°C for a period of over 5 minutes, and it is purged with rich exhaust gas ($\lambda < 1$). To increase the temperature, the same measures are used to regenerate the diesel particulate filter (DPF). As opposed to DPF regeneration, however, O_2 is completely removed from the exhaust gas by controlling the combustion process. Under these conditions, barium sulfate is converted back to barium carbonate.

The choice of a suitable desulfating process control (e.g., oscillating λ about 1) must make sure that the SO_2 removed is not reduced to hydrogen sulfide (H_2S) by a continuous deficiency of exhaust gas oxygen, O_2. H_2S is already highly toxic in very small concentrations and is perceptible by its intensive odor. The controlled conditions for desulfating must also avoid excessive aging of the catalyst. Although high temperatures (>750°C) accelerate the desulfating process, they also speed up catalyst aging. For this reason, optimized catalyst desulfating must take place within a limited temperature and excess- air-factor window, and may only have a negligible impact on drivability.

High sulfur content in fuel will speed up catalyst aging, since desulfating takes place more frequently, and will also increase fuel consumption. Ultimately, the use of storage catalysts is dependent on general availability of sulfur-free fuel at the filling station.

7.7.5 DE-NITROGENATION

Development of catalytic converters that can selectively convert oxides of nitrogen using HC or CO, even in the presence of excess oxygen is a challenging task. The following practical methods are possible for NOx reduction in the after-treatment (Fernando et al., 2006).

The same compounds that are used to store NOx are even more effective at storing sulfur as sulfates, and therefore NOx adsorbers require ultra-low sulfur diesel fuel. The durability of LNTs is linked directly to sulfur removal by regeneration and is a major aspect of technology development. Sulfur is removed from the trap by periodic high-temperature excursions under reducing conditions, a procedure called "DeSOx." The DeSOx regeneration temperatures are typically around 700°C and require only brief periods of time to be completed. However, the wash-coat materials and catalysts used in these technologies begin to deactivate quickly above 800°C and there-fore methods are being developed to reduce the desulfation temperature. Advanced thermally stable materials have allowed LNTs to achieve dura-bility over their full useful life.

7.7.6 APPLICATION OF NO$_x$ ADSORBER TECHNOLOGY

NOx adsorber technology has made significant progress and has recently been commercialized on a medium duty pick-up truck meeting EPA's (2010) on-highway emission standards. Several vehicle manufacturers have announced plans to commercialize LNT catalysts on diesel passenger vehicles prior to 2010. NOx adsorber technology is also being applied to gasoline vehicles powered by GDI engines and the results are impressive.

7.7.7 COMBINED LNT/SCR NO$_x$ REDUCTION TECHNOLOGIES

Engine and technology manufacturers are looking at novel approaches to address the need for alternative NOx control systems that do not require separate on-board reductant, like urea. The hybrid systems combine the catalyst functionality of lean NOx traps and ammonia SCR catalysts without the need for a second reductant on board the vehicle. These experimental systems typically incorporate a fuel reformer catalyst to generate a hydrogen-rich reformate from the onboard fuel which is then used to regenerate the lean NOx trap. The regeneration of the LNT forms ammonia which is then stored within the SCR catalyst. The systems primarily rely on the LNT for the bulk of the NOx reduction during lean operation but the SCR uses the stored ammonia to further reduce NOx, thereby extending the time between LNT regeneration and desulfations to reduce fuel penalties associated with these strategies.

An LNT plus SCR hybrid system is shown in Figure 7.12. This design shows the reformer, LNT, and SCR catalyst all in series within the exhaust stream without a bypass valve for LNT regeneration. The reformer processes the entire exhaust stream to generate the reductants used for LNT regeneration and ammonia formation.

Both systems rely on dual LNT and SCR catalysts where the SCR stores ammonia formed during LNT regeneration. The Mercedes E320 Blutec system uses independent LNT and SCR catalysts whereas Honda has announced a single catalyst with dual layer functionality.

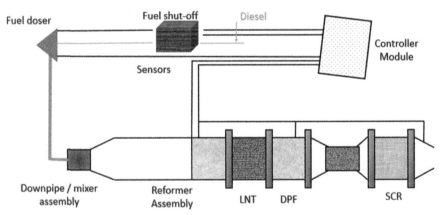

FIGURE 7.12   LNT/SCR combined catalyst: Series design.
Source: Modified from McCarthy and Taylor, 2007.

Compared to gasoline engines, carbon dioxide, carbon monoxide, and hydrocarbons produced in diesel engines are much lower. The issue of simultaneous reduction control of NOx and PM becomes more complex in diesel engines. Both aftertreatment and in-cylinder technologies to reduce emissions in CI engines have been reviewed. It is understood that various technologies to reduce emissions can just meet the present emission regulations. Better in-cylinder and/or after-treatment technologies have to be developed to meet future stringent emission norms. EGR seems to be a simple and most effective way of reducing NOx emissions, but suitable measures need to be taken to reduce soot. There is a need to modify the EGR system along with some in-cylinder solutions viz. various injection techniques, charge conditions, late intake valve closing to improve the combustion phenomena and to reduce the emissions. Simultaneous reduction in NOx and PM can be achieved with the modern combustion techniques such as LTC, HCCI, PCCI, etc. LTC can be achieved with high level of EGR and modified injection timings. LTC provides direct control over the combustion phasing through the injection timing, which helps in avoiding premature combustion that can normally occur with diesel-fuelled HCCI. Because of this, LTC is more favored than HCCI when the engine uses diesel-like fuels. However, achieving a wider operating range and controlling HC and CO emissions are the major challenges with this technique. An optimized combination of all in-cylinder solutions is required to overcome these issues.

KEYWORDS

- **carbon monoxide**
- **closed crankcase ventilation**
- **diesel oxidation catalysts**
- **light-duty diesels**
- **particulate matter**
- **soluble organic fraction**

REFERENCES

Johnson, T., (2008). Engine emissions and their control: An overview. *Platinum Metals Rev.*, *52*(1); https://www.technology.matthey.com/pdf/23-37-pmr-jan08.pdf

McCarthy, J. Jr., & Taylor, W. III, (2007). LNT+SCR aftertreatment for medium-heavy duty applications: a systems approach. *Diesel Engine-Efficiency and Emissions Research (DEER) Conference,* August 14, 2007.

Ravichettu, S. G., Rao, A. P., & Murthy, M. K. (2017). Effective abatement of diesel engine emissions with split injections International Journal of Automotive and Mechanical Engineering ISSN: 2229–8649 (Print); ISSN: 2180–1606 (Online); *14*(2), 4225–4242.

Shimizu, K., Sugino, K., Kato, K., Yokota, S., & Okumura, K., (2007). Atsushi Satsuma Reaction Mechanism of H2-Promoted Selective Catalytic Reduction of NO with C3H8 over Ag–MFI Zeolite. *J. Phys. Chem. C, 111*(17), 6481–6487.

Tavares, F., Johri, R., & Filipi, Z. (2011). Simulation study of advanced variable displacement engine coupled to power-split hydraulic hybrid powertrain. *Journal of Engineering for Gas Turbines and Power, 133*, 122803–11.

Twigg, M. V. (2011). Haren Gandhi 1941–2010: Contributions to the development and implementation of catalytic emissions control systems. *Platinum Metals Review, 55*(1), 43–53.

U.S. Environmental Protection Agency (EPA). (2010). An overview of technologies for emissions compliance: engines, 2010: Science and Research at the U.S. Environmental Protection Agency (EPA) Progress Report 2010, https://www.epa.gov/sites/production/files/2013–12/documents/annual-report-2010.pdf (accessed on 10 February 2020).

Yang, S., Deng, C., Gao, Y., & He, Y. (2016). Diesel particulate filter design simulation: A review. *Advances in Mechanical Engineering 8*(3), 1–14, Creative Commons CC-BY.

CHAPTER 8

Alternative Fuels

8.1 INTRODUCTION: NEED FOR ALTERNATIVE FUELS (AFS)

The innovators of internal combustion engines (ICEs) have improved the lifestyle by developing to fit in automobiles. The petroleum-derived automobiles have entirely changed the lifestyle of human beings and simplified the mobility requirements. The petroleum-derived gasoline and diesel fuels largely met the requirement in vehicles; automobiles since 1900s and though it led revolutionary propulsive power but at the same brought problems in the event of larger use in vehicles. The term "alternative fuel (AF)" has been used to describe any fuel as a substitute to widely used fuels such as gasoline and diesel fuels. Even Henry Ford envisioned many of today's concerns about fuel availability and environmental by investigating the use of ethanol as renewable, home-grown fuel whose production would benefit agriculture.

Over more than 150 years, the fossil fuel resources derived petroleum fuels have been exploited on a large scale to keep the wheels moving (majorly) and also for other purposes. The amount of energy consumed in terms of fuels has become a yardstick to rate a country as a developed, developing or under-developing one. Thermodynamic tests based on engine performance evaluations have established the feasibility of using a variety of AFs such as alcohol fuels, CNG, biogas, producer gas hydrogen and a host of vegetable oils. Methanol and ethanol are the two well-accepted safe AFs which possess the potential to be produced from the biomass resources. Even though host of AFs are developed, majority of them require blending with petro-derived fuels to use them in existing engines. Few of these fuels require ignition improvers for use in specific type of engine. It is not exaggeration to imagine a life without an automobile and thus to say that petroleum-derived development has brought large revenue to certain countries.

To obtain long term benefits from AFs, there is a great need for making modifications in the AF sources and treatment to suit to the existing engine to obtain better performance and less hazardous engine-out emissions. Few

AFs can directly be used with retrofitting for better economy, flexibility, effectiveness, and the simplicity of their use. There were times when the high impetus was given for the exploration and development of AF due to fossil fuel crisis again the idea was dropped due to sudden emergence of fossil fuels. In the initial days, AFs such as natural gas (in compressed or liquefied form), liquefied petroleum gas (LPG), methanol (from natural gas, coal, or biomass), ethanol (from food grains or sugar), vegetable oils, hydrogen, synthetic liquid fuels derived from the hydrogenation of coal, and various blends such as gasohol, have been considered for vehicular use. The nations and AFs respectively that adopted wide use are Brazil and Malaysia for Ethanol and Palm oil.

Many studies have also been performed for environmental assessment of AFs not only based solely on vehicle end-use emission characteristics but also pollutant emissions associated with the production, storage, and distribution of these fuels. In this direction, studies pertained to life cycle assessment of fuels have been emerged for assessment of emissions contributing to global warming.

8.2   NEED FOR ALTERNATIVE FUELS (AFS)

The petroleum-derived fuels have been serving as the main source for prime movers ever since its development. These fuels have been extracted from the earth crust but for various reasons, they could not replenish as the demand is heavy by various other sectors. The demand for alternatives due to fuel crisis of early 1970s (oil embargo) and pollution-related issues were the main driving forces for searching viable alternatives.

Both the vehicle manufacturing industries and researchers have given a serious thought for look out of AFs. The following points can cited as the need for AFs:

1.   **Dependence on OPEC Nations:** The countries which have no or limited fossil sources are heavily depending on Organization of Petroleum Exporting Countries (OPEC) for want of fuels to keep vehicles moving and meeting energy demands. In turn, such nations had to shell out their hard-earned revenue to import petroleum crude and other related products. The OPEC nations bestowed with large resources of fossil fuels and are majors in export of petro-derived products, since there is a mismatch in the distribution of natural

resources all over the world and thus OPEC nations are virtually ruling the world. At times due to heavy dependence on OPEC led heavy imports-very vulnerable to oil shortages and sudden price increases and fluctuating fuel prices. It is this dependence on petroleum on petroleum fuels that is prompting the lookout for AFs for transportation.

2. **Mismatch between Demand and Supply:** The fossil fuels are extracted from earth's crust and to meet large demands the sources were extracted at a great extent. For maintaining modern lifestyle and to meet other energy requirements, the fossil fuels have been indiscriminately exploited and thus leading to large scale gap between demand and supply.

3. **Sustainable Development:** Instead of importing resources from other nations, if a country can explore its own sources for meeting their energy demands, it is a positive sign for achieving sustainable development or energy independence. This also prompts for energy security.

4. **Cut Down Green House Gases:** Large scale use of petroleum-derived fuels for meeting the demand of exponential rise of vehicles have further worsened the global ecology and this has encouraged to search the good and eco-friendly AFs. To reduce harmful emissions (e.g., greenhouse gases) that is produced by the combustion of fossil fuels. The energy security issues began in 1990s forced the researchers to work intensively on the development of viable.

5. **SafeGuard of Aquatic Life:** While transport of petroleum sources via sea, eventual spillages caused a lot of damage to aquatic life and thus to have biodegradability, the search for fuels biodegradable AFs has emerged. Even with spillages on surface roads, the AF should be biodegradable.

Through experimentation with AF, it became clear that AFs had inherent environmental advantages as well. Each AF has some characteristic that gives it an environmental advantage over petroleum-derived fuel. Most are less damaging to the environment if spilled, and, in general, the emissions from AFs are less reactive. The initial research work on AFs focused on which one was best from the viewpoint of technical feasibility, production capability, and cost.

8.3   DEFINITION OF AN ALTERNATIVE FUEL (AF)

It is a fuel that substitutes to the largely exploited liquid fuels such as gasoline or diesel; the AF could be derived either from petroleum crude or renewable energy source. It is estimated that India imports about 80% of petroleum products to meet its demand in the transportation sector.

8.4   DESIRABLE CHARACTERISTICS OF A GOOD ALTERNATIVE FUEL (AF)

In order to identify or select an AF, one has to look for the desirable characteristics in a proposed fuel. The AF could be either for a SI or a CI engine.
The following characteristics may be met from a viable AF:

- **It Should Have High Anti-Knocking Tendency:** It can be seen that the thermodynamic efficiency of an engine strongly depends on the compression ratio (CR) of an engine. Beyond a certain CR of an SI engine, the engine starts detonating and hence for smooth operation of an engine, the fuel should able to resist knocking (in case of CI engines) and detonation (in case of SI engines). The anti-knocking tendency is thus a measure of ignition quality of a fuel-the ignition quality of SI and CI engine fuels are Octane Number and Cetane Number respectively. Also, it is an indication of resistance to knock e.g., Alcohol fuels, CNG, LPG, biogas.
- **It Should Have High Heating Value:** The fuel should posses high heating value on mass basis so that by burning unit quantity of fuel it should be able to release large amount of energy and also it can be stored in a smaller container. A fuel with high heating value thus gives rise to low specific fuel consumption or better mileage. Also small amount of fuel with high heating value is sufficient enough to produce similar power when compared a fuel having low heating value demanding large amount of fuel. Moreover, a fuel with low carbon to hydrogen ratio is preferable to release large amount of energy e.g., Hydrogen (120 MJ/kg) and Methanol (~20 MJ/kg).
- **It Should Possess Low Stoichiometric Requirement:** Stoichiometric requirement of fuel is nothing but chemically correct amount of oxidizer required to burn completely the given fuel. A fuel with low stoichiometric demands smaller space to accommodate oxidizer, i.e., swept volume of an engine. Therefore, an engine that employs a

fuel with low stoichiometric requirement demands small volume to fill air and thus a lighter engine.

e.g., Alcohols enjoys low stoichiometric air-fuel ratio (AF) but poor heating value fuels.

- **It Should Not Emit Harmful, i.e., Pollutant Emissions:** Fuels which emit less harmful pollutant emissions are preferred in the wake of greenhouse gas effect and help in reduction of global warming.
- **It Should Demand Minimum or No Modifications-Adoptability:** An AF should be easily be accommodated in an existing engine and thus shouldn't demand additional space or no new fuel filling stations.
- **It Should Not Leave Gummy Deposits Upon Combustion:** If any fuel after combustion leaves thick gummy deposits and thus sticks to intricate components and may interfere with smooth working of n engine-it should lead to ring sticking. The fuel injector may get frequently fouled. Also, the effective CR of engine gets altered.
- **It Should be Free from Sulphur:** A fuel should be free from sulfur to mitigate corrosion problems and should lead to formation low or no oxides of sulfur.
- **It Should Enable Ease of Starting:** The fuel should easily get vaporized and mix with oxidizer and enable to start quick start of engine.
- **It Should be Safe to Handle and Store:** The fuel should have safe limits of flash and fire point to avoid any hazards during storage or filling.
- **It Should be Available Abundantly:** The fuel should be easily available without causing difficulties of short supply and meeting regular demands.
- **It Should be Highly Biodegradable:** In case of any spillage, the fuel should be easily biodegradable.
- **It Should Not Demand High Maintenance and Running Costs:** The initial cost of a vehicle, with novel fuel run engine, should be low and its running and maintenance costs also should be low and the operation of such vehicle should be durable and reliable.
- **It Should be Economical:** The cost of such a viable fuel should be cheaper.

It is observed that no single fuel could posses all the above characteristics in practice. If one fuel is having, high heating values (ex. Hydrogen), it will be difficult to handle and if a fuel is having high anti-knocking tendency (ex. Alcohol fuels) then it may be having low heating value vice versa.

Apart from above, one should look at ease of handling, on board storage, cost of storage and safety of operation, in choosing right AF.

8.5   CLASSIFICATION OF ALTERNATIVE FUELS (AFS)

1. Petroleum derived:
    i.   Liquid fuels-for SI and CI engines: Kerosene.
    ii. Gaseous fuels-for SI and CI engines: LPG, CNG, LNG.
2. Non-petroleum derived (renewable fuels):
    i.   Liquid fuels-alcohols, vegetable oils (biodiesel), DME, DEE.
    ii. Gaseous fuels-hydrogen, biogas.

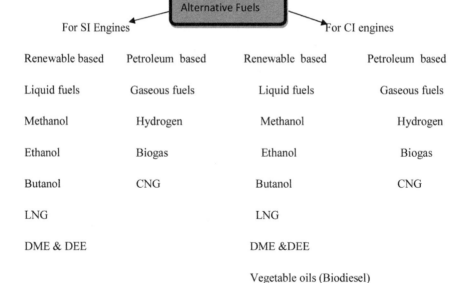

The following section deals with individual properties of each AF and emission characteristics.

8.6   COMPARISON OF PROPERTIES OF ALTERNATIVE FUELS (AFS) WITH PETROLEUM FUELS

The comparison of properties of alternative fuels (AFs) with petroleum fuels is shown in Table 8.1.

TABLE 8.1 Properties of Various Alternative Fuels

Parameter	Gasoline	Diesel Fuel	Methanol	Ethanol	Butanol	Natural Gas	Hydrogen	Biodiesel-Rapeseed
Formula	C_4 to C_{12}	C_8 to C_{25}	CH_3OH	C_2H_5OH	C_4H_9OH	CH_4	H_2	
Molecular Weight	100–105	200	32.04	46.07	74	16	2.02	294.8
Density, kg/L, 15/15°C	0.69–0.79	0.81–0.89	0.796	0.79	0.808	0.7–0.9	0.0013(g), 0.07(l)	0.882
Specific gravity, 15/15°C	0.69–0.79	0.81–0.89	0.796	0.794	0.8098	0.60–0.70	0.07(g), 0.07(l)	
Freezing point, °C	–40	–40 to –1	–97.5	–114	–90	–182	–275	
Boiling point, °C	27–225	188–343	65	78	117	–162	–253	
Vapor pressure, kPa @ 38°C	48–103	<1	32	15.9		Not applicable	Not applicable	
Specific Heat, kJ/(kg-K)	2.0	1.8	2.5	2.4	1.48	--	14.2	
Viscosity, Pa-s @ 20°C	0.37–0.44	2.6–4.1	0.59	1.19	1.782	0.01	0.009	4.63
Latent heat of vaporization, kJ/kg	349	233	1178	923	582	510	448	
Lower heating value, 1000 kJ/L	30–33	35–37	15.8	21.1	32.91 to 33.1	38–40	8.4	37.625
Flash point, °C	–43	74	11	13	35	–188	--	164.4
Auto ignition temperature, °C	257	316	464	423	345	540	--	
Stoichiometric air-fuel ratio, weight	14.7	14.7	6.45	9.00	11.13	17	34.3	
Flame spread rate, m/s	4–6	--	2–4	--	--	--	~3	
Octane number (RON)	88–100	--	108.7	108.6	96	--	--	
Cetane number	--	40–55	--	--	17–25	--	--	54.1

8.7   NATURAL GAS

Among the hydrocarbons, the first component of paraffin family is methane, CH_4 available as natural gas, is a combustible gas, found along with crude oil in subterranean deposits (also called "deep gas"). The major component of natural gas is methane.

Natural gas is one of the world's most abundant fossil fuels and currently supplies over 25% of the energy demand in the world as shown in Figure 8.1. Relative to liquid petroleum fuels, the ability to store sufficient amounts of natural gas for onboard vehicles has presented a significant barrier to its broad use as a transportation fuel.

Global Energy Consumption by Source, 2015

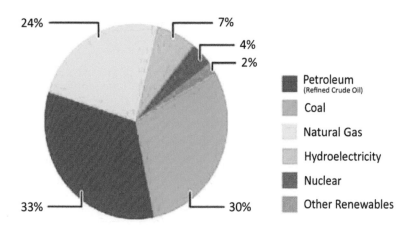

FIGURE 8.1   World energy consumption statistics.

There are two forms of natural gas; a gas, commonly known as compressed natural gas (CNG), stored under a pressure of about 200 bars and the other is the liquefied natural gas (LNG). CNG is more suitable for vehicular engines whereas LNG is not suitable as it is required to be stored in cryogenic containers. LNG requires only one-third of the storage volume of CNG, however, the storage of LNG requires a high expenditure of energy for liquefaction. Natural Gas is e sold at filling stations entirely in the form of CNG.

CNG has a very high octane number around 120 and with its high auto-ignition temperature because of the short-chain lengths and molecular sizes, and is thus more suited for use in SI engines like LPG. The natural gas low carbon/oxygen ratio with consequent reduction of CO_2 emissions upon combustion and is gaining significance.

Since it is a gas, it doesn't require atomization device and gets easily miscible with air for combustion, contains 85 to 99% methane and posses many desirable qualities as a fuel for spark-ignition engines. It's a clean-burning, cheap, and abundant in many parts of the world. The fuel is without any lubricity and sometime regarded as "dry fuel "with a heating value slightly more than gasoline, exhibits certain problem with the components with which comes in contact during combustion. For this reason, the valve seats gets frequent replacement and demands specialized materials for valve seats, etc.

The hydrogen/carbon ratio of NG is roughly 4:1 that of gasoline, on the other hand, is 2.3:1. As a result of the lower amount of carbon in NG, it produces less CO_2 and more H_2O than gasoline when it is burned. A SI engine can easily be converted to run on natural gas with retrofits since it releases 25% fewer CO_2 emissions than when run on gasoline (for equivalent power output). Most of the natural gas vehicles (NGVs) in operation worldwide are retrofits, converted from gasoline vehicles. The physical properties of natural gas make such a conversion relatively easy. Conversion costs typically due mostly to the cost of the on-board fuel storage system.

Natural Gas as Bio-methane has been emerging as an attractive replacement for vehicles as alternative for conventional transportation fuel with all the advantages of natural gas including a dense distribution, trade, and supply network. Studies are done to produce hydrogen through natural gas route.

8.7.1 COMPRESSED NATURAL GAS (CNG)

The major constituent of natural gas being methane and is first in the group of alkane family (the generic formula being C_nH_{2n+2}. The chemical formula for methane is CH_4 and has lowest carbon atoms among all hydrocarbons. Combustion chemistry is simple for methane (CH_4—a major constituent of natural gas) compared to conventional liquid fuels. Also, upon combustion of methane it has advantage of producing lower soot and lowest carbon dioxide. The natural gas contains 75% of carbon when compared with petrol and diesel which have 86–88% of carbon. This result in effective combustion which in turn lowers the HC, CO_2 and CO emissions in natural gas engine comparatively, per unit of energy released. Nowadays CNG-Diesel dual-fuel engines are used on a large scale due to its high fuel efficiency and ability in reducing the NO_x and particulate matter emissions compared to conventional diesel engines. However, CH_4 being one of the components of greenhouse gases, enough care is too exercised for arresting its leakage into atmosphere.

A carbureted air mixture and high octane gaseous fuel (e.g., natural gas), prepared in an external mixing device, is compressed and pumped into the engine cylinders. Due to the high ignition temperature, the compressed mixture cannot ignite automatically. Therefore, diesel as a pilot quantity is injected into the cylinders before the piston reaches the TDC in compression stroke. Thereafter the flame propagation begins with the combustion of pilot fuel, after a short ignition delay. As CNG is lean mixture, the cylinder temperature decreases which in turn mitigate the PM and NO_x emissions as the effective utilization of diesel has been dropped.

Studies have been done for optimizing the quantity of pilot injection diesel and injection timing to obtain advantage of its use in dual fuel mode of operation. Due to flame quenching, CO emissions were noticeably higher than the emissions under normal diesel operation mode, especially when the engines run on dual-fuel mode, even at high load, of the lean premixed natural gas-air mixture. As the fuel was combusted under lean premixed conditions, the local temperature got reduced which resulted in the reduction of NO_x emissions by 30% approximately. Due to the methane molecular structure and premixed combustion, the PM emissions were brought down, under dual fuelling mode (Liu et al., 2007).

It is found that PM emissions and pilot fuel quantity are directly proportional. They reported under dual-fuel mode the unburned HC emissions are higher compared to conventional diesel engine at low and medium loads. However, HC emissions were greatly reduced with increase in the pilot diesel quantity.

The researchers observed that when engine operated at high CNG composition and rated load, the gaseous and particulate emissions reduced to a greater extent due to enhancement in CR. The smoke opacity was reduced by 85 and 90%. Reduction in HC, CO, and NO_x emissions were found by varying the injection pressure on exhaust emission and performance. Due to jet, penetration and combustion spread in cylinder the gaseous emission were reduced. They found that in order to reduce the SO_2, CO_2, and O_2 the injection pressure should be high and low to reduce the PM and NO_x emissions.

The NO_x emissions can be reduced by increasing the intake swirl ratio in dual-fuel engines as heat losses increases with swirl. They inferred that engine operating at low mass ratio of natural gas at high engine load emits less NO_x emissions and vice versa for the lowering the PM emissions.

8.7.2 PRODUCTION

Natural gas is present in the earth and is often produced in along with production of crude oil. However, wells are also drilled for the express purpose of producing natural gas. It includes CO_2, N_2 and very small quantity of helium and hydrogen apart the main constituent of natural gas, i.e., methane. The composition of natural gas can influence the combustion greatly by varying its physical properties and heating value. It is generally to scrub out CO_2 from natural gas so that to reduce its dissociation effects.

Typically, storage system of CNG is less expensive than LNG. Natural gas exhibits superior properties among alternative vehicular fuels. Composition of natural gas differs with type of source. When natural gas is to be utilized in an engine Wobbe number plays a key role. Emissions advantages of natural gas make it a viable choice among AFs regardless of the method of storage and the cost.

8.7.3 ENGINE TECHNOLOGY AND PERFORMANCE

Natural gas engines can be grouped into three main types on the basis of the combustion system used: stoichiometric, lean-burn, and dual-fuel diesel. Majority of natural gas run vehicles operate in stoichiometric range and have been in wide use with retrofits with the engines originally designed for gasoline. Such type of engines may be either bi-fuel (able to operate on natural gas or gasoline) or dedicated to natural gas. Since the methane has higher auto-ignition temperature, normally for dedicated operation of CNG run engines, the engines need to be optimized for its CR and injection timing and such changes not usually done in retrofit situations because of the cost. Nearly all light-duty natural gas vehicles use stoichiometric engines, with or without three-way catalysts.

8.7.4   EMISSION CHARACTERISTICS

Cleaner combustion characteristics are the major advantage in natural gas engines. The major pollutants from natural gas combustion are NOx, HC, CO, and small amount of aldehydes and SO_2 are the major pollutant from the natural gas engines.

Natural gas is composed primarily of methane which dominates its emissions characteristics. Methane mixes readily with air and has a high octane rating which makes it a very good spark-ignition engine fuel. It has high ignition temperature that makes it unsuited for use in CI engines though it can be made to work in such engines. Methane barely participates in the atmospheric reactions that produce ozone, though it does contribute to global warming when released into atmosphere. Because of its high hydrogen-to-carbon ratio (the highest at 4 to 1), the combustion of methane produces about 10% less carbon dioxide than combustion of the energy-equivalent amount of gasoline or diesel fuel. (Carbon dioxide is the primary contributor to global warming. The emissions characteristics of both light-and heavy-duty vehicles are presented in the following sections.

Light-duty CNG vehicles are capable of very low gaseous exhaust emissions. CNG vehicles developed for auto industry to date, individual port fuel injection with a 3-way catalyst system to simultaneously oxidize exhaust hydrocarbons and carbon monoxide while reducing oxides of nitrogen. Because natural gas readily mixes with air, emissions of carbon monoxide are typically low when using natural gas, assuming that the AF is kept on the lean side of stoichiometric.

In addition to very low exhaust emissions, CNG vehicles also have the advantage of no evaporative emissions or running loss emissions caused by the fuel (Running loss emissions are emissions from the fuel tank during vehicle operation duel to heating of the fuel, losses from the evaporative canister, or losses due to permeation of fuel through elasto-metric fuel lines).

The heavy-duty engine manufacturers have taken a slightly different approach to natural gas engine development compared to light-duty engine manufacturers. Heavy-duty NG engines to date have been spark-ignition adaptations of diesel engines. To improve engine efficiency to be closer to that of diesel engines, the heavy-duty NG engines use lean-burn combustion. For emissions control, an oxidation catalyst is used to control methane and carbon monoxide emissions. Oxides of nitrogen are kept low through lean combustion and particulate emissions from natural gas are not a concern. Heavy-duty vehicles using NG favor storing NG as LNG instead of CNG.

HC emissions were formed from the un-burnt mixture of air and fuel when the exhaust valve is opened. Unburned HC tends to exist near the cylinder walls and inside fissure, volumes where the flame front is quenched which results in un-burnt emissions. HC formation is also a function of available air. The complete combustion occurs when there is sufficient amount of air, i.e., until the mixture become too lean and lead to misfire. As methane is slower to react the HC emission from natural gas engines are reasonably high. When the engine operates in lea condition, the HC emission s will be too high as the flame speed is too low to propagate during the power stroke. Due to its excellent capability of natural gas mixing with air during combustion at cold start, they emit very less smog. Incomplete combustion results in CO emissions.

It was found that the availability of air, i.e., air/fuel ratio and fruitful mixture of fuel and air decides the effective combustion which in turn decides the concentration of CO emissions. In most of the gasoline engines, the CO emissions are produced during the cold start condition, where the fuel is not completely vaporized and poorly mixed with the air. Due to its potential in mixing particularly during cold start NG-fuelled vehicles, emit 50 to 90% less CO emissions as compared to gasoline vehicles.

NO_x emission increases as temperature, availability of oxygen and combustion time at high temperature increases. NO_x formation reduces at very rich mixture, as combustion temperature decreases and lean mixture, as the oxygen availability is less. Formation of NOx reduces as natural gas engine operates at low flame temperature. However, the thermal efficiency increases with increase in CR and fuel's flame speed is compensated by incorporating advanced spark timing which in turn increase the NO_x formation. Due to these two opposite effects, NG-fueled vehicles emit either more or less NOx emission. NOx emissions can be controlled by exhaust gas recirculation (EGR) rate or lean-burn technology. When an engine is operating under very rich condition solid particulate carbon is formed as a combustion product which is not the case with properly-adjusted natural gas engine. The hydrocarbons in natural gas are also non-condensable under atmospheric conditions. Therefore, the only particulate emission from the spark-ignition natural gas engines is from the un-burnt lubricating oil on the cylinder walls.

From environmental point of view, even though the methane itself is greenhouse gas NG-fuelled vehicles still show benefits. As the carbon content in the natural gas is less compared to other fossil fuels, NG-fuelled vehicles emit less CO_2. Increase in concentration of OH is possible with decrease in CO emissions as CO plays a vital role in destroying OH radicals. This helps in decomposition of methane globally as methane and OH interaction results in methane removal. This also counteracts the increase in direct methane emissions (Figure 8.2).

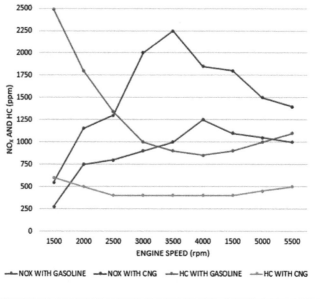

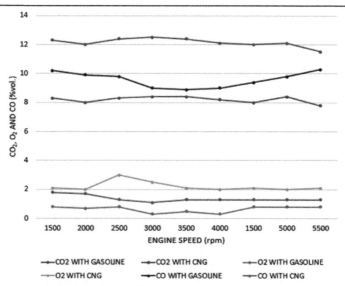

FIGURE 8.2   NOx, HC CO, CO_2, and O_2 emissions with CNG engine operation.

Two gaseous fuels-LPG and CNG have been exploited on a large scale as AFs as both exhibit high auto-ignition temperature. These two fuels are suitable for spark-ignited engines. Experiments were conducted on an air-cooled diesel engine modified into a spark-ignition engine with electronic ignition and speed dependant spark timing for use with LPG and CNG as fuels. CNG's major component is methane and its combustion chemistry is simple (Turns, 2005). Since the molecular weight of CNG is lower than air, the volumetric efficiency (VE) would reduce when used in engines. A resonator has been employed to supply more air and to compensate for loss in VE. Engine was optimized for stoichiometric operation on engine dynamometer. Ignition timing was mapped to work with gaseous fuels for different speeds. Ignition timing was observed to be the pertinent parameter in achieving good performance with gaseous fuels under consideration. Liquefied petroleum gas and natural gas fueled engines could be operated lean with an equivalence ratio as low as 0.7 resulting lower in-cylinder temperatures that reduce NOx emission levels. The engine-test results showed that AFs exhibit longer ignition delay, with slow burning rates (Amhed et al., 2013). It was observed that with dual-multi mapped ignition timing, the existing diesel engine was successfully converted into spark-ignited engine and ignition timing is required to be advanced for use with LPG and CNG in the same engine or needs higher CR. The tests on chassis dynamometer revealed that CO_2 emission is lower with CNG operation. Also, low CO and HC + NOx emissions were observed with CNG fuel operation. It was felt essential to adopt wear-resistant high-grade material needed for intake and exhaust valve and valve guide for dry gaseous fuel application to improve life of engine (Syed and Amba, 2010).

8.8 LIQUEFIED PETROLEUM GAS-LPG

Among the petroleum-derived fuels, next to CNG, LPG is the lightest gaseous component and has been preferred for domestic use over the years. It has excellent combustion properties and is widely used for automotive applications. In fact, the major components of LPG either being propane or butane and propane-air mixtures have been tried and tested extensively in earlier days of spark-ignition engine. Suitably, SI engine uses LPG having greater octane number because it has less size of molecules and short chain lengths. It is stored at pressure range of 5 to 10 bar and is gaseous in nature.

At present including India, majority of vehicles in several countries are powered by LPG which become more predominant of gas fuels. Additionally, it has an advantage of carrying easy to aboard the vehicle as being a natural gas with similar advantages. With varying proportions, it is a mixture of many gases containing mainly butane (C_4H_{10}) and propane (C_3H_8), with low proportions of many butenes (C_4H_8), propene (C_3H_6), iso-butane, and ethane (C_2H_6). The variation of its composition quite large.

Propane is preferred as an automotive fuel rather than butane because of its higher-level knock-resistance. Cold start enrichment is not needed with LPG as it has no cold starting problem. The propane has high octane rating as 112 compared to gasoline allow its use in engines with high CRs of (11 to 12):1. Lean burn combustion is convenient with the use of LPG in SI engines and reduces emissions improving efficiency. The use of higher CRs are permitted by the lean mixtures as they are more knock resistant.

LPG is obtained during the extraction of crude oil and during different refinery processes. It can be liquefied at room temperature under comparatively low pressure. Because, it has lower carbon content than gasoline, approximately 10% less CO_2 is produced during combustion. The ON is approximately 100–110 RON.

8.8.1 LIQUEFIED PETROLEUM GAS (LPG) USAGE AS ENGINE FUEL

Studies with LPG-Diesel fuel blending investigated on the effects of pollutant emissions and performance of DI diesel engine and informed that with by mass fraction increase in LPG in that fuel bending resulted in peak cylinder pressure of lower values. Equivalent *BSFC* declined under fuel blending mode at lesser engine loads which is antithetical to diesel operation. However, equivalent *BSFC* values were nearer to the recorded values under diesel operation at greater engine loads. The report by them states that there was a NOx emission dropping as the LPG fraction's mass fraction increased in fuel blending, when compared to diesel operation yields low emissions of NOx. As the mass fraction of the LPG fraction increased, CO emissions were reduced at greater engine loads.

Under low engine load operation, modest increase in CO emissions was observed. With increase in the mass fraction of the LPG fraction in the fuel blending, marginally HC emissions are increased. So for controlling engine-out (smoke and NOx) emissions, LPG-Diesel fuel blending is a most encouraging technique. Investigated with various engine parameters and LPG fuel blends on the dual-fuel engine performance running on diesel which is a pilot fuel. The LPG fuel mixtures were Butane to Propane with volume ratios of 0:100, 45:55, 30:70, 75:25, and 100:0. The volume ratio of 45:55 exhibits highest noise levels with lowest efficiency.

Accompanying with the higher combustion noise of engine, all fuel types inclined to have greater efficiency as there was an increase in mass flow rate of gaseous fuels. With the volume ratio of 100:0, the fuel recorded the highest pressure rising rates. On the other hand, the overall efficiency of the engine was yielded with increase in pilot fuel's mass flow rate while there is a slight reduction in noise levels of combustion. Under all varying parameters, the lower noise levels with comparison to engine performance to others were created by the fuel with the volume ratio of 100:0.

A study examined five gaseous fuel effects with various proportions of LPG in a dual fuel engine's exhaust emissions and performance on diesel operation as pilot fuel. The experiments showed that when using various LPG compositions with higher butane content, the dual-fuel engine's exhaust gas temperatures, the exhaust emissions and fuel conversion efficiency were affected leading to less NOx levels while CO levels reduced by higher propane content. Based on observations, a better trade-off amidst NOx and CO emissions suggested by them would be attained within 5 to 15% EGR rate. Comparing with the conventional diesel engine, SO_2 and NOx emissions for #3-diesel fuel blend declined by 69%, 27% at 35%, 51% at 25% load and full load respectively. At 25% load, CO emissions declined up to 100% and at full load, it inclined about 15.7%.

Moreover, fuel #3 (70% propane, 30% butane) of diesel fuel, replaced 40% mass fraction was the best LPG composition in the dual-fuel operation because except at part loads, the overall performance of engine was in equivalent with conventional diesel engine. With 5% EGR rate at part loads, the fuel conversion efficiency of #3-diesel fuel blend was improvised. Hence, up to 90% substitution of diesel fuel, capacity of smooth running of test engine was concluded and compared to a conventional engine for maintaining high thermal efficiency, $(M_{propane}/(M_{diesel} + M_{propane})) = 40\%$ was the best ratio.

8.8.2 PERFORMANCE AND TECHNOLOGY OF ENGINE

Engine technology for natural gas vehicles is similar to that for LPG vehicles, with an exception that because of its poor resistance for knocking, LPG is not commonly used in dual-fuel diesel applications (as it has low Cetane number) (Hutcheson, 1995). Both lean-burn and stoichiometric LPG engines have been improved with better results. In majority of the countries, LPG is used in existing vehicles with retrofits mostly using mechanically converted systems. As compared to natural gas, the costs of gasoline conversion to LPG operation are greatly less mainly because of less cost of fuel tanks.

8.8.3   EMISSIONS

Many emission characteristics of natural gas are similar with LPG. On an energy basis compared to diesel/gasoline fuel, LPG has lower carbon content. LPG, when used in SI engines produces moderate HC, near-zero particulate and very little CO emissions. Due to concentration variations of different HCs in LPG, the reactivity of HC exhaust emissions and species composition was affected. As olefins (like butene and propene) are much more reactive in contributing to ozone formation compared to paraffins (like butanes and propane), olefin content increase of LPG is likely to be resulted in increased potential of ozone formation of exhaust emissions. Because requirement of tight seals of gas on the fuel system, evaporative emissions are very small. During refueling, LPG emissions are significant and to avoid overfilling, vapor vent is needed on the tank to be opened.

Compared to gasoline, exhaust CO and NMHC emissions of LPG are lesser. Because of the higher octane quality and lower carbon-energy ratio of LPG, the CO_2 emissions are lesser than those for gasoline. To those from gasoline vehicles, NOx emissions are similar and using three-way catalysts, they can be effectively controlled. Primarily, because of the photo-chemically more reactive hydrocarbon emissions and higher emissions of carbon monoxide, benefits of less air quality is provided by LPG than CNG. Of late, ECE+EUDC cycle based tests revealed reduced emissions with improved vehicle-engine combinations.

Emissions can further be reduced with three-way catalyst employed modern spark-ignition LPG-fuelled engines. In fusion with an oxidation catalyst, very less emission results can be achieved by lean burn engines. With both lean-burning LPG and stoichiometric engines, very low levels of particulate emissions can be attained less NOx emissions. In response with energy consumption and CO_2 emissions, typically 20 to 30% more energy is consumed by heavy-duty vehicles of LPG-fuelled. As compared to diesel, LPG has lower carbon mass fraction and higher energy content per kilogram, so the CO_2 emissions of LPG in heavy-duty vehicles used are hardly comparable with diesel. As an advantage over CNG, LPG can be stored at relatively less pressure in lighter tanks, in which a lower energy penalty is imposed.

8.8.4 STORAGE OF FUEL

On the vehicle, LPG is stored as a liquid under pressure containing an internal pressure of 40 to 20 atm which are usually cylindrical tanks with rounded ends and are more stronger compared to the storing tanks used for diesel/gasoline fuel (much lower than those used for CNG). Like any liquid propane could be supplied from one to another tank, but the requirement for maintaining pressure needs a tight seal for gas. Using as vehicle fuel, LPG can be distributed in much the similar kind as a diesel/gasoline fuel except for the need for a normalized tight connection of gas. As a by-product in petroleum refining or in the extraction of weighty liquids from natural gas, LPG is produced. Compared with other HCs, the price of LPG is less. So in most petroleum-refining countries, the demand in LPG supply is exceeded.

8.8.5 SAFETY

Higher safety risk than CNG is posed by LPG. LPG vapors are weighty than air unlike natural gas so that at ground level, it leaks from fuel system tending to "pool," where there might be ignition source contact. LPG vapor in the air has flammability limits broader than those of natural gas. LPG is non-toxic like natural gas. Additionally, on the vehicle, LPG is stored in sealed pressure vessels like natural gas that are stronger than gasoline fuel tanks.

8.8.6 EMISSION TESTS AS PER EURO NORMS ON TWO-VEHICLE ENGINES FUELLED WITH CNG AND LPG

Two Euro 5 compliant bi-fuel spark-ignition vehicle engines with (petrol/LPG) and (petrol/CNG) combination were tested on a chassis dynamometer. Along with heated oxygen, sensors and typical three-way-catalysts employing in closed-loop control, both SI engines are equipped with MPI feeding systems. Featured multipoint gas injection (MPGI) systems were tested by the vehicles. The vehicles were tested for analyzing the emissions such as THC, NMHC, CO, NOx, and CO_2. By using special additional equipment mounted onto the existing petrol fuelling system for fuelling with CNG or LPG, the vehicles had been adapted by their manufacturers.

Vehicle engines are checked whether bi-fuel operated vehicles can comply with EURO VI emission norms.

The team (Bielaczyc et al., 2016) worked with Engine Research Department of BOSMAL Automotive Research and Development Institute Ltd in Bielsko-Biala, Poland, within a research program probing the AFs impact on exhaust emissions from both CI and SI engine light-duty vehicles (LDVs).

Due to the influence on CO_2 emissions (greenhouse effect), these fuels are interested from the ecological point of view. Among gaseous fuels like Propane and Butane inherently that posses a less carbon and more hydrogen than diesel/gasoline fuel, and more CO_2 is produced by the higher carbon fractions and also particulate matter emitted upon combustion is also less (Figure 8.3).

FIGURE 8.3 BOSMAL emission testing laboratory—the climatic chambers internal view.

Featured MPGI systems are tested by vehicles. After 1–2 minutes and few seconds, the engines automatically switched over to LPG or CNG operation always the first startup using gasoline if the vehicle is in LPG mode and is in CNG mode respectively. Consisting of a three-way catalytic converters (TWC) especially organized for bi-fuel cars (fuelled with LPG or CNG and gasoline) are tested by the aftertreatment car systems.

The exhaust emissions analysis from the bi-fuel vehicles with SI engines fuelled with LPG or CNG and later during the NEDC cycle fuelled with petrol was executed in the Emissions testing laboratory of BOSMAL Automotive

R&D Institute in Bielsko-Biala, Poland on a chassis dynamometer. The automotive-type vehicles are tested rigorously tested using UDC, EUDC, and NEDC and emission traces are obtained (Figure 8.4).

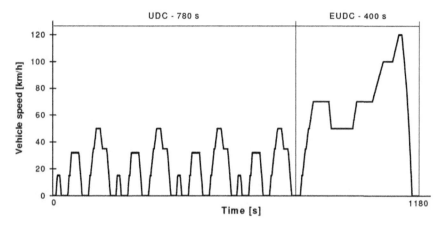

FIGURE 8.4   The new European driving cycle (NEDC).

This study focused to compare and assess the performance of Euro 5 vehicles emissions is a representative of the European market that operating on LPG or CNG and petrol for collation to the Euro 6 limits established in September 2014 (Figure 8.4). A few unregulated exhaust emissions (CH_4 and CO_2) were also analyzed and measured.

On a chassis dynamometer facility, sequences of tests were conducted on the test vehicles which are fuelled with CNG or LPG and with petrol (in turn). In order to find the gaseous fuels (CNG and LPG) impact on emissions in juxtapose with petrol testing and in juxtapose with Euro 6 limits, test have been conducted. For comparison ease, the results obtained using all fuels are put-up side-by-side. The mean emissions in (g/km) of NMHC, CH_4, THC, NOx, CO_2, and CO for both the phases were presented in Figure 8.5. For example, the EUDC and UDC and the total NEDC (EUDC+UDC) from the test vehicles. HC emissions, i.e., NMHC, THC, and CH_4 (Figure 8.5) from vehicle A fuelled with LPG and gasoline were equal for both the fuels. The total HC emissions (Figure 8.6) for vehicle B fuelled with CNG and gasoline over the whole NEDC were greater by 25% when running on CNG, which is caused by very high CH_4 emissions over the UDC phase, i.e., 7 fold higher than those of petrol.

CH_4 is a compound that is very hard to be oxidized in TWC and as its composition is greater than 96% in vehicle B fuelled with CNG, is responsible for high emissions of methane. Although, presently CH_4 emissions are not subjected to the EU direct regulations, because of concerning methane's role as a greenhouse gas (GHG), they are modulated in the United States which is 21 times as more absorptive of IR radiation than CO_2. Since there is a non-methane HC limit and a THC limit in which the difference being CH_4 itself and for the indirect regulation in the EU, methane could be considered. As a HC of natural gas vehicle (NGV) exhaust gas, CH_4 (mainly but not solely) is a relatively inert component (no saturated molecular structure or longer carbon chains and double carbon bonds) which leading it to have a greater light-off temperature for conversion than other HCs (e.g., those which are with longer chains or unsaturated). Even though CH_4 is not present in gasoline, by cracking processes in gasoline combustion CH_4 is formed.

Of both vehicles A, B fuelled with all fuels, emissions of NMHC (Figure 8.5) of the Euro 6 limit were lesser than 30%. When running on CNG, NMHC emissions for vehicle B were 45% lesser over the whole NEDC. Relative to gasoline, the results by using CNG leads to lesser NMHC emissions but greater THC emissions is a phenomenon which relates directly to the 2 fuel type's composition.

CO emissions from both vehicles fuelled with both fuels of Euro 6 limit were lesser than 50%. For vehicle, B when running on CNG, CO emissions over the whole NEDC was greater by 24%. Contrary to the vehicle a when running on LPG, CO emissions over the whole NEDC, lower with a 43% difference. It is worth to keep in mind that even if LPG or CNG fuel was chosen, in gasoline mode only the vehicles will be started and engines are automatically after 1–2 minutes (if the vehicle is in LPG mode) or after a few seconds (if the vehicle is in CNG) switched over to LPG or CNG operation. This goes some way towards explaining the small difference in CO emission for both vehicles during the UDC phase (Figure 8.5). During, the EUDC phase, i.e., the 2nd phase of the NEDC cycle, since when these are in LPG or CNG mode gasoline is not consumed by the vehicles and these differences were at times much higher.

NOx emissions of the Euro 6 limit (Figure 8.5) were lesser than 30% from both vehicles fuelled with all fuels. For vehicle B, when running on CNG NOx emissions were 10% lesser, but for vehicle A when running on LPG were higher with 35% difference over the whole NEDC. NOx emissions can be greatly impacted by the precise chemistry of the TWC and the engine calibration and these trends in turn could be contrary and it may be possible that for other vehicles fuelled with CNG or LPG and with gasoline. Due to this, the changes that are observed might not be directly attributable to the fuels chemistry itself in full, but to the fuel chemistry's interaction with TWC chemistry and also operation, variables of engine like spark timing, etc.

Due to the fact that fuel economy and CO_2 emissions are related intimately, with respect to global warming CO_2 emissions are important even though currently to direct limits in automotive emissions legislation are not subjected by them. As compared to gasoline, CNG, and LPG have lesser carbon/hydrogen ratio (Figure 8.5). So during combustion, lesser CO_2 emissions are produced by them per unit energy are released. From the total NEDC cycle in both phases rather than for gasoline, for vehicle B of CNG fuel, these emissions were lesser about 25% and for vehicle A of CNG fuel were lesser about 10% can be observed.

Without any further modifications, during the emissions test of NEDC, after testing of the obtained results found that both vehicle tested with an integrated CNG/gasoline or LPG/gasoline ECU (electronic control unit) and LPG or CNG multipoint injection gas already meets the Euro 6 emissions limits. Particularly for the light-duty vehicle (LDV) category, Euro 6 limits were met by CO, NOx, THC, and NMHC emissions.

So emissions limit in this area where for all fuels (CNG, gasoline, LPG) and both vehicles rather than the NMHC limit stipulated by these regulations, were met comfortably as the observed THC emissions were lesser. When running on CNG Euro 6 limit are 7 times greater than the NMHC emissions. When vehicle B was fuelled with CNG, during the NEDC cycle THC emissions are inclined comparing with gasoline, but this inclination was very low for problems causing with emissions limits of Euro 6. For vehicle A fuelled with gasoline and LPG, THC emissions were same for both fuels.

For vehicle A fuelled with LPG, CO emissions were considerably less by 43%. When vehicle B was fuelled with CNG during the NEDC cycle, there was an increase of 24% CO emission, even though CO emission limits are low from both fuel types.

NOx emissions from vehicle A when fuelled with LPG were about 35% higher, but from vehicle B when fuelled with CNG were about 10% lower in comparison to petrol.

When vehicle B was fuelled with CNG, CO_2 emissions were declined by 25% and when vehicle A was fuelled with LPG declined by 10%, CNG, and LPG are commonly used gaseous fuels in SI engines since their power trains are easily converted from liquid to gaseous fuels. Compared to diesel/ gasoline, these fuels are preferred since they are cheap. When using CNG or LPG as a fuel, Euro 6 emissions standards can met by certain pre-Euro 6 technologies were shown by this paper. In addition to that, the CO_2 emissions on gaseous fuels (CNG or LPG) from vehicles operation are lesser.

The CNG and LPGs better performance CNG and LPG have worthy environmental potential are confirmed in terms of low CO_2 and particulates emissions, even if certain emissions are sometimes greater than when running on petrol. The emissions and performance of bi-fuel vehicles changes when operating on fuel other than gasoline (e.g., CNG or LPG). The main influencing factors of tailpipe emissions are the ambient temperatures, fuel, and lubricant oil properties, the individual engine calibration strategy, driving conditions and aftertreatment technology and engine.

From Figure 8.5, it can be concluded that gaseous are far superior in terms of emissions even in the case of CO_2 emissions with CNG operation is better. Except NOx emissions, all other emissions are lower relatively.

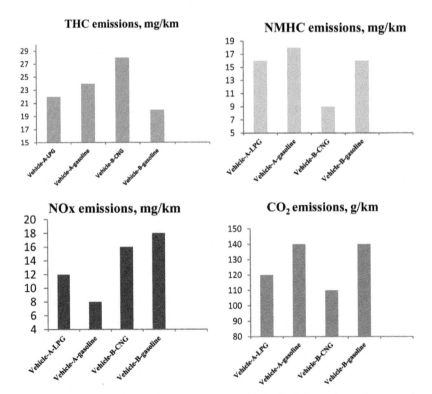

FIGURE 8.5   Comparison of different emissions with LPG, CNG, and gasoline operation.

8.9 ALCOHOLS AS FUELS-METHANOL AND ETHANOL

Lead and MTBE have been explored as additives for gasoline to improve ignition quality in the context of detonation. However, lead has been phased out and MTBE posed adverse health and environmental effects, straight use of alcohols as fuels or additives has increased. Addition of small amounts of alcohols enhances octane rating of gasoline and thereby allows the petroleum to combust completely with reduced harmful air pollutants (Venkateswara et al., 2011). Moreover, ever since the fuel crisis is recognized in the early 1970s, the researchers started exploring alcohols as engine fuels. Alcohols belong to "ol" group of alkane family and have "hydroxyl" in its molecular structure. This is an attractive feature as external oxygen supply is reduced. This is justified as for methane combustion stoichiometric requirement is far more than methanol combustion since both contain the same number of carbon and hydrogen atoms. Among alcohols, lower carbon atom alcohols have gained interest; methanol and ethanol.

Stoichiometric Reaction	Stoichiometric A/F Ratio
$CH_4 + 2O_2 + 7.52N_2 \rightarrow CO_2 + 2H_2O + 7.52N_2$	$[7.52 \times 28 + 2 \times 32]/[12+4] = 274.56/16 = 17.16{:}1$
$CH_3OH + 1.5O_2 + 5.64N_2 \rightarrow CO_2 + 2H_2O + 5.64N_2$	$[5.64 \times 28 + 1.5 \times 32]/12 + 4 + 8] = 205.92/24 = 8.58{:}1$

From the above equations, it can be observed that there is a reduction of 50% external supply of air for stoichiometric combustion due to the presence of oxygen in molecular structure of fuel.

8.9.1 METHANOL

The Methanol, a lowest weight among alcohols is promoted widely in many countries as a "clean fuel." CH_3OH has many combustion and emissive characteristics which desirable, including lean-combustion properties, low flame temperature (helps in reduction of NOx emissions), and low photochemical reactivity. CH_3OH can be produced from natural gas, coal, crude oil, biomass, or cellulose and even from municipal solid waste (MSW). Methanol is popularly called as "wood alcohol." It is rejected from fuel pumps in same manner that of gasoline. In the early days, the cost of methanol was on the higher side and now it has become *price-competitive with conventional fuels.* With a 112 octane number and good lean combustion properties, CH_3OH is most effective fuel for lean burning spark igniting engines. Its lean combustions are similar to that of natural gas, while it's less energy density shows low flame temperature relative to hydrocarbon fuels, and in lower NOx emissions. Methanol's low vapor pressure helps in reduction in evaporative losses.

Alcohols especially lower alcohols have lower heating value compared to petroleum fuels poses energy release issues. The low energy density of CH_3OH indicates that a large amount (roughly twice the mass of gasoline) is required to attain the same amount of power output. This inborn disadvantage is part decreased by the high value octane number of CH_3OH and it charges cooling capability, so that 1.5 liters of CH_3OH gives the same distant mileage as 1 L of gasoline in usage vehicle. The higher heat of vaporization of CH_3OH mixed with huge amount, makes it difficult to make sure of complete vaporization, and requires special consideration in designing of intake manifolds and cold-start methods. With the alcohol fuels, the vehicle/engine components suffer with corrosion related issues with metallic fuel system. Hence, it is required to redesign certain vehicle/engine components with which the alcohols come in contact. The points to consider in the re-design of components have been presented thoroughly by Venkateswara et al. (2011) in their review paper.

To get fuel advantages of alcohols, they can be used in blended form. CH_3OH (M100) has large latent heat of vaporization and when used in SI engines pose start problems, hence requires special fuels or additional heating techniques. Problems like low temperature starting and other problems with pure CH_3OH have led the researchers of light-duty CH_3OH vehicles to indicate an 85% CH_3OH 15% gasoline blend (M85) for the present running generation of methanol vehicles. The added gasoline raises the vapor pressure of the blend, enhancing starting, and making very rich headspace mixture to burn. Most of the emissive benefits of CH_3OH (such as low evaporative emissions) are lost by switching to M85, however. If the fuel mixture consists even a little amount of methanol, the vapor pressure increments dramatically.

CH_3OH shows, a number of handling and safety problems which leads to consideration over its possible wide area usages. Unlike HC fuels, methanol burns with a light non-luminous flame that is not possible to view in daylight. This has introduced consideration over the effects on passers-by and fire fighters in the event of a fire. Methanol's vapor pressure is also such that at normal ambient temperature it can give a flammable mixture in the headspace of its fuel tank. The toxic nature of methanol and due to its tastelessness or odorless can be a great deal of problem compared to gasoline. Methanol can as well be produced from MSW in addition to steam reformation etc. It is used as feedstock in the preparation of MTBE, and the heavy rise in the demand of MTBE for reformulated gasoline affected the methanol prices. MTBE is used as an additive to improve anti-knocking tendency of gasoline.

8.9.2   FLEXIBLE FUEL VEHICLES (FFVS)

FFVs are those vehicles which are capable of running on any combination of methanol (approx. 85%) and gasoline. These have been developed, and widely verified. There is a semantic behavior between these vehicles and that of the advanced technology gasoline vehicles with respect to engines and emission control systems, and the overall energy efficiency and emission properties are semantic as well.

The amount required for the maintenance of gasoline fuelled engines could be lower as compared to that of methanol fuelled engines. Certain Methanol fuelled engines tested till date illustrates that life of the engine these days is likely to be shorter and also Methanol combustion products are fairly corrosive. Other designs have experienced no considerable added wear and also something that is to be shortened is oil change intervals which counteract the increased corrosiveness of the products that are combustible.

8.9.3   ENGINE EMISSIONS WITH METHANOL

There are benefits of using neat methanol (M100); however, the power and torque would reduce. To take advantage of methanol properties, it has been blended along with gasoline. Lower blends have well tested concluding that a reduction of power, torque due to its lower heating value but better thermal efficiency. For better performance, engine's spark timing needs to be optimized. The experiments for performance and emissions by Liu Shenghua et al. (2007) revealed that lower HC and CO emissions can be obtained when the two-way catalytic converter is used, however, with increased proportions of methanol, formaldehyde emissions would increase. It is also observed that could start of SI engine have been improved significantly. While the potential emissions benefits of neat methanol (M100) remain a matter of concern for various reasons, the use of M85 offers some advantages. Due to the absence of complex organic compounds in methanol, its use as fuel also offers two more benefits such as it develops the potential to form ozone in the lower layers and reduction in emissions of a number of toxic air contaminants (polycyclic aromatic hydrocarbons such as benzene especially). However, about five times more levels of emissions of formaldehyde compared to the use of gasoline are formed with methanol (a primary incomplete combustion product of methanol. The control of formaldehyde is of utmost importance since it is a toxic and carcinogenic compound, especially its emission in confined spaces such as garages and tunnels, emissions are more harmful. Due to cold start difficulties, methanol run engines bound to emit high CO, HC, and formaldehyde emissions. Methanol is less photochemical in nature than most hydrocarbons; it was always believed that urban ozone problems can be reduced by using methanol engine vehicles. Recent studies, however, illustrate that formaldehyde which is highly reactive in nature offsets methanol which is less reactive in nature, in order to attain small net ozone benefits. Reduction in emissions of formaldehyde and 1,3-butadiene can be achieved by the use of M85, but emissions of acetaldehyde (not significant with neat methanol) approach the levels associated with ethanol-gasoline blends (OECD, 1995).

Since the heating value of methanol far lower than gasoline and diesel, the methanol is used in blended form with gasoline in M10 and M85 forms.

Heavy-duty vehicles which use methanol engines produce emissions which significantly produce lower NOx and PM than semantic heavy-duty bus diesel engines, while NMHC, CO, and formaldehyde emissions, tend to be much higher. However, Catalytic converters have been successful in controlling these emissions.

8.9.4 *ETHANOL*

After methanol, the next primary alcohol higher in molecular weight is ethanol which resembles methanol in all of its physical properties and combustion nature. The difference is that ethanol is considerably cleaner, exorbitant, less corrosive, and toxic and higher volumetric energy content. Ethanol is also known as "grain alcohol" can be produced by processing sugar cane or corn. However, this alcohol is much preferred among other alcohols, as it is produced on a large scale from bio-derived sources. Ethanol is most manufactured in either of the two forms: hydrous or hydrated ethanol which comprises of a 95% mixture with water, known as hydrous or hydrated ethanol, and anhydrous or absolute ethanol comprises of a 99.5%

mixture with benzene. Ethanol if handled improperly can cause a fire hazard which is considered as moderate when flame is exposed to it and is also more explosive than gasoline though less volatile. Vapors that form above a pool of ethanol are potentially explosive. The preference of ethanol depends on choice of feedstock for its production. It also depends on how much amount of petroleum products are utilized for its production is the amount of petroleum used to produce the crop and then prepare it for fermentation. This, in turn, becomes a factor in judging whether it is renewable or not. Depending on the area of the country, the crop grown, and the agricultural practices used governs the use of petroleum use for ethanol production.

There are three primary ways that ethanol can be used as a transportation fuel:

1. As a blend with gasoline, typically 10% and commonly known as "gasohol" (Gasoline+ Alcohol);
2. As a component of reformulated gasoline both directly and/or transformed into a compound such as ethyl tertiary butyl ether (ETBE); and
3. Used directly as a fuel, with 15% or more of gasoline known as "E85."

In Brazil, ethanol has been extensively used. Fuel grade ethanol is manufactured by distillation, and contains several parts water (by volume). Ethanol blended with ordinary gasoline up to 22% results in mixture known as gasohol, may be burned in ordinary spark-ignition automobile engines.

In Brazil, South Africa, and the United States, the alcohol that is extensively used for blending of gasoline is pure ethanol. The presence of even a small amount of water in gasohol mixture can result in phase separation even though ethanol is completely miscible in gasoline; therefore, only anhydrous ethanol is preferred to overcome problems with separation and eventually stalling of vehicles. To make use of alcohols even lower temperatures, researchers have added surfactants such as biodiesel, stearyl alcohol (saturated), 1-hexanol, oleyl alcohol (unsaturated), methyl oleate (unsaturated with ester group) and tetrahydrofuran (Molea et al., 2017).

With the formation of azeotropes that cause an undesirable increase in vapor pressure together with a reduction in front-end distillation temperature. Since neat ethanol cannot be used similar to methanol due to lower heating value and other difficulties, it is preferred to be used in blended form with gasoline like gasohol which exhibits a higher RON and almost as equal MON. Road performance of gasohol is semantic to or better than gasoline with the same octane quality (gasohol is inferior to gasoline) only under low speed and acceleration conditions.

Majorly, ethanol is produced from fermentation molasses and in a sense; it's a biofuel as it is obtained from food crops/grains, etc. Gay Lussac's law depicts the formation reaction of ethanol (fermentation of carbohydrates).

$$C_6H_{12}O_6 \rightarrow 2C_2H_5OH + 2CO$$

Through this, about 50% of alcohol is produced from carbohydrates. Other routes of production of ethanol have been reviewed by (Venkateswara et al., 2011).

Anhydrous ethanol contains heat energy which is equal to that of 65% of that of gasoline for when equal quantities are considered. The former has more thermal efficiency than latter. Up to a proportion of about 20% anhydrous ethanol, there are no noticeable mileage penalties in substituting gasohol for gasoline except for engine type and driving conditions. Dynamometer test results and road trials on non-catalyst cars show no change in fuel economy catalyst-equipped cars, it is the other way around.

8.9.5   ENGINE EMISSIONS WITH ETHANOL

Brazil has been in the forefront as far as the use of ethanol in its fleet, it has adopted about 20% ethanol as an octane enhancer by blending with gasoline. Emissions from ethanol-powered vehicles are believed to be high in unburnt ethanol, acetaldehyde mostly formaldehyde which can be controlled with catalytic converter. However, these vehicles generate 20 to 30% less CO and roughly 15% less NOx compared to gasoline-fueled vehicles. Evaporative emissions from ethanol powered vehicles are significantly lower. However, low vapor pressure could cause cold-starting problems in colder climates resulting in higher cold start exhaust emissions.

8.9.6   EMISSION CHARACTERISTICS OF METHANOL AND ETHANOL FUELS

Ethanol by itself has a very low vapor pressure, but when blended in small amounts with gasoline, it causes the resulting blend to have an undesirable increase in vapor pressure. For this reason, there is interest in using fuels such as ETBE as reformulated gasoline components because ETBE has a small blending vapor pressure, which will reduce the vapor pressure of the resulting blend when added to gasoline. The primary emission advantage of using ethanol blends is that CO emissions are reduced through the "blend-leaning" effect that is caused by the oxygen content of ethanol. The oxygen in the fuel contributes to combustion much the same as adding additional air.

E85 produces acetaldehyde instead of formaldehyde and Ford unit found that the level of acetaldehyde was the same as for M85. An advantage of acetaldehyde over formaldehyde is that it is less reactive in the atmosphere which contributes less to ground-level ozone formation. The specific reactivity of ethanol exhaust emissions (grams ozone per gram emissions) has been measured to be significantly lower than conventional gasoline (about 30%) and lower than from specially formulated gasoline.

Another feature of alcohol is its low sulfur content and even in blend of E85, a benefit in reducing catalyst deterioration compared to vehicles using gasoline. Insufficient data have been gathered to date to determine whether this effect is significant. Similar to methanol operated engines, redesigning of the engine is essential either neat or in or blended form, especially with higher alcohols, to improve fuel characteristics, provided to get the advantages of alcohols for engine.

The vehicle technology to use E85 is virtually the same as that to use M85, allowing auto manufacturers to quickly build E85 vehicles since they have already engineered M85 vehicles. The low emission characteristics of higher blends of methanol and ethanol are a promising to satisfy ultra-low emission vehicle standards (ULEV).

A spark-ignition engine with a bore of 65.1 mm and a stroke of 44.4 mm is used in this study. The engine is 1-cylinder, 4-stroke with a 7:1 CR, air-cooled, no catalytic converter unit, and a carburetor fuel system. One may claim that carburetor is hardly current engine technology but carbureted engines are still widely used and developed. In addition, the carburetor fuel system is very appropriate for use with fuel blends. This is due to its high quality of mixture preparation and mixing of different fuels. The engine is connected with an air-cooled Dynostar Model ECB500 eddy current engine dynamometer with 7000 r/min maximum engine speed. An electronic ignition control unit (EICU) was used in the engine setup for defining the proper ignition at different loads.

The highest emissions of CO and UHC and the lowest emissions of CO_2 at all engine speeds (2600–3450 rpm) are neatly presented by gasoline. However, the lowest emissions of CO and UHC and the highest of CO_2 are shown by the methanol-gasoline blends, e.g., the best emission results among all test fuels. The ethanol-gasoline blended fuels show lower CO_2 and higher emissions of CO and UHC than those of M at all rates (3, 7, and 10 vol. %). A moderate level of emissions between E and M test fuels is shown by EM. In general, a significant reduction in CO and UHC emissions when compared to neat gasoline fuel is obtained by using blended fuels containing ethanol and/or methanol with gasoline. The reason lies with the blended fuels contain oxygen, which can enhance the combustion process significantly which will be discussed later in further detail. The relative decreases in the CO emissions of E3, E7, and E10 are about 15.5%, 31%, and 42%, respectively; the CO emissions of M3, M7, and M10 are decreased by about 17.7%, 51.5%, and 55.5%, respectively when compared to neat gasoline. But the CO emissions of EM10, EM7, and EM3 are decreased by about 46.6%, 35.5%, and 17.5%, respectively.

UHC emissions for test fuels are compared and shown also in Figure 8.6. When compared to neat gasoline; a reduction of about 3.5%, 14%, and 21.5%, for the UHC emissions of E3, E7, and E10 respectively; while a reduction of about 19.6%, 16% and 26% for UHC emissions of M3, M7, and M10, respectively while a reduction of about 10.7%, 15.3% and 23.2%for EM3, EM7, and EM10, respectively. When we observe clearly by comparing with neat gasoline, M is having very low value of UHC, then for EM and followed by E blends which is an indication of combustion quality, e.g., the combustion is enhanced when it is lower. CO_2 emissions for test fuels are compared as shown in Figure 8.6. When compared to the CO and UHC emissions the changes in CO_2 emissions have an opposite manner; the CO and UHC emissions decrease while the CO_2 emissions increase.

This can be related to the fact that CO_2 emissions depend on CO and UHC emissions concentration. M can have the maximum value of CO_2, then for EM, E, and followed finally by the G fuel. In particular, when compared to the neat gasoline the CO_2 emissions of M10, M7, and M3 are higher than that for gasoline fuel by about 9.2%, 8%, and 3%, respectively; the CO_2 of EM10, EM7 and EM3 are higher by about 7.1%, 5.1%, and 3%, respectively; the CO_2 of E10, E7 and E3 are higher by about 4%, 1.7%, and 1%, respectively. The effect of M on CO_2 emissions is minute significant than those of EM and E as seen from these experimental values. In general, due to sufficient amount of air in the air-fuel mixture and plenty of time in the cycle for completion of combustion process, the emission of CO_2 is a product of complete combustion. So that, the process of CO-CO_2 as well as UHC-CO_2 will be enhanced with sufficient oxygen/air and time and, in turn, maximizing the emissions of CO_2, and it will be discussed in details subsequently.

The performance and emission characteristics of the spark-ignition engine running on ethanol and methanol blended with gasoline were evaluated and compared with neat gasoline fuel. Apparatuses used in the present study were an engine, a dynamometer, and an exhaust analyzer. A four-stroke, single-cylinder, carbureted, SI non-road engine (type Bernard Moteurs 19A), was chosen. There is a lot of difference between non-road gasoline and automotive engines in terms of several technical specifications. Because of these design variations, the influence of alcohol/gasoline blended fuel changes on emission and performance characteristics for non-road gasoline engines are different from the conventional gasoline engines. This engine had a 58 mm stroke and a 56 mm bore (total displacement 143 cm^3) and rated power of 2.2 kW. The ignition system was composed of the conventional coil and spark plug arrangement with the primary coil circuit operating on a pulse generator unit. The engine was coupled to a hydraulic dynamometer. Exhaust gases were evaluated on line by Bosch analyzer, which were sampled from the outlet.

Engine emission characteristics to investigate the effect of different ethanol/methanol gasoline blended fuels on exhaust emissions, results of the engine test at 2000 rpm with full throttle valve opening were selected for comparison (Figure 8.6).

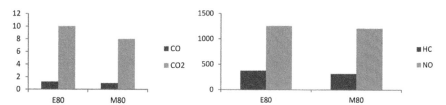

FIGURE 8.6 The effect of various ethanol/methanol gasoline blend fuels on CO, CO_2, HC, and NO emissions.

The incomplete combustion results in toxic gas named carbon monoxide. The combustion of engine becomes efficient, when methanol and ethanol-containing oxygen is combined with gasoline which in turn reduces the CO emissions (Stump et al., 1996; Yasar, 2010). As seen in Figure 8.8, the values of CO emission are about 3.654% (3.637%), 3.161% (3.145), 2.842% (2.825), 2.337% (2.306), 1.851% (1.824%) and 1.275% (1.248%) for E5 (M5), E10 (M10), E20 (M20), E40 (M40), E60 (M60) and E80 (M80) fuels, respectively. CO_2 is not harmful but contributes to the greenhouse effect. With full-throttle valve, opening using methanol and ethanol-gasoline blends the CO_2 concentrations were reduced, at 2000 engine speed. As the carbon, atoms were less in ethanol and methanol the CO_2 emissions were less (Knapp, 1998; Celik, 2008). The value of CO_2 emission is about 13.88% for gasoline fuel, while the values of CO_2 are about 13.12% (12.96%), 12.95% (12.78%), 12.25% (12.12%), 11.73% (11.68%), 10.42% (10.39%), and 9.78% (9.57%) with E5 (M5), E10 (M10), E20 (M20), E40 (M40), E60 (M60), and E80 (M80) fuels, respectively. The HC concentration in the exhaust gas emission at 2000 rpm with full throttle valve opening, for gasoline fuel, was 345 ppm, while the HC concentration of E5 (M5), E10 (M10), E20 (M20), E40 (M40), E60 (M60), and E80 (M80) fuels was 341 (304), 301 (297), 282 (223), 265 (234), 273 (261), and 380 (372) ppm, respectively. The HC concentration at 2000 rpm using E5 (M5), E10 (M10), E20 (M20), E40 (M40) and E60 (M60) was decreased by 8.98% (11.88%), 12.75% (13.91%), 18.26% (35.36%), 23.19% (32.17%) and 20.86% (24.35%), respectively, while the HC concentration of E80 (M80) fuels was increased by 10.14% (7.83%), respectively in comparison to gasoline. The above results illustrate that the primary alcohols such as ethanol and methanol can be treated as hydrocarbons (whose emissions decrease to a certain extent when added to gasoline) which are partially oxidized when added to the blended fuel. The volumes of methanol and ethanol content blends reduce the cylinder temperature as the heat of vaporization of primary alcohols is higher when compared to gasoline. The lower temperature causes partial burn in the regions near the combustion chamber wall. This results in increase of emissions and decrease of power of engine. This behavior has been reported by other investigators on various types of engines and conditions (Celik, 2008; Najafi et al., 2009). It shows that as the percentage of ethanol/methanol in the blends increased, NOx emission was decreased. The NOx concentration in the exhaust gas emission at 2000 rpm with full throttle valve opening, for gasoline fuel was 2247 ppm, while the NOx concentration of E5 (M5), E10 (M10), E20 (M20), E40 (M40), E60 (M60) and E80 (M80) fuels was 1957 (1945), 1841 (1828), 1724 (1574), 1498 (1379), 1366 (1338) and 1223 (1207) ppm, respectively. The NOx concentrations at 2000 rpm using E5 (M5), E10 (M10), E20 (M20), E40 (M40), E60 (M60) and E80 (M80) fuels were decreased by 12.91% (13.44%), 18.07% (18.65%), 23.27% (29.95%), 33.33% (38.63%), 39.21% (40.45%) and 45.57% (53.72%), respectively in comparison to gasoline. The higher heat of vaporization of primary alcohols blends temperature at the end of intake stroke decreases resulting in decrease of temperature of combustion resulting in decrease of engine-out NOx (He et al., 2003; Celik,

2008). The fuel blends containing high ratios of ethanol and methanol had important effects on the reduction exhaust emissions. These results show that E40 and M20 are most suitable in terms of HC emission. CO, CO_2, and NOx concentrations of E80 and M80 were the lowest.

Even though the cetane number of ethanol is very low, it is used in diesel engine. Researchers observed that there is reasonable reduction in smoke and NO_x emissions by adding cetane improvers while performing the experiment (Lu¨ Xing-Cai et al., 2013).

8.9.7 USE OF HIGHER ALCOHOLS AS FUELS IN ENGINES

Many researchers started exploring the use of higher alcohol in recent times. Butanol is used in non-modified spark-ignition engines because of its solubility (sparingly) in water miscibility with most solvents. Butanol can be produced using fossil fuels or biomass, in which case it is called bio-butanol. However, both of these have same chemical properties.

Butanol has been attracting attention in recent times and has longer hydrocarbon chain, lower oxygen content, and higher heating value which makes it semantic with that of gasoline. The former is anyhow a promising fuel candidate. Butanol has several benefits such as high tolerance to water contamination as compared to methanol and ethanol. Recently, the fuel oil that is attracting attention is a by-product of alcohol production after fermentation during the distillation process. Fuel oil is a natural source of amyl alcohols (Vinod et al., 2017).

Among the liquid alternate fuel due to lower heating value alcohols are produced through different feedstock attain significant power and torque. Alcohol content is preferred over conventional fuels for lower sulfur, olefins, and aromatics to get emission advantages with the use of alcohols (Gajendra Babu, 2013). Moreover, due to lower cetane number of ethanol, effect of advanced injection timing has also been studied and observed lower levels of CO, smoke levels when used in blended form with diesel in CI engine. However, the NOx levels have increased with increased proportion of ethanol in the blend (Cenk and Kadir, 2008).

8.10   BIO FUELS-BIOMASS FUELS

Over about 150 years, the fossil resource derived fuels have dominated the transportation sector-especially the automotive sector. The large scale exploitation of such fossil fuels led not only to doubts about its long term availability but also raised concern about environmental degradation. To achieve sustainability in fuels, there should be a shift in resource for fossil fuels. In this context, the biomass derived fuels or biodegradable fuels have attracted many. The biomass derived fuels such as alcohols, biodiesel (obtained mainly from vegetable oils), biogas gained lot of popularity among researchers, and extensive studies have been on the use of these fuels. In the above sections, already details of methanol and ethanol are discussed. Energy is recovered from biomass feedstocks is attained by direct combusting or thermo-chemical conversion indirectly. Direct combustion is burning of solid biomass. In indirect, methods wood to a liquid or gas, conversation is happened. Liquid or gaseous fuel thus obtained is then burned to generate heat and by-products of combustion. The use of biofuels can reduce the emission of CO_2 and other gases associated with global climate change. As plants grow, they capture CO_2 from the atmosphere. When the fuels from the plant sources are consumed, new plants absorb the CO_2 released during combustion process, thus effectively recycling the carbon. As far as a significant feature of biofuel chemistry is concerned, they contain oxygen molecule in its structure and thus require lesser external oxygen for its full combustion and also the biofuels are almost sulfur-free fuels.

Biofuels have started becoming important part of transportation fuels in many countries. In biofuels, widely used fuel is ethanol at present and Brazil is the forerunner in its use. Whereas vegetable oil-based oils from both edible and non-edible, plant oils have attracted by researchers (Agarwal and Das, 2001). Edible oils plants such as Palm trees have been used to derive seed oils as substitute for petro-diesel fuel. Biofuels are good option for agricultural dominated economic countries.

The biofuels are mostly obtained from bio-derived sources and when it is converted to fuel like liquid, combusted (for power and transport, etc.) yielding CO_2 will be absorbed by plants for photosynthesis on the contrary to combustion petro-derived fuels (Figure 8.7).

It is observed from the majority of the research works have done on both edible and non-edible type vegetable oils (seed-based oil). A few of the researchers felt that oils are moving from kitchen tables to garages. Looking at the scale of exploration of these vegetable oils, it is feared that the large scale use of biodiesel plants may interfere with food plantations and with land-related issues. These issues prompted the research in the direction of using waste materials to be converted into biodiesel.

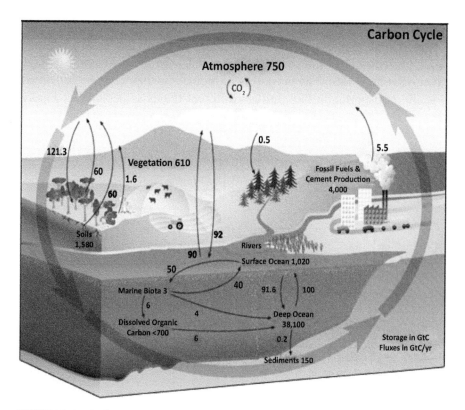

FIGURE 8.7 Carbon cycle https://articles.extension.org/pages/27533/biofuels-and-greenhouse-gas-reductions-creative commons.

There appears to be a large potential to explore biodiesel like fuel from the waste cooking oil available from large restaurants, waste tires, waste tallow, and plastic oil. These are now acting as feedstock for preparation of biodiesel.

Biofuels are classified into different categories as the first generation, second generation and third-generation biofuels, which is based on the feed stocks and their respective conversion technology adapted for their production: (i) First-generation biofuels are produced from bioenergy crops cultivation (i.e., food-crop feedstock), sugar, vegetable oil, starch are main source to make them. (ii) Second-generation biofuels come from non-food feedstock which is bio-based product. This is lignocellulosic biomass such as forestry and agricultural-based. They consist of residues from agricultural and food processing (discarded biomass). Third-generation biofuel is generated from aquatic cultivated feedstock (i.e., algae). The huge improvement of biofuel gives the issues of production sustainability and food vs. fuel usage, availability of land, environmental hazardous impacts, indirect land-use change (ILUC) effects. The work on algae biodiesel is in the development stage (Adeniyi et al., 2018). The versatility of biomass is that it can be converted to useful transportation fuel through different routes as depicted in Figure 8.8.

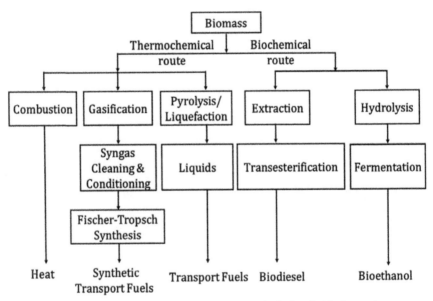

FIGURE 8.8   Various routes of conversion of biomass in fuels suitable for engines.
Source: Reprinted from Sikarwar et al., 2017. Open access.

8.10.1   *VEGETABLE OILS AND BIODIESEL*

Among the biofuels, seed-based or vegetable oils have gained a lot of interest among the research community as even in their raw form (untreated), the physico-combustion properties are close to petroleum-derived diesel fuel. Thus, the seed-based oils are good replacement to diesel in compression-ignition engines (CI engines). Vegetable oils, with their physical and combustion properties close to petro-diesel fuel, can stand as an immediate and viable substitute candidate for petro-diesel fuel as diesel fuel is meeting about 80% of the fuel requirement of developing countries. Vegetable oils are produced from plant seeds, also called seed oils or energy plant oils. It is not now new to use vegetable oils in engines, the use dates back to 1917 when the great inventor of Diesel engine Sir Rudolph Diesel demonstrated his engine with peanut oil at the French exposition.

Vegetable oils are plant or animal origins which are insoluble in water, which consists mostly of glyceryl esters of fatty acids, or triglycerides. Structurally, a triglyceride is the condensed output form of one glycerol molecule with three fatty acids molecule to give three water and one triglyceride molecules. However, they have poor atomization of fuel, ring stickings, fuel injector blockage, and contamination of lubricating oil, which due its high viscosity which makes them usable as blends with diesel fuels in up to 50%. Vegetable oils tend to emit higher emissions of CO, PM, and HC as relative to pure diesel fuels when used in blends.

Studies show that it is possible for usage of vegetable oils in combination with alcohols in diesel engines with satisfactory performance. It creates problems in diesel engine if untreated vegetable oil is used. Vegetable oils have considerable amount of nitrogen, which can affect emission pattern spectra. They consist of HC's of high aromaticity. Low volatility and high viscosity are the two properties, which leads to undesired problems in engine. Vegetable oil's viscosity introduces strong parameter in shaping of the fuel spray. Poor atomization, high spray jet emission, and large droplets are some of the effects caused due to high viscosity. This results in partial mixing of fuel to burn which is undesirable. Many vegetable oils have kinematic viscosity of 40–70 cSt at 30°C whereas for diesel is 3–5 cSt. From the literature as well as experimentation activities, investigators observed the problems faced during engine tests could be differentiated as durability and operational problems. The former relates to starting ability, performance ignition, and combustion. The latter relates to the deposit formation, ring sticking, carbonization of injector tip, and lube oil dilution/degradation. Researchers worked out on this problem either by suiting fuel to engine or vice versa.

The engine modifications included:

- Operation in dual-fuel mode;
- By high fuel injection pressures (FIP);
- Supercharging;
- Heated fuel lines;
- Concept of IDI engine;
- Blending with low viscous fuels;
- Hot combustion chamber concept—LHR engine; whereas the fuel modifications covered;
- Transesterification;

- Pyrolysis/cracking;
- Micro-emulsification.

Transesterification is acceptable over a wide area, one among it is by making it close to diesel oil by reducing the viscosity of vegetable oils. This process is a chemical combination of natural oil or fat with a methanol or ethanol along with an alkali hydroxide catalyst, it forms alkyl esters of component fatty acids and glycerol. Biodiesel is well known as alkyl ester of vegetable oil. The process of biodiesel production is illustrated in Figure 8.9. Host of vegetable oils can be grouped into two categories as edible type and non-edible type. Sunflower, peanut, ground nut, rapeseed (canola), soybean palm oil are few among the edible type while linseed oil, cottonseed, karanji, neem oil, honge oil, and Jatropha Curcas oil are examples for the non-edible type. As there is already a great demand for edible type, research should be focused on the development and commercialization of vegetable oil as a fuel from non-edible category (Figure 8.10).

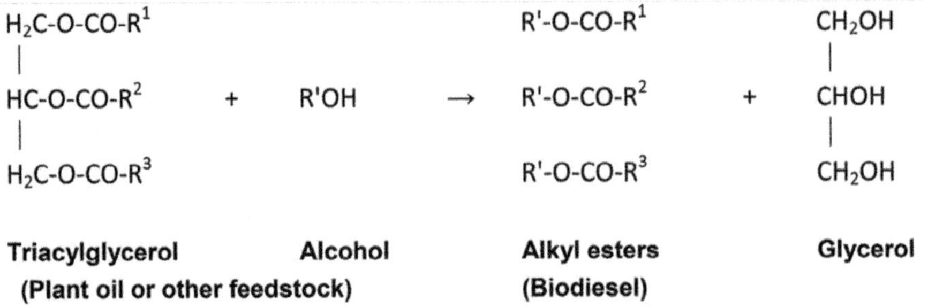

$$H_2C-O-CO-R^1 \qquad\qquad R'-O-CO-R^1 \qquad\qquad CH_2OH$$
$$|$$
$$HC-O-CO-R^2 \quad + \quad R'OH \quad \rightarrow \quad R'-O-CO-R^2 \quad + \quad CHOH$$
$$|$$
$$H_2C-O-CO-R^3 \qquad\qquad R'-O-CO-R^3 \qquad\qquad CH_2OH$$

| **Triacylglycerol** | **Alcohol** | **Alkyl esters** | **Glycerol** |
| (Plant oil or other feedstock) | | (Biodiesel) | |

FIGURE 8.9 The transesterification that produces biodiesel. R1, R2, and R3 represent different fatty acid chains.
Source: Knothe, G., & Razon, L. F., (2017). *Progress in Energy and Combustion Science, 58,* 36–59.

Even though there are many AFs, except biodiesel, all other fuels require some way or other blending with petroleum-derived fuels such as gasoline or diesel. In that sense, by single fuel the biodiesel could be a better option for automotive engines.

The idea of using vegetable oils as fuels was begin within the agriculture community to run agricultural machines. The most popular types of crops for extraction include sunflowers, soybeans, peanuts, Chinese tallow trees, and rapeseed.

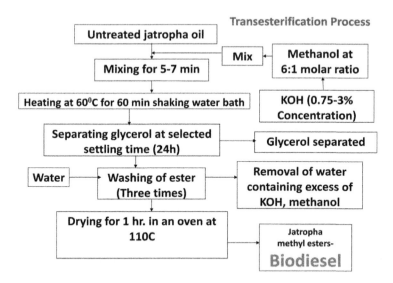

FIGURE 8.10 Preparation of biodiesel from seed oils.

When oil is reacted with methanol or ethanol, it gives esters which have much closer to the fuel. These vegetable oils (esterified) are generally termed as "biodiesel." Producers of biodiesel are targeting transit bus fleets to use a mixture ratio of 20% biodiesel with diesel fuel, known also as "B20." When used in the typical transit bus favorable emissive properties of this blend smoke, gaseous emissions and particulates are reduced (Figure 8.11).

Besides lower gaseous emissions, biodiesel contains no toxins (other than perhaps very small amounts of unreacted methanol in methyl ester biodiesels) and biodegrades quickly when spilled on the water or ground.

In general, the biofuels, apart from serving as resource to engines, offer multitude of benefits to the world:

- Sustainability;
- Regional development;
- Security of supply;
- Reduction of greenhouse gas emissions;
- Social structure and agriculture.

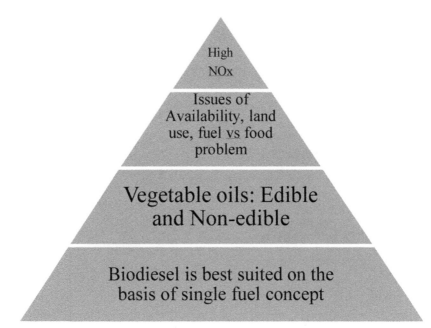

FIGURE 8.11   Pyramid representing issues related to the use of biodiesel.

As described in the above sections, there is a potential for use in CI engines but is raising issues related fuel versus food, availability, and its use in engines (without any modifications) leading to high NOx emissions. However, the biodiesel is more preferred than other AFs, as it is safe and doesn't require blending unless it is warranted. That's why this can be most sought after fuel on the basis of single fuel concept.

8.10.2   *BIODIESEL FUELLED VEHICLE EMISSION CHARACTERISTICS*

It is observed through many studies that CO_2-equivalent from the different combinations of alternative fuels, especially bio- fuels, in heavy duty vehicles, emissions of GHGs are reduced by 60% in relation with the diesel/gasoline fuelled engine emission levels (Beer et al., 2000). This observation is also in-line with the Commonwealth Scientific and Industrial Research Organization in 2000 studied on emissions of fuel-cycle.

The analysis of wheel and fuel-cycle includes exhausted emissions as a result of conversion, extraction, production, oil refinery, and transportation of the nonconventional oil. The other study by Hill et al. (2001) on biodiesel of soybean when compared with conventional diesel produces a declination of 41% in GHGs emissions for fuel combustion and production. It shows that if feedstock with less agricultural input is produced, huge benefits of biodiesel can be made like with less pesticides and artificial fertilizers used low value agricultural land and requires low-input energy in feedstock to biodiesel conversion stage. However, the argument by Searchinger et al. (2008) was that because when calculating net CO_2 emissions of land-use changes and prior studies did not take into account of upfront release of CO_2 isolated on land. The carbon would be liberated into atmosphere through fire or decomposition which is stored in soils and plants if the biofuel is produced through clearing of rainforests.

PM emissions reductions in biodiesels were contributed by absence of aromatic and sulfur compounds. The exhaust SOx emissions reduction is achieved with inherently low sulfur content in biodiesel. Thus minimizes the formation of H_2SO_4, H_2SO_3, and acid rain with availability of SOx combined with atmospheric water vapor is very less. Also, expected that the emissions of soot from biodiesel combustion will develop as sulfur in fuel is main responsibility of sulfates absorption on soot particles.

In 2001, International Agency for Research on Cancer, categorized the diesel emissions of different inorganic and organic compounds as possible carcinogen or mutagen to humans, in solid and gaseous phases wiz; polycyclic aromatic hydrocarbons (PAHs), monocyclic aromatic hydrocarbons (MAHs) and aldehydes.

Because of lack of aromatic compounds in biodiesel, the concentration of tail pipe aromatics will be less, in turn, will decline carcinogenic potency. Many of exhaust particles of diesel are categorized as ultrafine (PM1.0) or fine (PM2.5). comparing with medium-sized particles (PM10), these ultrafine and fine particles exert higher adverse effects on human health as smaller particles has more surface area for fixing chemicals that stay airborne larger and deeper penetration into many of the lung parts. A US Environmental Protection Agency produced an analysis of the estimated soybean biodiesel effects on the mean values of emissions from 43 heavier duty highway engine vehicles under steady-state or transient test cycles showed quite acceptable reductions in emissions of UHCs, CO, and PM except for NOx. The significant reductions are UHCs, CO, and PM emissions reduce the problems of minor irritations to the eyes and throat, CO poisoning.

Relative to automobile sector, the high scale biodiesel penetration into the fuel market would have been straight forward. Therefore, exhaust-out emissions of NOx, CO, UHCs, and PMs can be strongly reduced by the usage of biodiesel in which health risks showed by those species are limited indirectly. However, technical challenges in areas involving carbon deposition, low temperature operation, corrosion, and storage are hindered extraordinarily. With many degrees of success, these are addressed with the simple addition of additives of fuel.

From any oil plants, Biodiesel (Demirbas, 2007) can be produced by over 350 oil-bearing crops being carried out for the biodiesel making.

Based on price, chemical, and regional availability, the feedstock for biodiesel is selected. So only certain crops are used like cottonseed, palm, rapeseed, soybean, and sunflower seed. Due to the fact that vegetation suitability is greatly dependent on local climates, the biodiesel feedstock, however, will be varying among all the countries. Soybean is the predominant feedstock in the United States whereas rapeseed is the important feedstock in Europe.

Jatropha, coconut, and Palm are the usually seen feedstocks in India, the Philippines, and Malaysia, respectively. There are other bio-lipids besides vegetable oil that can be made use for the biodiesel production in addition with non-edible oils, animal fats, and WVO (Demirbas, 2007). The costs of edible oil plantations are usually higher than that of non-edible oil during cultivation because of greater need on irrigation system and soil nutrients (Gui et al., 2008). In relation with edible oils, the cost of non-edible oil plantations is less and translated into the lesser cost per unit oil. Palm oil is exception case where despite having the huge plantation cost per unit area; it has one of the minimal costs per unit oil because of its yielding superiority.

At present, biodiesel derived from palm oil was contrarily opposed and the producers were accused because after rainforest clearing only, the plantations were built and the feedstock crops has grown by them (Stewart and Webster, 2007). As a whole, this results in deforestation negating emissions and causes biodiversity by biodiesel usage. Predominantly, using biodiesel on its own as an AF does not cause deforestation, but the way of its production will do deforestation. With better agricultural control and practice, unnecessary deforestation and indiscriminate burning can be avoided. For oil plantation expansion of palm, the land needed can really be made without deforestation by using fallow land, pastures, and grassland.

Oxygen content: relative huge concentrations of oxygen in biodiesels that are developing soot oxidation process have been found as the main donator for the PM emissions related advantages. Complete combustion is promoted by huge oxygen content and thus contributed to the CO emission reductions. With the NOx emission formations which are rigidly depending on burned gas temperature and oxygen concentration, huge content of oxygen from biodiesel is the important cause for greater NOx emissions; namely, because of its huge concentrations, reaction of oxygen with nitrogen is easier during the combustion and thus causes an inclination in NOx emissions.

Compared to mineral diesel, the biodiesel's CO emission formations is because of small lesser carbon/hydrogen ratio of biodiesel fuels. Because of higher oxygen also, CO emission declines when biodiesel replaces the mineral diesel. It is also been proven that engine load have an important influence on CO emissions. With inclining engine speed, there is a sole conclusion that CO emission of biodiesel declines.

The HC emissions also typically decline when biodiesel is utilized instead of mineral diesel. This is partially based on properties with biodiesel source, especially on saturation level or the chain length of biodiesels. The proceedings in combustion and injection of biodiesels for mechanically controlled injection systems favoring lesser emissions of HC. However, there are inconsistent conclusions related to the engine load effects on emissions of HC.

When biodiesel is replaced by mineral diesel, the increase of NOx emissions determined by the experimental results. This increase is partly due to relatively high oxygen content in biodiesel. Furthermore, the huge content of unsaturated compounds and differences of injection behaviors and having greater biodiesel's cetane number also have a worthy influence on NOx emissions. NOx emissions may be declined using EGR of biodiesel, but those EGR rates should not be similar with the mineral diesel; for utilizing biodiesel, they have to be optimized. With increase in engine load, the NOx emission levels of biodiesel also will be inclined. PM emissions of biodiesel might be destroyed using EGR, although the calibrated levels are still much lesser than those obtained with mineral diesel. However, a note should be made on biodiesel that PM emissions might be abnormally inclines in case of mini temperature conditions.

By considering into account, the calculated maximum firing temperature and maximum heat release rate it is quite interesting that the NOx emission of biodiesel should be lesser than normally measured experimental values. Maximum NOx formation in cylinder temperature is analyzed by some combustion and indicated and the maximum heat release rate may not be as keen as the advancement in the start of injection timing, in a mechanically controlled injection system is influenced by the greater bulk modulus and compressibility of biodiesel. The advanced combustion and injection process lead to greater in-cylinder temperature at combustion process starting. Peaks will prolong earlier with favorable conditions for formation of NOx and consequently to the total NOx emissions.

Rather inconsistent results are revealed by the CO_2 emission measurements of biodiesel. For example, the CO_2 emission declines as a result of the low carbon to hydrocarbon ratio when biodiesel is replaced by mineral diesel actually indicated by some researchers and some others displayed that the CO_2 emission stays same or rises due to the huge effective biodiesel combustion. If whole life cycle of CO_2 is taken into consideration, then the CO_2 emission of biodiesel is less rather than mineral diesel.

Emissions were scarcely measured because the engine durability, reliability, and power are aimed by early testing of biodiesel and vegetable oils.

Rather than typical No.2 diesel, cetane rating of Soy methyl ester (straight biodiesel) is much greater, viscosity is high, and heating value is low and nearly contains 10% mass of oxygen. If fuel system is not sequenced, then maximum power is affected by lower heating. Conducting transient-mode engine test in the diesel engines pre-chamber, a declination in nitrogen oxides (NOx) emissions is showed by straight biodiesel, lesser formation of particulate matter mutagenicity, and particulate emissions are decreased.

When mixing of biodiesel fuel with diesel, somehow the emissions will change. A significant decrease in CO and HC emissions is predominant, a small rise or no change in NOx emissions, and particulate emissions are significantly decreased. In accordance with diesel fuel substitution percentage, toxins emissions would be decreased.

Rather than diesel fuel, biodiesel is largely liable to polymerization and oxidation since it has a problem with engine oil dilution. The reason for thick sludge formation is biodiesel presence in engine oil may be responsible factor for the oil becoming too thick to pump and is a major result. For minimization of dilution effects with biodiesel, remarkable formulations of engine oil are being carried out.

By using other sources and since biodiesel is gaining importance in recent times and the production of biodiesel from fresh feed stock such as edible and non-edible oils is raising issues of land availability and food-fuel crisis. To make use of other sources for production of biodiesel, researchers have explored waste cooking oil, Plastic oil, waste tire oil, and animal Tallow.

In some studies, they considered 2 types of biodiesel like type B and type A with (30% waste cooking oil methyl ester and 70% chicken tallow) and (20% canola oil methyl ester and 80% tallow) respectively. It was found that with increasing blend ratio, the performance (majorly brake power and torque) of both type fuels is reduced which can be considered to lesser energy biodiesel quantity. As expected, comparing with diesel fuel Specific fuel consumption for both biodiesels will rise. Few emissions of GHGs were observed to be lesser than diesel petroleum and a few were greater. As a whole, compared to Biodiesel B and diesel, lesser emissions across the board were produced by Biodiesel A (Figure 8.12).

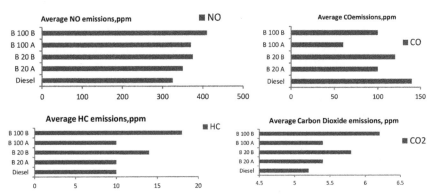

FIGURE 8.12 Comparison of emissions with different diesel and blends of diesel and biodiesel.

NOx emissions for diesel Biodiesel B, and Biodiesel are compared in Figure 8.12. Nitric oxide emissions of Biodiesel A displayed declining trend with rising blend ratio and Biodiesel B emissions also rises. NOx emissions can rise or fall depending on various items like test procedure usage, engine, and biodiesel type. In comparison to diesel, the US EPA reports a 10% rise in NOx emissions for B100.

Up to 51% CO emissions fall of biodiesel have been displayed by Studies and this can decline or raise CO_2 emissions, with the percentage change from −7% to +7% range based on the biodiesel types. The literature is clearly agreed in terms of CO_2 and CO emissions for Biodiesel A, however, a greater rise in CO_2 emissions comparing with literature findings was observed for Biodiesel B. As CO_2 emissions are unregulated, this variation can be called as not so significant (Figures 8.12).

The use of biodiesel is tried out in both DI and IDI engines and experiments are performed on a multi-cylinder IDI engine with palm oiled methyl esters that are preheated. They concluded that through etherification, one could not bring the viscosity of ester equivalent to diesel fuel and hence diesel-like performance cannot be obtained. They called the combustion of preheated blends as "hypergolic combustion."

Supercharging effect on the DI engine performance confirmed that super-charging is beneficial in reducing smoke from the trans-esterified (Amba Prasad Rao and Rammohan) cottonseed oil. Palm has been extensively exploited as a feed stock for biodiesel in countries like Malaysia. While extracting oil from palm seeds, the waste material is leftover called Palm Stearine and has been used to prepare biodiesel and employed in engines (Babu et al., 2017). Even though B20 has been established as the better option yielding lower emissions without penalty on the performance of engine. Its performance is further improved and NOx emissions have been reduced with the oxygenates addition like diethyl ether (DEE) and ethanol (USV Prasad et al., 2011).

Jun Cong Ge et al. (2015) performed experiments on c engine adopting pilot injection timing and EGR techniques in an attempt to reduce NOx emissions without engine efficiency effect, with canola oil biodiesel-diesel (BD) blend of B20 (Figures 8.13 and 8.14).

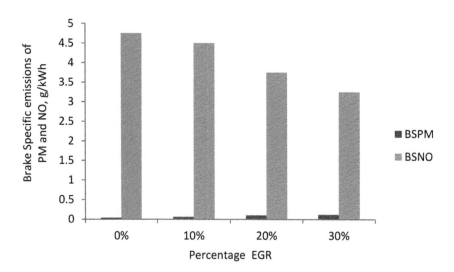

FIGURE 8.13 Trade-off between BSPM and BSNO emissions at pilot injection timing of 20 for various EGR levels.
Source: www.mdpi.com/journal/energies [Open Access-Creative Commons].

For their experiments canola, oil-based biodiesel was prepared. With the widely accepted BD blend; B20 (volume mixture of 80% diesel fuel and 20% canola oil) is allowed using EGR rate and pilot injection timing at 2000 rpm engine speed with B20 fuelled. With the advancement of injection timing, brake specific fuel consumption (BSFC), the combustion pressure, and peak combustion pressure (P_{max}) are medially differentiated. Clearly, CO and PM emissions reduced at BTDC 20° compared to BTDC 5°, but NOx emissions are lightly raised. With retarded injection timing and higher EGR, levels the NOx emissions decreased significantly. Pilot injection timing of 10 BTDC and 30% EGR levels, drastic reduction in NOx emissions can be seen from Figure 8.12. Similar results were observed with adoption of retarded injection timings with biodiesel prepared from a mixture of crude rice bran oil methyl ester (CRBME) (Saravanan et al., 2014) and with palm oil-based biodiesel also (Teoh et al., 2015).

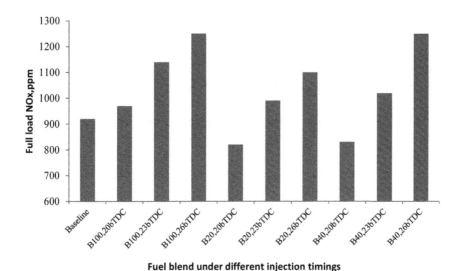

FIGURE 8.14 Comparison of NOx emissions for various blends of fuel with various timings of injection under full load condition.

Puneet et al. (2018) investigated the use of diesel-like fuel produced from waste tires as alternatives to diesel. They observed parameters affecting production method and the influence of reactor operating parameters (such as reactor temperature and catalyst type) on oil yield. These have a major effect on the performance and emission characteristics of diesel engines when using tire-derived fuels. BSFC and BTE values with oil produced from waste tire oil were observed to be inferior to neat diesel fuel operation. They also noticed mixed results for NOx, CO, and CO_2 emissions with waste tire-derived fuels. It also reported an increase in HC emission and attributed to lower cetane number and higher density, whereas some studies have inferred reduced HC emissions. It has been found that the higher aromatic content in such fuels can lead to increased particulate matter emissions.

A recent experimental study by Kiran and Anand (2018) observed the variation in properties by altering the palm biodiesel composition and on a 20%, 15%, and 10% volume basis; its fuel properties effects are studied. Palm biodiesel was mixed with different neat methyl esters. Out of the 34 biodiesel blend samples prepared, the properties like heating value, kinematic viscosity, surface tension, and density were determined. They observed decrease in viscosity with decrease with the degree of unsaturation and methyl oleate concentration declination. The minimum viscosity was conducted by a mixture of 20% methyl octanoate + 80% palm. The heating value of the sample was observed to be effected by the methyl esters' changes in the chain lengths and the mean number of double bonds and the mean number of single bonds. A 20% methyl + 80% palm octanoate mix exhibited the least surface tension of the fuel blends.

Many authors explored the use of different seed oils which contain sufficient oil in them. One such experimental study was performed by using biodiesel produced from false flax seeds along with butanol and petro-diesel in two different proportions (Mustafa, 2016). He evaluated the properties of 70% diesel–20% biodiesel–10% butanol and 60% diesel–20% biodiesel–20% butanol by volume and BD 20% biodiesel–80%diesel and 100% biodiesel by volume. He observed the engine performance and emission characteristics of engine with these blends. Experiments showed that 10% alcohol addition to diesel and biodiesel fuels caused a decrease in torque value up to 8.57%. When butanol ratio rose to 20%, torque value decreased to an average of 12.7% and power values decreased to an average of 13.57%. To an average of 10.63% and 12.80% with 10% and 20% butanol addition, respectively. An important observation made from his study that torque was reduced with an increase in butanol addition and also the specific fuel consumption increased. Also, he noticed that NO emissions reduced with butanol addition compared to neat biodiesel operation (B100). It was attributed to the reason that oxygen content improved the oxidation during the combustion process. In addition to these emissions, he observed variations in CO and CO_2 emissions with variations in blends.

For the promotion of higher biodiesel blends and diesel fuel, Prasad (2012) has adopted two blends B20 and B40 with biodiesel prepared from jatropha-a non-edible oil. He made use of techniques such as retarded injection timing, the addition of cetane improvers (DEE) and ethanol as oxygenated additive in different proportions. He observed lowest NOx emissions with both blends with retarded injection timings (the retarded injection of chosen engine is 23BTDC), the lowest values observed for 20BTDC, as shown in Figure 8.15.

However, with addition of cetane improver DEE, the NOx levels performed better than retarded injection timing. A comparison between the two techniques is depicted in Figure 8.15.

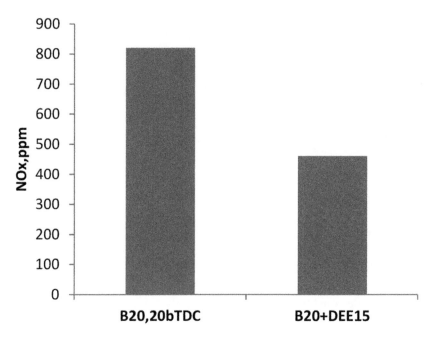

FIGURE 8.15 Comparison of NOx levels for B20 with DEE and retarded timing under full load condition.

In attempt to promote biodiesel as large scale fuel, it is required to establish its impact on environment. He evaluated ecological efficiency proposed by Christian Rodriguez Coronado et al. (2010); he compared the ecological efficiencies from different combinations tested. It is observed that B20+DEE10 outperformed when compared to other combinations tested. It is illustrated in Figure 8.16.

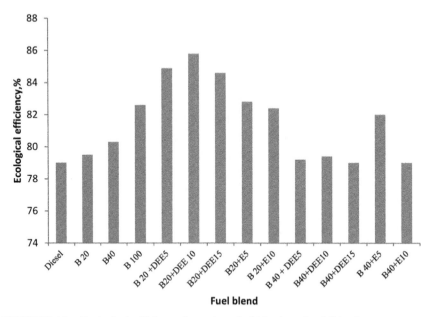

FIGURE 8.16   Ecological efficiency for various fuel blends under full load.

8.11   BIOGAS

Among the renewable gaseous fuels, biogas has occupied a prominent place. It has been long known gas obtained by anaerobic digestion of organic matter (in the absence of air). The main constituents of biogas are methane (CH_4), carbon dioxide (CO_2), hydrogen sulfide (H_2S), and moisture. Among the combustible constituent, methane is the only component that helps in the release of energy with the combustion of biogas. The more the proportion of methane in biogas, the better is the quality of biogas used for energy purposes. On the other hand, the presence of H_2S makes the gas smelly) foul and unbearable smell). The presence of carbon dioxide makes the biogas poor and it is required to be removed from it. Moreover, the presence of CO_2 absorbs the heat released by the main fuel like gasoline or diesel and CO_2 dissociates by absorbing the heat released. It is essential to remove CO_2 from the biogas. By the process of scrubbing, the carbon dioxide is removed and the enrichment of biogas is done. In few of the applications, the enriched gas is stored in compressed form in cryogenic containers. In India, biogas is produced with the anaerobic digestion of any degradable organically material mostly animal dung or kitchen waste mixed in equal proportions with water and mixture is left in a closed container for about 30–45 days (called gestation period). There are plants that produce gas using paper waste, poultry excreta, and degradable organically material. Around this period, a foul smell starts emanating indicating production of biogas. The gas thus produced is taken for end-use application either to run engine in dual-fuel mode of operation. Such dual-fuel engines reduce the burden on petroleum fuels commonly used in agricultural pump sets and lift water from the ground (Figure 8.17).

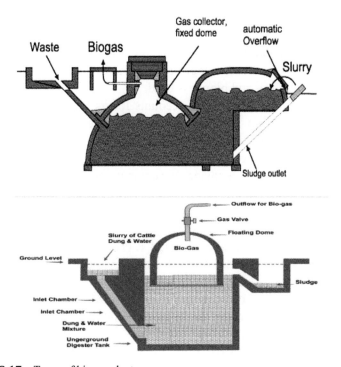

FIGURE 8.17 Types of biogas plants.

Farmers who were having their own land were encouraged to have their biogas plant since the waste or residue from agriculture can be used for the purpose of biogas production. Moreover, the leftover after the production biogas called slurry can be used as a manure which is rich in nitrogen prosperous and potassium required for organic farming (N-P-K). Instead of feeding directly the organic material to the crops, the slurry is much more useful in meeting the requirement of manure. The details of biogas plant are shown in Figure 8.17.

The biogas is produced in a plant called digester is available in fixed dome or floating drum type. Each has got its own advantages and limitations. The floating drum type is permissible for repairs; however, the drum undergoes corrosion with continuous exposure to water and other constituents. The top of digester is exposed to open air, wherein once the top surface gets heated; thermophyllic reaction takes place inside the digester. The production of biogas is faster in summer.

The auto-ignition temperature of Biogas is higher and hence it cannot be used as a neat fuel and is required to be blended with diesel for use in a CI engine in dual-fuel engines. The dual-fuel engine is a modified diesel engine in which usually a gaseous fuel called the primary fuel is inducted with air into the engine cylinder. For this purpose, biogas is extensively in SI engine. In few instances to get an advantage of high CR, the CI engines are modified into spark-ignited form and biogas is used.

Also, its gaseous nature which accounts for its low volumetric density implies that apart from the basic modification needed to accommodate the fuel, the engine might need further alterations to get the best from this relatively low cost and readily available fuel. Various modes of enhancing performance particularly methane enrichment, pre-chamber combustion, alteration of ignition parameters, increasing CR and addition of hydrogen to improve performance and emissions were drawn from previous works to validate its efficiency as a viable substitute fuel in SI engines.

Studies have been done with varying CR (18, 17.5, 17, and 16) in a dual fuel diesel engine run on raw biogas. A single-cylinder diesel engine with a power output of 3.5 kW in which direct injection (DI) is used and is water-cooled with a flexibility of variable CR is modified into a duel fuel engine which runs on biogas and the modification is done by attaching a venturi gas mixer at the inlet manifold. By fixing the standard injection timing at 23 prior to TDC, experiments have been carried out. At CRs of 16, 17, 17.5 and 18 the break thermal efficiencies of the diesel engine on dual fuel mode are found to be 16.42%, 17.07%, 18.25%, and 20.04%, respectively at 100% load and at the same load conditions the diesel engine operating in diesel mode showed an efficiency of 27.76% at a CR of 17.5. At the CRs of 16, 17, 17.5, and 18, the maximum replacement of the diesel fuel is found to be 72%, 74%, 76.1% and 79.46% at 100% load.

On increasing the CRs from 16 to 18 in the duel fuel mode, the average reduction in CO and hydrocarbon emissions is observed to be 26.22% and 41.97% but the NO_x emissions and CO_2 emissions increased by 66.65% and 27.18% respectively. It is observed that from all the tests carried out, CO_2

and hydrocarbon emissions are noticed to be more in duel fuel mode operation than that in diesel mode which can be attributed to the reduction in VE of the former. The experimental evidence suggests operating the dual fuel diesel engine at high CRs.

An experimental study was carried out on a constant speed SI engine that is powered by biogas in which the CO_2 concentration reduction was done. Its effect on the performance, emissions, and combustion process were studied. The reduction in CO_2 concentration was achieved by using a lime water scrubber which allowed a reduction from 41% of CO_2 in biogas to 30% and 20%. All the equivalence ratios from rich mixture to lean mixture were covered in the tests carried out at a constant speed of 1500 rpm by incorporating a masked valve which boosts the swirl process by applying a CR of 13:1. With a reduction in the CO_2, level there was a significant improvement in the performance and reduction in emissions of hydrocarbons (HC) particularly with lean mixtures. An extension in the lean mixture range for combustion process was also attained. Heat release rates indicated enhanced combustion rates, which are mainly responsible for the improvement in thermal efficiency. When the CO_2 Concentration was reduced by 10%, the hydrocarbon levels found to be reduced and there is no significant rise in the NO concentration levels. This has to be carried out by spark retardation of about 5.

A SI engine with biogas is run with removal of CO_2 from biogas is obtained when the engine is operated with lean fuel-air mixtures in the range of equivalence ratios between (0.8 and 0.95). An equivalence ratio of around 0.95 and full throttle, the engine performance and HC and CO_2 emissions have been lowered with reduced CO_2 in biogas. Moreover, faster combustion and reduction in the cycle by cycle variations in combustion are also obtained (Porpatham et al., 2008).

Though the hydrogen sulfide (H_2S) is present in biogas, even though it doesn't contribute to the engine's performance can affect the engine overhaul life since its corrosive effects can attack valves and bearing materials. It is required to avoid use corrosive materials and make use of high TBN engine oil can prevent the corrosive effects (Nirendra et al., 2006).

8.12 HYDROGEN

Among other attractive AFs, hydrogen occupied the highest position as it a substance with no carbon and hence no particulate matter and no carbon compounds and upon combustion. As an element, it is lightest among all elements. It's a good energy carrier as it has very high energy content (120 MJ/kg) it has many characteristics that make it the "ultimate" AF to fossil energy fuels. Hydrogen can be combusted directly in ICEs or it can be used in fuel cells to produce electricity with high-efficiency (30–50% over the typical load range). These days, use of hydrogen is gaining much importance through fuel cell mode. When hydrogen is oxidized in fuel cells, the only emission is water vapor. When hydrogen is combusted in ICEs, water vapor is again the major emission, though some oxides of nitrogen may be formed if the combustion temperatures are high enough. Thus, it can be said when hydrogen is produced from electrolysis of water, the source of production of hydrogen is the product of combustion. Therefore, the use of hydrogen as a transportation vehicle fuel would result in few or no emissions that would contribute to ozone formation.

8.12.1   HYDROGEN PRODUCTION

There are abundant sources ways to produce hydrogen as an energy carrier.

One of the most common methods of production of hydrogen is from coal. In this method of steam reformation from the underground coal gasification gas consisting of hydrogen, carbon monoxide and methane can be produced with the following reactions.

$$C(s) + H_2O + heat \rightarrow CO + H_2 \tag{1}$$

$$CO + H_2O \rightarrow CO_2 + H_2 + heat\uparrow \tag{2}$$

The first reaction is endothermic and the CO is converted to CO_2 and H_2 through the water gas shift reaction.

Other methods of production of hydrogen are from splitting of water by water electrolysis, photo-electrolysis, photo-biological production, and high-temperature water decomposition. The methods differ in availability of source and application as well. The best method to produce hydrogen is the one which has simplest process, easily to get the main sources, low cost and environmentally safe.

Even though the use of hydrogen through combustion mode in IC engines is well understood but now--days, the hydrogen use is explored in fuel cell mode. Fuel cell converts directly the chemical energy of fuel into electricity and is the most preferred method in electric vehicle or fuel cell vehicles through the use of proton-exchange-membrane (PEM). Though Hydrogen is regarded as renewable but the source of obtaining hydrogen may not be completely renewable.

8.12.2 HYDROGEN AS A FUEL IN INTERNAL COMBUSTION ENGINES

Since auto-ignition temperature of H_2 is high, it is more suitable spark-ignition engines. Hydrogen can be used in SI engine by three methods;

1. By manifold induction;
2. By direct introduction of hydrogen into the cylinder; and
3. By supplementing gasoline.

Emission production Hydrogen can be produced form any kind of energy source and it is combusted without emitting carbon dioxide or soot, it is considered as an ideal AF to conventional hydrocarbon fuels. The only potential emissions are the NOx as pollutants from hydrogen combustion; hence, it becomes crucial to minimize the (NOx) emissions from the combustion of hydrogen. The best reliable way to use H_2 as a fuel in vehicles is by reacting it with O_2 in a fuel cell which generates electric energy that can be used to propel a hybrid-electric vehicle. Because of the high production costs involved, lack of availability of hydrogen reserves and larger demand of H_2 to be used as fuel in vehicles, it is quite improbable that H_2 will be a cost-effective fuel in the near future. In order to make the hydrogen fuel economically viable, new production and storage techniques are required. The details of hydrogen use and production are figuratively explained in Fayaz et al. (2012).

Hydrogen may have a very high weight-related energy density (approximately 120 MJ/kg and thus almost three times as high as that of gasoline), but its volume-related energy density is very low on account of the low specific density. When it comes to storage, this means that the hydrogen has to be compressed either under pressure (at 350–700 bar) or by liquefaction (cryogenic storage at- 253°C) in order to acceptable tank volume.

Though H_2 has the ability to be a clean source of energy, its properties makes the use of H_2 in vehicles for propulsion a question. It is because no carbon atoms are present in its molecular structure the unique ability of hydrogen to be a cleanest source of energy is possible. As there is no carbon atom, the only pollutant from hydrogen upon combustion is NO_x (without considering the HC, CO, and PM emissions that were caused due to the use of lubricating oil). The combustion of hydrogen causes no direct emissions of CO_2 whereas the emission of CO_2 could be indirect in nature based on the energy source used to produce hydrogen. In the long run, when the CO_2 emission norms were very stringent, the use of hydrogen as a fuel that is produce from renewable energy sources could be a probable solution.

The magnitude of the NOx emissions that was noticed from the existing prototype hydrogen vehicles was found to be similar to that of the gasoline powered vehicles.

When hydrogen is used in engine in dual-fuel mode with gasoline in SI engines, the only significant emission is water vapor. The effect of blending of hydrogen with gasoline is analyzed in this current paper. Both HC and NOx concentration decrease with the increase in hydrogen fractions. The results were analyzed by collecting data on different crank angles and at 5%, 10%, 15%, and 20% of hydrogen introduction into combustion chamber along with gasoline as base fuel. Though CO should not be seen but small traces are seen due to burning of lubricating oils due to high temperature prevalent during combustion.

The use of ECU was adapted to the existing engine unit and converted the engine to operate on dual-fuel mode of operation. Since hydrogen is a high energy carrier and was used along with diesel fuel in water cooled naturally aspirated engine with CR of 14.7 and was made to run at 1500 rpm. The hydrogen was used in pilot spraying mode in different propotions: 0, 15, 30, and 45%. An appropriate software and hardware along with ECU are designed and developed to operate a diesel engine with diesel + hydrogen dual fuel mode. Modest improvement in thermal efficiency was observed at 45% addition of hydrogen. As far as emissions are concerned, except NOx emissions, THC, CO, and UHCs, were significantly reduced. They attributed that high flame temperatures with higher percentage fraction of hydrogen was responsible for formation of high NOx emissions. They finally concluded that existing diesel engines with mechanical fuel system can be easily converted to common-rail fuel system with a properly designed ECU and thus engines can be made to run on dual-fuel mode easily and it increases the flexibility in changeover of fuel.

8.13   HYTHANE (HCNG)

HCNG (hydrogen enriched compressed natural gas) has been studied in great detail in the past 20 years and is proven to be a promising alternative energy to traditional fuels used in ICEs. Results of these studies generally show that HCNG engines have advantages on emissions and fuel economy over traditional CNG engines. Hydrogen addition to methane can reduce HC, CO, and CO_2 emissions due to its high hydrogen-to carbon ratio. Hydrogen enrichment enables operation with leaner mixtures and retarded ignition timing, so significant reductions in NOx emissions can also be obtained. Furthermore, hydrogen addition can significantly strengthen the engine's lean burn capability, which is beneficial for higher thermal efficiencies. Through optimizing ignition timing and EGR ratio, high-efficiency and low emission spark-ignition engine can be realized fuelled with natural gas-hydrogen blends.

The influence of the spark timing and CR on a H_2 enriched CNG engine was studied. By altering the spark-ignition timing and by varying CRs, the experiment was carried out and the data was collected. The engine was kept running at a constant speed of 1200 rpm with a constant excess air ratio of 1.6 and a constant manifold absolute pressure of 50 kPa. From the obtained outcomes of the experiment, it was observed that higher CRs leading to higher indicated thermal efficiency. The rise in CR caused greater brake torque and lesser break specific fuel consumption. The enhancement in the above-mentioned properties was reduced between higher CRs. With a high CR, the peak pressure was found to be higher. The rate of heat release was improved and the coefficient of variation of indicated mean effective pressure seems to be lowered at higher CRs.

8.14 SYN-GAS

Upon partial burning of biomass, produce gas is obtained which mainly consists of CO and H_2 and thus produces combustibles sufficient enough to use in engines. In some cases, a large amount of CO_2 would be seen.

8.14.1 SIMULATED PRODUCER GAS OR SYNGAS

In the process of alleviating the PM and NO_x emissions, researchers unriddled the feasibility of CO_2 usage in simulation of EGR. Zhao et al. used CO_2 in simulating the effects of EGR in one of their works.

The researchers found that the soot and NO_x emissions were reduced and EGR cooling system cost was brought down with the incorporation of CO_2. In a much later study, four different mass fractions of CO_2 such as 17.82, 12.33, 7.18, and 3.13% were used in diesel fuels by Jin et al. (2018) in order to study the characteristics of jet flame with the influence of CO_2

concentration. They inferred that the flame structure was greatly affected by the atomization and dilution of dissolved carbon dioxide component in the fuel. At constant injected pressure, the flame length at low temperature was increased whereas the total flame length and flame length at high temperature was decreased, with increase in the concentration of CO_2. At constant injected pressure, increase in concentration of CO_2, first increased and then dropped the mean temperature of flame.

Qu et al. investigated the influence of CO_2 in-cylinder injection on PCCI combustion. They found that for CO_2 in-cylinder injection, with little compromise in thermal efficiency the NO_x emissions were mitigated to a greater extent either by increasing the quantity of cyclic injection or by assisting injection timings of CO_2. Besides increase in PM emission, the intake pipe injection had same effect to EGR, such as depletion in NO_x emission unlike CO_2 in-cylinder injection which had a small effect on smoke. Furthermore, the effect of CO_2 as diluent was studied by introducing CO_2 in the ratio of 6, 4, and 2% in the intake manifold of diesel engine. For moderate injection, the NO_x emissions are inversely proportional to the CO_2 concentration. They inferred that as the ratio of molar heat capacity of CO_2 to that of atmospheric air was greater than 1, the combustion enthalpy was absorbed by CO_2 which in turn reduced the in-cylinder temperature and formation rate of NO_x. They found that along with NO_x emissions reduction many other parameters got decayed. The BSFC, power, BMEP, and energy torque were decayed by 3.3, 5.5, 6, 5.9%, respectively, emissions of smoke was increased by 60%, CO emissions were increased by 8.5 times from its origin level and NO_x emissions were mitigated by 50%, with admission of 6% CO_2.

Using biogas as primary fuel in a two-cylinder, stationary dual-fuel and four stroke DI power generation engine, the influence of pilot quality and mixing system on performance was calculated. They inferred that for all loads biodiesel and biogas as power sources can totally replace the diesel. Usage of super charged mixing system along with biodiesel as pilot fuel increases the pilot fuel substitution and thermal efficiency and simultaneously mitigates the CO and CH_4 emissions.

8.14.2   SYN-GAS DUAL-FUELLING

As the syngas production is available all over the world with the biomass feedstock, researchers have planned to run either Otto or diesel engine with syn gas. By varying the type of biomass and operating conditions, the concentration of syngas produced can be varied. Imitated syngas operations range versus pure diesel fuel (Table 8.2).

TABLE 8.2 Producer Gas Constituents

Constituents of Producer Gas from Biomass Gasification						
Gas Components	CO	H_2	CH_4	CO_2	N_2	Water Vapor
Percentage in Volume (%)	18–22	15–19	1–5	7–12	45–55	4

Table 8.2 lists the percentage range of the components of producer gas produced, on the volumetric basis, from biomass downdraft gasifier. This chapter deals with replacement of syngas in IC engines. Syngas was used in duel fuelling mode in CI engine as it can operate at high self-ignition temperature. In this process, the primary fuel was syngas and pilot fuel was diesel, these two were mixed with air, carbureted at engine intake and combusted in cylinder.

A small amount of diesel is required to combust the syngas-air mixture similar to the normal diesel operation, at the end of compression stage. With the combination of two fuels the stability of ignition and energy sustainability are obtained. The operation system can be easily switched either for syngas dual fuelling or normal diesel operation. Besides showing better performance with lean mixture re, syngas dual fuelling has some drawbacks, with lean mixture could lead to better performance. The syngas dual fuelling in CI engines has some limitations, where there is requirement for ignition source in order to retain the power output of the engine. In syngas, dual fuelling system a regular inspection for the nozzles is required for every 500 hrs in order to rectify the problem of overheating of injection jet at low diesel flow rate of the normal flow. Figure 8.18 shows theoretical diagram of dual fuelling system for CI engine.

Syngas duel fuelling operation with SI engine results in instable combustion at high load condition due to fluctuations in syngas. Apart from this high power, deterioration was expected due to low energy density of syngas compared to gasoline. This system operates efficiently with CI engine, but the performance and emissions were to be determined. The syngas obtained from the coffee husks gasification were used in the diesel engine operation.

With the application of dual concept 31% diesel restoration was possible. The CO and NO_x exhaust emission concentrations were observed by Sridhar et al. for dual fuelling a CI engine. In duel fuelling operating engine, besides lower NO_x emissions, CO emission levels raised due to the inefficiencies in ignition.

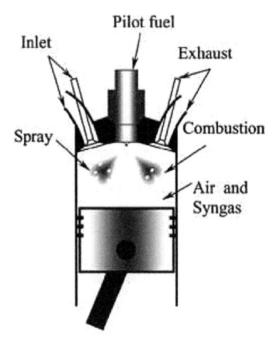

FIGURE 8.18   Conceptual diagram of dual fuel CI engine.

An experimental work to study the effect of variation in load and H_2: CO ratio on the performance of syngas dual fuelling in a diesel engine.

The performance, emissions, and combustion characteristics of a single cylinder, water cooled DI diesel engine were investigated for the different syngas compositions such as were 100:0%, 75:25% and 50:50%. At low engine load conditions, there was small change on the brake thermal efficiency varying H_2: CO ratio. However, beyond medium load conditions for dual fuelling with increase in the hydrogen content in syngas the brake thermal efficiency increases. In this system, the high VE, cylinder pressure, and rate of diesel substitution were observed at 80% load and 100% hydrogen syngas. With the complete utilization of H_2 the combustion temperatures were raised which in turn increased the NO_x emissions and mitigated CO and HC emission compared to the one from other compositions at all load conditions. They found that the engine load and emissions levels are inversely proportional.

The previous studies were discussed about the possibility of syngas as an AF in both Otto and diesel engines. They obtained the results by varying the compositions syngas at constant engine speed.

8.14.3   PERFORMANCE OF SYNGAS DUAL FUELLING IN A CI ENGINE

This section discusses about the influence of syngas composition on the performance characteristics of syngas dual fueling through developing a gaseous fuel supply system in a naturally aspirated CI diesel engine. Brake thermal efficiency, engine brake power, fuel equivalence ratio, BSFC, VE, and exhaust temperature were the different parameters evaluated, to obtain engine performance.

In syngas duel fuelling system, it was found due to dilution effect of nitrogen in air-syngas mixture the NO_x emissions got reduced and they also observed that the reduction concentration of O_2 gave the same result. Therefore, in this study, syngas dual fuelling with composition A results in lowering the concentration of NO_x, because of its lower heating valve which lead to low combustion temperature. Similarly, due to composition C the NOx levels reached to maximum, due to high combustion temperature. It can be inferred that CO and UHC concentrations greatly depend upon the hydrogen content in syngas composition. Besides mitigating the CO and UHC concentrations, syngas dual fuelling with Composition C, produced lowest brake specific fuel with highest brake power output and with the maximum diesel replacement ratio.

8.15   ETHERS AS FUELS-DIMETHYL ETHER AND DIETHYL ETHER-DME & DEE

Dimethyl ether, also called DME, is one among the newer candidates for the replacement of present conventional fuels in transport, for the energy generation, and for domestic cooking and for heating markets. It has been used for many years as an aerosol propellant. DME in the atmosphere decomposes to CO_2 and water in the time frame of a day or as and, therefore, does not react with ozone in the upper atmosphere as fluorocarbons previously used for aerosol propellants did. It is also attractive in this respect due to its vapor pressure, very low toxicity, and high activity as a solvent. In Japan since 1995, a large research and development work has been performed. It is reported in the DME Handbook (JDF, 2007), published by the Japan DME

Forum. The reader in need of more detailed information than that which can be presented here. DME can be treated as a multisource-multiuse chemical, of which fuel is a major application. Figure 8.19 depicts the raw materials applicable to DME production and the many uses.

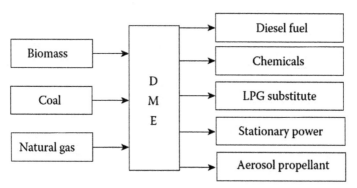

FIGURE 8.19   The flexibility of DME as a multisource-multipurpose fuel/chemical.

Among the ethers, DME is the simplest and among organic compounds, consist of an oxygen molecule bonded to two organic radicals. DME is a colorless gas, with a slight odor at room physical conditions. It is important to keep DME in closed, low pressure containers for normal use and distribution. DME, with a declined carbon-hydrogen ratio having the chemical formula CH_3-O-CH_3. The storage, handling, and transport necessity for DME are same as LPG due to their similar properties. At atmospheric conditions, DME is an invisible gas. By increasing the pressure to 0.5 MPa at standard temperature, it passes from the vapor phase to the liquid phase. DME gas is heavier than dry air, and the density of DME in liquid phase is 660 kg/m³ at normal temperature and normal pressure.

8.15.1   DME/METHANOL PRODUCTION

The production of DME and methanol are closely related. DME is formed through the methanol by a dehydration process, is the direct method whereas when it is produced through methanol from syngas is the indirect method. These processes can occur in separate reactors or in a common reactor. Methanol is formed first and the dehydration reaction occurs thereafter in the same reactor.

8.15.2   DME AS INTERNAL COMBUSTION ENGINE (ICE) FUEL

In order to meet the strict particulate emission standards that had been proposed for city buses in the United States for 1991, the first use of DME as an ignition improver was observed (Brook et al., 1984; Murayama et al., 1992) (Figure 8.20).

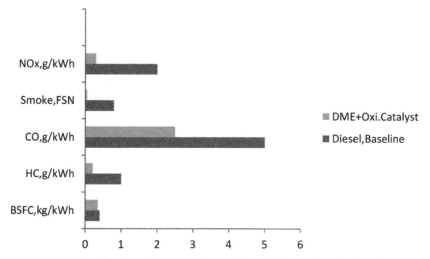

FIGURE 8.20   Road load test data comparing engine emissions using diesel and neat DME with oxidation catalyst.
Source: Adapted from: Jurchiş, B., Burnete, N., Iclodean, C., & Burnete, N. V., (2017). Study of emissions for a compression-ignition engine fueled with a mix of DME and diesel IOP Conf. *Series: Materials Science and Engineering, 252,* 012065. doi: 10.1088/1757-899X/252/1/012065.

The road load tests conducted by Jurchiş et al. (2017) revealed that the effect of DME with addition of oxidation catalyst is much superior to baseline tests conducted with diesel fuel under similar conditions of operation. The smoke, NOx, and HC emission levels much lower than compared to neat diesel fuel operation. This establishes the fact that DME is a good substitute for diesel fuel.

8.15.3   DME IMPACT ON POLLUTANT EMISSIONS

Troy et al. (2006) presented the pros of DME compared to conventional diesel; include reduced emissions of NOx, carbon monoxide and hydrocarbons that

DME doesn't gives soot. There are some papers showing that NOx emissions increase if DME is added in a blend with diesel. Fe researchers reported that with the addition of DME, there is an increase in NOx emissions, the EGR is necessary to reduce the NOX emissions during DME combustion. DME enhances the burning process by producing less soot due to its oxygen content. Soot emissions decreases if the amount of DME is increased for all engine speeds and NOx emissions are higher by adding DME compared to diesel for all engine speeds.

It has been noticed that the change in the injection pressure has reduced the NOx emission but not enough compared to diesel, so exhaust gas aftertreatment systems such as the gas recirculation valve should be used to engines that are fuelled by a mix of DME and diesel. Moreover, DME can improve the engine's economical and reduced emissions when used as an additive in SI engine.

8.15.4   DIETHYL ETHER (DEE)

DEE have high cetane number, it is non-corrosive and have high energy density than ethanol relatively. Any proportionate mixture with diesel can be done. Common fuel injection systems can be used without any design modification. Ethanol can be converted by dehydration process to DEE. The vapor pressure of DEE lower than DME. Hence, vapor lock problems in the fuel pump are minimized. In this work, DEE was mixed with diesel fuel. DEE (CH_3CH_2-O-CH_2CH_3) is a great solvent and highly flammable. An experimental analysis is done to study the effects of DEE blends with diesel and kerosene on combustion, emission, and performance characteristics of a direct injecting diesel powered engine. Initially, 2%, 5%, 8%, 10%, 15%, 20% and 25% DEE (by volume) were blended into diesel. The DEE-diesel blends shows decline in trade-off between PM and NOx of diesel engine. Optimum performance is obtained at blend of DE15D. It was seen that the density, calorific value and kinematic viscosity of blends decreases, while the cetane number oxygen content of the blends rises with the amount of DEE addition. The test results showed that DEE-kerosene-diesel blends gives less brake thermal efficiency, high BSFC, high smoke at full applied load, low smoke at little load, overall less NO, almost similar amount of CO, high HC at fully loaded and low HC at partially loaded as compared to DE15D blend.

DEE (C_2H_5-O-C_2H_5) is one such cetane and oxygenated enhancer fuel that is attracting attentions of researchers as an AF for compression igniting engines. The experimental observation has shown that kerosene with diesel up to 15% and blends of DEE with diesel up to 25% by volume were possible. The laboratory tests concluded that as DEE is completely miscible with diesel fuel it can be mixed in any ratio in diesel and kerosene. The density, calorific value and kinematic viscosity of the blends falls down while the content of oxygen and cetane number of the blends rises with the amount of DEE in the blends. It shows that the mixture of fuels can reproduce the desirable physical properties of diesel fuel but includes the cleaner burning capability of DEE. The smoke emissions were partially eliminated with rise in DEE percentage in the DEE-diesel blends. However, engine tends to emit high smoke content at fully loaded condition at higher blends. And coming to DEE-kerosene-diesel blends have shown huge smoke at fully loaded conditions and low smoke at partially loaded condition as compared to that of DE15D blend. The NO emissions released by DEE-diesel blends were sharply reduced than compared to neat diesel, with the reduction going higher the DEE fraction in the blends is. Thus, it known that the fuel-bounded oxygen is very much effective than that of external oxygen supplied by air relatively in the formation of NOx. The CO released by all DEE-diesel blends was lower while HC emission was larger than the corresponding neat diesel. The DEE-kerosene-diesel blends have shown that overall emissions of NO is less, which is almost same as CO emissions, and higher HC emissions at fully loaded conditions and at partially loaded conditions gives low HC emissions as compared to DE15D blend. The complete test results show the trend between NOx and PM of diesel engine was decreased by using oxygenated DEE-diesel blends and DE15D is the most optimistic ratio. The mixing of kerosene with DE15D blend further declines the overall performance of the engine, hence it is not suggestible.

Since DEE has a very high cetane number, for additional improvement the characteristics of combustion BD blends (B20), small amounts of DEE in proportions of 5, 10, and 15 have been added and studied by USV Prasad et al. (2011). The tests were performed on a single cylinder water cooled constant speed CI engine under varying load conditions. It was observed that the HC emissions and BSFCs are lower with B20 and DEE 5. While B20 with DEE15 gives, lower NOx emissions. Also, the mixing of oxygenating materials has improved the process of combustion and emissions are reduced to. The present analysis concludes that compared to neat blend; blends with higher cetane rating and oxygenated additives are superior. The emission tests revealed that the drop in NOx and HC emissions is observed by the addition of higher percentage of DEE to B20 when compared to neat biodiesel operation (Figure 8.21).

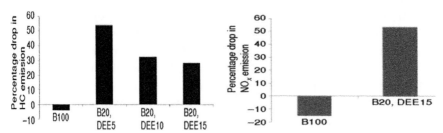

FIGURE 8.21   Drop in HC and NOx emissions for different B20 + DEE blends at 100% load power.
Source: Adapted from Satya, et al., 2011.

8.16   REFORMULATED GASOLINE

The ability of reformulated gasolines to decrease emissions has attracted a vast area of attention worldwide, in Europe and USA it is one of the major sector of the research in oil and automobile industries.

Through reducing volatility, adding oxygenated blend stocks, reducing sulfur (to improve catalyst efficiency) most significant change in emissions is observed in gasoline. Reduction of emissions in fuel injected catalyst-equipped vehicles can be seen by change in midrange distillation, but this effect is relatively small and it is demonstrated in vehicles having carburetors.

The overall effect of reducing distillation temperatures in these vehicles is to increase HC emissions by those changes which lead to large increase in hot-soak evaporative emissions from older vehicles.

Research by AQIRP on exhaust emissions by gasoline reformulation of catalyst-equipped older, current, and advanced (U.S. tier 1) cars shown that non-methane hydrocarbon (NMHC) emissions are 12–27% lower with reformulated gasoline than with industrial-average gasoline; further CO emissions were declined by 21 to 28% and NOx emission by 7 to 16%.

Such reformulations bring the greatest effect and also the greatest cost-effectiveness—when applied to the fuels. This fuel is leaded gasoline in most instances. But in Modern emission-controlled automobiles generally require unleaded gasoline, whose sensitive is less to changes in gasoline parameters, and have comparatively lower emissions to begin with, so that even a large reduction in emissions can be insignificant. This suggests that a "two gasoline" might be very advantageous, with the fuels (whether leaded or a specially unleaded blend) used by the older, noncontrollable vehicles being specially formulated for reduction of emissions, and unleaded gasoline is intended for emission-controlled automobiles subject to low stringent restrictions.

Some of the combinations can be controlled at low cost beyond the single parameters controlling cost, while the other combinations have higher controlling cost. This is because of the synergy that comes into occurrence among gasoline properties when one is controlled and others properties are allowed to get change.

In Europe, the total effect of reformulated gasolines, attained by adding 10 to 15% oxygenates (by volume) and reducing level of aromatics, has been a car weighted average reduction of 11% in both NOx and HC emissions and 23% decrement in CO emissions. These declines pertain primarily to cars without carbon canisters or catalytic converters. The additional cost of manufacturing these cleaner gasoline engines is estimated around 0.5 to 1.0% of the retail price.

The main pros of gasoline reformulation are to reduce emissions (especially evaporative emissions) from automobiles with little emission control. Reformulated unleaded gasoline intended for modern emission-controlled automobile is not considered to be cost-effective, relative comparison with most other potential emission control measures, so it should be considered only where many of the cost-effective measures are insufficient.

8.17 FISCHER-TROPSCH DIESEL FUEL

In 1923, German scientists Hans Tropsch and Franz Fischer established Fischer-Tropsch synthesis. In this process, a synthetic liquid fuel obtained from biomass, natural gas, and coal, oil shale, sources through the process of synthesis of Fisher Tropsch. The production process of synthetic fuel is hydrogenation wherein hydrogen is added. Low hydrogen residues which are carbon-containing are produced from the sources of hydrogen which are intramolecular. Hydrogenation can be either direct or indirect. In direct hydrogenation, exposing raw materials to hydrogen gas at high-end pressures is involved. In indirect hydrogenation, reaction of the feedstocks with steam, and hydrogen gas is generated inside the system.

The Fischer-Tropsch diesel is manufactured by the liquefaction of synthesis gas (mixture of CO and H_2) produced from gasification of biomass and fossil fuels. Depending upon the synthesis process, several types of fuels can be produced. If diesel from gasification of biomass is produced through the Fischer-Tropsch synthesis process, is called green diesel.

Without any design, changes in automobile engine Fischer-Tropsch diesel can be substituted directly into conventional diesel. The output of the Fischer-Tropsch process is around 50% high-quality, high cetane synthetic diesel, sulfur-free, 30% naphtha; and 20% other products (Figure 8.22).

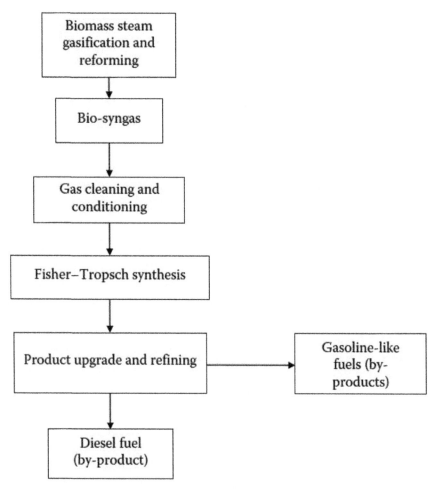

FIGURE 8.22   Production of F–T diesel by using Fisher-Tropsch synthesis process. *Source:* Adapted from Babu and Subramanian, 2013.

Fischer-Tropsch diesel fuel was selected for the explanation on the lower exhaust emissions while enhancing fuel economy by Noboru Uchida et al. (2011).

They examined three Fischer-Tropsch fuels and a most basic diesel fuels to study the effect of the detailed effects of fuel properties on their emissions and combustion characteristics with three diesel engines having different engine displacements. Another objective is to observe combustion process of both fuels on a single cylindrical engine with optical access.

By the above findings, to bring changes in both emissions and fuel consumption by the use of neat Fischer-Tropsch fuels some modifications are made in a test engine. The conversion efficiency of catalyst has been enhanced. Absence of poly-aromatic hydrocarbon contents and High cetane number are fully used to improvise the conventional diesel combustion limitations to show the affinity to reduce emissions with simultaneous improvement in economy of the fuel. Hence, it shows not only the superior emission characteristics of F-T fuels, but also evidence that lower excess air ratios and higher EGR will be key concepts of both engines and improvement in optimized fuel consumption.

In another study by Atkinson et al. (1999), investigated on a transient cycle to observe emissions by conducting with 12 speeds on full load conditions. The chosen 12 speeds are taken as 12 points. The study revealed that the regulated emissions on FT fuel are almost entirely lower than those of Diesel No.2 (with the exception of HC emissions at some speed and load points). On average, for the 12 speed-load points, HC was reduced by 14%. On average, CO was reduced by 21% for the FT fuel over Diesel No.2. The reduction in CO ranged from a low of 3.1% at Point 12 to a maximum of 46% at Point 1. On average, CO_2 was reduced by 4.9% with a range of 3.8 to 6.2%. NOx is reduced, on average, by 20% with the range from 7.14% for Point 11 to 29% for Point 1. The PM was reduced between 13% and 42% with an average reduction of 31% for the FT compared to the Diesel No.2.

KEYWORDS

- **compressed natural gas**
- **ethyl tertiary butyl ether**
- **liquefied natural gas**
- **liquefied petroleum gas**
- **monocyclic aromatic hydrocarbons**
- **proton-exchange-membrane**

REFERENCES

Adeniyi, Oladapo Martins, Azimov, Ulugbek, & Burluka, Alexey, (2018). "Algae biofuel: Current status and future applications," *Renewable and Sustainable Energy Reviews,* Elsevier, *90*(C), 316–335.

Agarwal, A. K., & Das, L. M. (2001). Biodiesel development and characterization for use as a fuel in compression ignition engines. *Journal of Engineering for Gas Turbines and Power, 123*(2), 440–447.

Ahmed A. Taha, Tarek M. Abdel-Salam, & Madhu Vellakal (2013). Alternative Fuels For Internal Combustion Engines: An Overview Of The Current Research International Energy And Environment Foundation www.IEEFoundation.org, Alternative Fuels Research Progress Editor: Maher A.R. Sadiq Al-Baghdadi.

Atkinson, C. M., Thompson, G. J., Traver, M. L., & Clark, N. N., (1999). "In-Cylinder Combustion Pressure Characteristics of Fischer-Tropsch and Conventional Diesel Fuels in a Heavy Duty CI Engine," SAE Paper 1999–01–1472.

Ayhan Demirbas (2007). Progress and recent trends in biofuels. *Progress in Energy and Combustion Science, 33*(1), 1–18.

Babu, A. R., Amba Prasad Rao, G., & Hari Prasad, T. (2017). Experimental investigations on a variable compression ratio (VCR) CIDI engine with a blend of methyl esters palm stearin-diesel for performance and emissions, *International Journal of Ambient Energy, 38*(4), 420–427, DOI: 10.1080/01430750.2015.1132768.

Beer, T., Grant, T., Morgan, G., Lapszewicz, J., Anyon, P., Edwards, J., Nelson, P., Watson, H., & Williams, D. (2001). *Comparison of Transport Fuels*, Report EV45A/2/F3C, CSIRO Atmospheric Research.

Brook, D. L., Cipolat, D., Rallis, C. J. (1984). Methanol with dimethyl ether ignition promotor as fuel for compression ignition engines. *Volume 2*; *Conference: Intersociety Energy Conversion Engineering Conference*, San Francisco, CA, USA.

Celik, M. B. (2008). Experimental determination of suitable ethanol–gasoline blend rate at high compression ratio for gasoline engine. *Applied Thermal Engineering, 28*, 396–404.

Cenk Sayin & Kadir Uslu (2008). Influence of advanced injection timing on the performance and emissions of CI engine fueled with ethanol-blended diesel fuel. *Int. J. Energy Res. 32*, 1006–1015.

Christian Rodriguez Coronado, Aparecida de Castro Villela, & José Luz Silveira (2010). Ecological efficiency in CHP: Biodiesel case. *Applied Thermal Engineering, 30*, 458–463.

Esmail Khalife, Meisam Tabatabaei, Ayhan Demirbas & Mortaza Aghbashlo. (2017). Impacts of additives on performance and emission characteristics of diesel engines during steady state operation. *Progress in Energy and Combustion Science, 59*, 32–78.

Fayaz, H., Saidur, R., Razali, N., Anuar, F. S., Saleman, A. R., Islam, M. R. (2012). An overview of hydrogen as a vehicle fuel. *Renewable and Sustainable Energy Reviews, 16*, 5511–5528.

Gajendra Babu, M. K., & Subramanian, K. A. (2013). Alternative Transportation Fuels Utilisation in Combustion Engines, CRC Press Taylor & Francis Group.

Gui, M. M., Lee, K. T., & Bhatia, S. (2008). *Feasibility of edible oil vs. non-edible oil vs. waste edible oil as biodiesel feedstock, 33*(11), 1646–1653.

He, B.-Q., Wang, J.-X., Hao, J.-M., Yan, X.-G., & Xiao, J.-H. (2003). A study on emission characteristics of an EFI engine with ethanol blended gasoline fuels. *Atmospheric Environment, 37*, 949–957.

Hill, J., Nelson, E., Tilman, D., Polasky, S., & Tiffany, D. (2006). Environmental, economic, and energetic costs, and benefits of biodiesel and ethanol biofuels. *Proceedings of the National Academy of Sciences of the United States of America, 103*, 11206–10. 10.1073/pnas.0604600103.

Hutcheson, R. C. (1995). *Alternative Fuels in the Automotive Market,* CONCAWE Report No. 2/95. Brussels.

JDF (Japan DME Forum). (2007). DME Handbook. Minato-Ku, Tokyo: Japan DME Forum. (Cross reference: A. S. Ramdhass – Alternative Fuels for Transportation. CRC Press Taylor & Francis Group 6000 Broken Sound Parkway NW, Suite 300 Boca Raton, FL 33487–2742 © 2011 by Taylor and Francis Group, LLC, CRC Press is an imprint of Taylor & Francis Group, an Informa business.

Jerzy Merkisz, Jacek Pielecha, Piotr Bielaczyc, & Joseph Woodburn (2016). Analysis of emission factors in RDE tests as well as in NEDC and WLTC chassis dynamometer tests. *SAE Technical Paper 2016–01–0980.*

Jinxing Zhao, Qingyuan, X., Shuwen Wang, & Sen Wang (2018). Improving the partial-load fuel economy of 4-cylinder SI engines by combining variable valve timing and cylinder-deactivation through double intake manifolds. *Applied Thermal Engineering, 141*, 245–256.

Jurchiş, B., Burnete, N., Iclodean, C., & Burnete, N. V., (2017). Study of emissions for a compression-ignition engine fueled with a mix of DME and diesel IOP Conf. Series: *Materials Science and Engineering, 252*, 012065. doi: 10.1088/1757–899X/252/1/012065.

Kiran Raj Bukkarapu & Anand Krishnasamy (2018). A study on the effects of compositional variations of biodiesel fuel on its physiochemical properties. *Biofuels,* DOI: 10.1080/17597269.2018.1501638.

Knapp, K. T., Stump, F. D., & Tejada, S. B. (1998). The effect of ethanol fuel on the emissions of vehicles over a wide range of temperatures. *J, Air Waste Mgmt. Assoc., 48*(7), 646–653.

Liu Shenghua, Eddy R. Cuty Clemente, Hu Tiegang, Wei Yanjv (2007). Study of spark ignition engine fueled with methanol/gasoline fuel blends. *Applied Thermal Engineering 27,* 1904–1910.

Molea, A., Visuian, P., Barabás, R., Suciu, C., & Burnete, N. V., (2017). Key fuel properties and engine performances of diesel-ethanol blends, using tetrahydrofuran as surfactant additive. *IOP Conf. Series: Materials Science and Engineering 252,* 012077, doi:10.1088/1757–899X/252/1/012077.

Murayama, T. C., Guo, J., & Miyano, M. (1992). A study of compression ignition methanol engine with converted dimethyl ether as an ignition improver. SAE 922212.(Cross reference: A. S. Ramdhass, Alternative Fuels for Transportation, CRC Press Taylor & Francis Group 6000 Broken Sound Parkway NW, Suite 300 Boca Raton, FL 33487–2742 © 2011 by Taylor and Francis Group, LLC CRC Press is an imprint of Taylor & Francis Group, an Informa business.).

Mustafa Atakan Akar, (2016). Performance and emission characteristics of compression ignition engine operating with false flaxbiodiesel and butanol blends. *Advances in Mechanical Engineering, 8*(2), 1–7.

Najafi, G., Ghobadian, B., Tavakoli, T., Buttsworth, D. R., Yusaf, T. F., Faizollahnejad, M. (2019). Performance and exhaust emissions of a gasoline engine with ethanol blended gasoline fuels using artificial neural network. *Applied Energy* (Elsevier), *86*(5), 630–639.

Nirendra N. Mustafi, Robert R. Raine, & Pradeep K. Bansal (2006). The use of biogas in Internal Combustion Engines: A review. *Proceedings of ICES2006ASME Internal Combustion Engine Division.* Spring Technical Conference May 8–10, 2006, Aachen, Germany.

Porpatham, E., Ramesh, A., Nagalingam, B. (2008). Investigation on the effect of concentration of methane in biogas when used as a fuel for a spark ignition engine. *Fuel 87,* 1651–1659.

Puneet Verma, Ali Zare, Mohammad Jafari, Timothy A. Bodisco, Thomas Rainey, Zoran D. Ristovski & Richard J. Brown (2018). Diesel engine performance and emissions with fuels derived from waste tyres. *Scientific Reports, 8,* 2457, DOI:10.1038/s41598–018–19330–0.

Recommendation of the Council of the OECD on Improving the Quality of Government Regulation, (1995). http://www.oecd.org/officialdocuments/publicdisplaydocumentpdf/?d oclanguage=en&cote=OCDE/GD(95)95 (accessed on 11 Febraury 2020).

Saravanan, S., Nagarajan, G., & Sampath, S. (2014). Combined effect of injection timing, EGR and injection pressure in reducing the NOx emission of a biodiesel blend, *International Journal of Sustainable Energy, 33*(2), 386–399.

Satya, V.P.U.; Murthy, K.M.; Rao, G.A.P (2011). Effective utilization of B20 blend with oxygenated additives. *Thermal Science, 15*(4), 1175–1184.

Searchinger, T., Heimlich, R., Houghton, R., Dong, F., Elobeid, A., Fabiosa, J., Tokgoz, S., Hayes, D., & Yu, T.-H. (2008). Use of U.S. Croplands for Biofuels Increases Greenhouse Gases Through Emissions from Land-Use Change. *Science, 319*, 10.1126/science.1151861.

Sikarwar, V.S., Ming, Z., Fennel, P.S., Shah, N., and Anthony, E.J. (2017). Progress in biofuel production from gasification. *Progress in Energy and Combustion Science, 61*, 189–248. https://creativecommons.org/licenses/by/4.0/

Stump, F. D., Knapp, K. T., Ray, W. D., Siudak, P., & Snow, R. (1996). Influence of ethanol blended fuels on the emissions from three pre-1985 light-duty passenger vehicles. *J. Air Waste Mgmt Assoc, 46*, 1149–1161.

Syed Kaleemuddin & G. Amba Prasad Rao (2010). Conversion of Diesel Engine into Spark Ignition Engine to Work with CNGand LPG Fuels for Meeting New Emission Norms, *Thermal Science, 14*, No. 4, pp. 913–922.

Teoh, Y. H., Masjuki, H. H., Kalam, M. A., & How, H. G. Effect of injection timing and EGR on engine-outresponses of a common-rail diesel engine fueled with neat biodiesel, *RSC Adv.*, *5*, 96080.

Troy A. Semelsberger, Rodney L. Borup, Howard L. Greene (2006). Dimethyl ether (DME) as an alternative fuel. *Journal of Power Sources 156*(2), 497–511.

Turns, S. R. (2000). *An Introduction to Combustion: Concepts and Applications,* McGraw Hill.

Uchida, N., Sakata, I., Kitano, K., Okabe, N., & Sakamoto, Y. (2012). Simultaneous improvement in both exhaust emissions and fuel consumption by means of Fischer-Tropsch diesel fuels. *International Journal of Engine Research, 15*, 20–36. 10.1177/1468087412456528.

Venkateswara Rao Surisetty, Ajay Kumar Dalaia, & Janusz Kozinski (2011). Alcohols as alternative fuels: An overview. *Applied Catalysis A: General 404,* 1–11.

Vinod Babu, M., Madhu Murthy, K., & Amba Prasad Rao, G. (2017). Butanol and pentanol: The promising biofuels for CI engines – A review. *Renewable and Sustainable Energy Reviews, 78*, 1068–1088.

Xingcai, L., Jianguang, Y., Wugao, Z., & Zhen, H. (2004). Effect of cetane number improver on heat release rate and emissions of high speed diesel engine fueled with ethanol–diesel blend fuel. *Fuel 83*(14–15), 2013–2020.

Yasar, A. (2010). Effects of alcohol-gasoline blends on exhaust and noise emissions in small scaled generators. *Metalurgija, 49*(4), 335–338.

CHAPTER 9

Alternative Combustion Concepts

9.1 INTRODUCTION

The chapters presented above dealt with in cylinder measures, aftertreatment of exhaust gas and adoption of viable alternative fuels (AFs) approaches to address pollutant emission-related issues. In an attempt to mitigate harmful emissions from existing engines, the researchers have developed different techniques/alternative combustion concepts. In this chapter, brief details about above techniques have been presented. Internal combustion engine (ICE), as a prime mover, has served the mankind for over 150 years finding versatile applications in both industrial, power, and transportation sectors, has simplified the lifestyle of mankind. In its conventional form, IC Engines are broadly are of two types; either compression engines or spark-ignition. The fuel characteristics for these two engines are entirely different with high auto-ignition temperature for the former case and low auto-ignition temperature for the latter case. Also, the compression ratios (CRs) are different. The SI engine types (with gasoline as fuel) are CR limited ones due to on-set of detonation and suffers from reduced part-load efficiency with high NOx and CO emissions. Whereas, the ignition delay is the major cause of abnormal combustion (with diesel fuel) even though it works with lean mixtures and also release higher levels of PM and NOx emissions. The SI engines are homogeneous type and CI engines are heterogeneous type combustion engines. In spite of these limitations, both types have revolutionized the automotive sector and consequent exponential growth in automotive population brought large wealth and associated air pollution-related issues. The environmental impact of harmful tail pipe emissions and ever-tightening of emission regulations are enticing both researchers and manufacturers to develop engines that are environ benign significantly without compromise on performance. Over the years, the researchers are tackling these issues with either adopting alternate fuels, aftertreatment devices or modifications in engine designs. The researchers confirmed that if any step is taken to make a difference in combustion process, will be great step in minimizing the harmful emissions.

In this regard, number of alternative combustion concepts has been evolved. These have been simultaneously were developed with the convention engine combustion by observing limitations with regards to conventional concept. The modern concepts would include either modifying engine combustion process or making use of different thermodynamic cycles. The approaches include:

1. Homogeneous charge compression-ignition engines (HCCI) or its variants such as PCCI or RCCI concept.
2. Stratified engine concept.
3. Lean burn concept.
4. Low heat rejection or thermal barrier coatings (HHR or TBCs).
5. Miller cycle.
6. Multiple or split injection.
7. Variable CR engines.
8. Cylinder deactivation.

9.1.1   HOMOGENEOUS CHARGE COMPRESSION-IGNITION (HCCI) ENGINES: HCCI CONCEPT

Considering the best characteristics of conventional SI and CI engine combustion concepts, a new concept that combines the best features of compression-ignition (CI) and spark-ignition (SI) engines, called HCCI Engine has been evolved. HCCI combustion includes the auto-ignition of the uniform mixture made either inducted or inside into the chamber for combustion by compression. Since there is no specific source of ignition, the spontaneous ignition of uniform charge thus prepared takes place in explosive manner. Therefore, the combustion is volumetric in kind in HCCI as there is no spark plug for flame transmission and the uniform mixture tends to automatic-ignition temperatures at the completion of compression stroke. This volumetric automatic-ignition feature of HCCI engines facilitating it to function with lean mixtures at low temperatures, called low temperature combustion (LTC). This particular feature of HCCI is raising hopes among combustion to extensively adopt to simultaneously reduce PM-NOx emissions. Lean mixtures functioning ability, reducing the ISFC to a noticeable degree and lower temperature combustion is accountable for lower NOx about 90–98%.

HCCI model is conversely not a recent discovery. In the late nineteenth century, heated bulb engines functioned with a HCCI-like combustion. The type of HCCI combustion, as volumetric auto ignited combustion, is stated by Christensen et al. (1997) and Thring et al. (1989) and many other scholars due to the presence of the dense lean fuel-air mixture and was credited to the profits of HCCI engines. Many scientists have stated the restrictions of HCCI engines such as narrow operating range, low specific output, and long start-up time, lack of control over the ignition process and high levels of CO and UHC secretions. The CO and UHC secretions can be after dealt utilizing catalytic converters. Few scientists like Onishi et al. (1979), Ali et al. (2001), Yu et al. (2006), and Krishna et al. (2011) made experiments to envisage the procedure of combustion in both CI and SI engines participating in HCCI modes utilizing particulate image velocimetry (PIV), Laser Doppler methods, and Schlieren photography. From their experimental outcomes they stated that very distinct transmission of flame was discovered in SI function mode, but no visible propagation of flame was found in HCCI type of combustion; verifying volumetric combustion of HCCI engines.

HCCI engines gave hope to the scientists as up-coming combustion technology to encounter the stringent emission norms. The researchers are focusing both on experimental as well as computational methods to make HCCI a viable option for future engines. In HCCI engines, once conducive environment prevails inside the combustion chamber, the mixture burns volumetrically in the absence of spark but only due to compression. Comparison of SI, CI, and HCCI combustion can be seen in Figure 9.1.

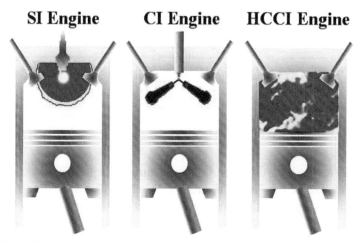

FIGURE 9.1 Pictorial comparison of SI, CI, and HCCI combustion phenomenon.

HCCI engine utilizes uniform air and fuel mixture as in the SI engines and uses higher ratios of compression to let the mixture to auto-ignite as seen in the diesel engine. HCCI combustion comprises the automatic ignition, by compression of the uniform mixture, made either inducted into or inside the chamber of combustion. Fundamentally, HCCI engine runs on the fundamentals of having a dilute, premixed charge that reacts and burns volumetrically, without the aid of spark, throughout the cylinder as it is compressed by the piston. This particular feature of HCCI engine helps its functioning with lean mixtures at lower temperatures. Therefore, lean mixture functioning capacity decreases the indicated specific fuel consumption (ISFC) to a noticeable level while lower temperature combustion is accountable for extremely low NOx (about 90–98% decrease). The brake thermal efficiency was similar with that of conventional CI engines and much greater to SI engines. In spite of attractive features of HCCI concept, due to spontaneous combustion, there would be a sudden rise of peak pressure, it suffers certain limitations/ challenges. Of late, the studies are concentrated on realizing HCCI concept while addressing challenges being posed by this concept.

9.1.2   CHALLENGES ASSOCIATED WITH HCCI ENGINES

Though the HCCI concept appears to be so attractive, but it suffers from few limitations. The difficulties posed in realizing the HCCI are compiled herewith as shown in Figure 9.2.

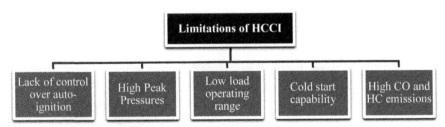

FIGURE 9.2   Limitations of HCCI engine.

- **Deficiency of Control Over Automatic Ignition:** A larger issue with HCCI engines is the lack of control over auto-ignition as it is not having any direct mechanism to start the ignition like in SI and CI engines. The volumetric type of combustion in HCCI engines by CI heads to the deficiency of power over automatic ignition. The prime combustion in HCCI relies on the temperature, pressure, low temperature chemical reactions, and reaction between the species this directed to a deficiency of power over the ignition.

- **Reduced Load Bearing Capacity:** As HCCI engines are trading with slender mixtures the load, bearing capacity is extremely poor. The resultant dilution restricts the quantity of fuel that can be combined for a steady charge mass resulting to a loss in engine capacity.
- **Cold- Start Capacity:** While in cold-start the slender mixtures in the HCCI engine couldn't attain the automatic ignition temperatures. HCCI engine cold-start will be problematic lacking some ignition mechanism while in cold weather circumstances.
- **Excessive CO and HC Emissions:** HCCI engines release excessive CO and HC emissions due to the uniform mixture confined in cleft areas being not burnt. Due to the lower temperature combustion, the unburned fuel from cleft areas while in the expansion stroke couldn't be expended much resulting in a meaningful rise in both CO and HC emissions. Also, the low combustion temperatures hinder the oxidation of CO to form into CO_2, which generally happens at higher temperatures.

It is noticed that the scientists have equipped different processes to gain HCCI type of combustion to gain NOx emissions and low soot. It seemed that there are two processes to allow an engine to run in HCCI type by injection and premixed techniques. There noticed to be a great aptitude for HCCI engine to be the future prime mover for various applications. Classification of processes to gain the HCCI type of combustion can be seen in Figure 9.3.

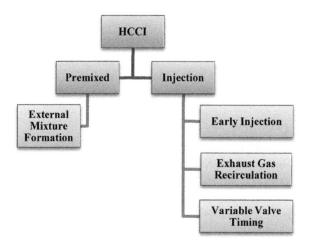

FIGURE 9.3 Classification of methods to achieve HCCI mode of combustion.

Many of the scientists used the Injection mode as it needs very nominal changes to the present engines. But there are question about this method one can't say that the charge inside the combustion chamber is completely homogeneous under different conditions. For high speed engines, this method is not suitable. Ideal homogenous conditions cannot be achieved by this method.

Premixed method is gaining the attention of the researchers to achieve ideal HCCI mode of combustion at all engine running conditions. Premixed type of HCCI combustion can be attained by external mixture formation technique. Wherein the fuel vaporizer is utilized to vaporize the fuel and then the mixture of air and fuel vapors enters the chamber of combustion wherein it gets compressed to higher pressures. Since there won't be any spark plug, it is so designed that the temperatures obtained due to compression results to automatic ignition of the charge like in CI engines. Therefore, HCCI engines gain the merits of both CI and SI engines. The utilization of higher CRs aids the use of very slender mixtures, thereby lower ISFCs are achievable with HCCI engines. The utilization of very slender premixed charge results in lower temperature volumetric combustion bringing lower soot and NOx emissions. The low temperatures are also accountable for higher HC and CO emissions hindering the oxidation of HC and CO. Further, there are issues linked with external mixture formation techniques like, ISFC being high when related with the conventional engines and also high HC emissions. Making use of this technique, few researchers carried out tests to gain HCCI combustion. They successfully gained HCCI type of combustion. The control over the faster rise in heat emission rate at higher load functioning and faster falls in combustion periods at rich mixture conditions.

Large number of scientists has worked their finest to gain HCCI phenomena and inferred low soot and NOx emissions, but have faced with higher CO and HC emissions. Studies also showed that external mixture formation technique could be installed in attaining HCCI type of combustion. Engine functioning can be developed by changing operating and geometrical parameters. Also, emissions can be eradicated by these parameters besides installing EGR, swirl, etc.

Majority of studies to realize HCCI concept have been done in CI engine. For this, either too early injection or too late injection has been considered. Moreover, the way the air and liquid fuel form the uniform mixture is the main issue in the HCCI work. Karthikeya Sharma 2016 did consider the premixed mode to achieve HCCI combustion did systematic analysis and extensive computational work through the use of ECFM-3Z combustion model developed in STAR-CD is employed. His work focused on the use of swirl for overcoming challenges. For this purpose, 3D- CFD based combustion software STAR-ice ECFM code was adopted with an external mixture formation technique proposed by Ganesh et al. (2010). Thus, for a systematic computational study a premixed charge concept for achieving HCCI mode and examined the effect of geometrical and operational parameters, as detailed below. The study revealed that if swirl is adopted, few of important challenges could be overcome with HCCI concept:

- Swirl;
- Compression ratio;
- Boost pressure;
- Exhaust gas recirculation (EGR);
- Speed;
- Piston Bowl shape;
- Piston Bowl geometry;
- Equivalence ratio;
- Fuel type (Fuel ignition quality).

Widespread numerical trials were carried out and operation is estimated in terms of turbulent kinetic energy (TKE), piston work, emissions, and velocity magnitude (VM). HC, NOx, CO_2, CO for the chosen engine specifications. Simulation results showed that induction induced swirl motion has meaningful effect on the HCCI engine operation and emissions as velocity magnitude (VM) and TKE raised with raise in intensity of the swirl. Reduction in in-cylinder pressures, temperatures and piston work, NOx, and CO_2 emissions were attained with raise in swirl ratios. With raise in swirl ratios raise in CO emissions and wall heat transfer losses were attained. Effective, efficient reduction and combustion of emissions were noticed by improving the air movement inside the chamber of combustion with optimized geometry and piston bowl shape. Also, the computational study recognized that with the addition of swirl and any other parameter would overcome specific problems linked with realization of HCCI phenomena.

It is observed that added effect of swirl and other input parameters pacify the challenges like decrease in piston work, increasing in emissions and, automatic ignition and poor load-bearing capability. The added effect of swirl and equivalence ratio enhanced the engine operation and decreased emissions to a significant level. Load-bearing capacity of HCCI engines enhanced by equipping boost pressure. A raise of 151.00% in piston work was noticed with raise in boost pressure from 1 bar to 2 bar with swirl ratio 1.

A sum of 11.12% decrease in NOx emissions with swirl ratio 4 and 42.59% raise in CO emissions with swirl ratio 1 have developed with raise in boost pressure from 1 bar to 2 bar. Figure 9.4 shows the relation of maximum piston work attained by conventional HCCI and CI types with swirl ratio 1.

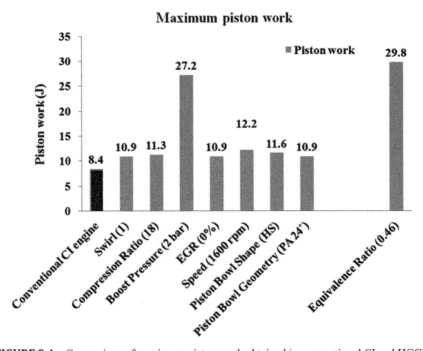

FIGURE 9.4   Comparison of maximum piston work obtained in conventional CI and HCCI modes with swirl ratio 1.

It was found that lower ratios of compression with higher swirl for an equivalence ratio of 0.46 and re-entrant angle of 24° is seen to be optimal among the various parameters studied. For swirl ratio extending 4, it leads to combustion irregularities and increased wall heat transfer losses.

The premixed technique is effective in gaining HCCI type of combustion with induction induced swirl. By equipping the plan, it leads to low CO and HC emissions and low NOx emissions. The study showed that there lies a trade-off between piston work and NOx emissions. Since HCCI has been seen as an alternative combustion concept and is effective in achieving simultaneous reductions in PM and NOx emissions with leaner mixtures and low combustion temperatures. Sometimes this concept is also called low-temperature combustion concept.

Calculations performed by Kitamura et al. (2003) for n-heptane at an ambient pressure of 60 MPa and a residence time of 2 ms; provide the regions of soot and NOx formation in the Ø-T plane shown in Figure 9.5.

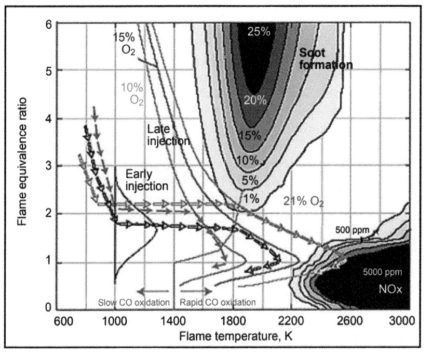

FIGURE 9.5 Equivalence ratio vs. local temperature with the typical paths of different combustion concepts.

From a theoretical point of view, HCCI would permit a radical decrease of NOx and particulates emissions to near-zero levels on the basis of two basic methods: first, a uniform combination has to be created, and second, this combination automatically-ignites because of compression heat. Nevertheless, these important features are also the prime problems. If an entirely uniform combination is formed, the temperature and pressure rise while in the compression stroke will result in instantaneous ignition which varies from the classical diesel automatic-ignition in the nous that it doesn't happen at a certain place in the spray, but concurrently across the chamber of the combustion. Successively, if automatic-ignition happens concurrently in the complete cylinder, no large temperature flame front will emerge as in the case of SI engines. The non-appearance of a higher temperature flame front will result in a rationally insignificant creation of nitrogen oxides (NOx), and because of the uniform slender mixture, fuel rich zones are not present and thereby soot formation is also avoided.

HCCI is one of the very widely recognized abbreviations to classify new combustion strategies depending on the improvement of premixed combustion modes.

The basic procedure contains the formation of a uniform (or greatly premixed) combination under automatic-ignition conditions. This combination will experience concurrent and dispersed oxidation so that no flame front is there, local temperatures are kept lower, and NOx formations can be shunned. Moreover, by keeping local air-fuel ratios (AFs) value low, typical of conventional non-premixed flames, soot formation, can also be also avoided.

Many studies have been undertaken as HCCI has given a promising solution to simultaneously reduce the PM and NOx emissions. This is established through both theoretical and experimental investigations. Since the mode of combustion is heterogeneous in CI engines, HCCI concept has a large potential to reduce NOx and soot emissions in diesel-fuelled CI engines, while maintaining its low fuel consumption. HCCI is sustaining under low load conditions but is giving large scope for its development for practical application especially to meet EURO VI and Super ULEV emission norms.

Prolonging the functioning limits needs prime enhancements in terms of combustion control, and the employment of sophisticated technologies such as advanced boosting, adjustable valve actuation, or adjustable CR, or newer enhancements in injection systems. On this last apprehension, adoptions of present injection systems for new combustion modes should trade with hardware necessities rather than with injection function.

HCCI is a crossbreed of the SI and CI engine concepts. As in a SI engine, a uniform air-fuel combination is formed in the inlet system. While in the compression stroke, the temperature of the mixture raises and tends to the point of automatic ignition, just as in a CI engine (or diesel). One problem with HCCI engines is the requirement for better timing control of the combustion. Automatic ignition of a uniform combination is very subtle to functioning circumstances. Even the little variation of the load can alter the timing from too early to too late combustion. Thus, a faster combustion timing control is essential because it sets the performance restrictions of the load control. HCCI is an assuring substitute combustion strategy having higher thermal efficiency while sustaining the soot and NO_x emissions below the present emissions directives.

The way fuel and air mix and enter the combustion chamber makes a lot of difference in achieving HCCI mode of combustion. Many authors/researchers adopted too early injection of fuel and few used injection at the TDC itself (Zhao, 2007). However, the works of Ganesh and Nagarajan (2010) involved in external mixture formation. Due to this, chemically, and pre-mixed mode and reactivity-controlled combustion, the HCCI is alternatively are called PCCI and RCCI combustion notion. For achieving simultaneous reduction of PM and NOx emissions, many researchers adopted cooled/hot EGR and thus allowing low temperature combustion and lean mixtures. Whatever may be the mode of mixing and attaining homogeneity, the HCCI combustion remained pose many challenges with regards to combustion phasing (combustion instability limit), combustion noise limits, high in-cylinder peak pressures, poor high load efficiency, and high HC/CO emissions. As there is specific mode of combustion in HCCI Combustion concept, it all happens due to auto-ignition of fuel and air. The sudden outburst of combustion through auto-ignition is a cause of concern for high peak-pressure in HCCI engines. Prominent works in this concept have been compiled in a book edited by Zhao.

HCCI combustion is kindled by virtue of the compression solely through in-cylinder fuel and ai mixture chemical kinetics. This characterises HCCI operation from conventional SI and CI engine combustion process. Ming and Maozhao (2007) have chosen iso-octane fuel studied and designed a skeletal chemical kinetic mechanism including 38 species and 69 reactions and predicted reasonable ignition timing, burn rate and the emissions of CO, HC, and NOx for HCCI multi-dimensional modeling. With the help of a single-zone model, their results were observed to in good agreement with that Curran et al. (2002) but the computing time was decreased highly.

As HCCI, engines pose problems with respect to load-bearing capacity and give audible noise. Ringing intensity (RI) has been taken as a measure to identify the smoothness of engine working with hydrogen fuelled HCCI working and is predicted through ANN model precisely by Maurya and Saxena (2018) and experimentally by Mack et al. (2016) with butanol isomer combustion in HCCI mode.

Assanis and Fiveland (2001) simulation of the CI engine has provided a firm construction and execution of the governing equations for the HCCI model. Grounded on the work done and concrete expectations through zero-dimensional model work, additionally such a outline has been employed for HCCI concept for estimating performance and combustion of a four-stroke engine has been established for studies. Using models for energy, species, and mass in a user-outlined chemical kinetic mechanism employing CHEMKIN libraries to link a rigid kinetic chemical solver fit for mixing inside an entire engine cycle replication, presenting models of gas exchange, wall heat transfer, and turbulence. For determining the model capacities with natural gas and Hydrogen chemistry, the simulations have been made. Whole 23 reactions and 11 species for hydrogen, while for natural 325 reactions and 53 species, comprising NO_x chemistry were taken into consideration via GRI mechanism 3.0. The computer simulations made for HCCI displayed the significance of coupling specified chemistry for illustrating physical models of the HCCI engine processes.

Mingfa, Zhaolei, and Haifeng (2009) HCCI combustion has been drawing the noticeable consideration because of higher efficiency and low particulate matter (PM) and NOx emissions. Nevertheless, there are still difficulties in the fruitful functioning of HCCI engines, like high unburned hydrocarbon and CO emissions, controlling the combustion phasing, and extending the operating range. Significant progress in the control of HCCI combustion has happened to make it as practically viable with massive research throughout the world.

- First, many studies have been performing numerical simulations to check HCCI and to improve control strategies for HCCI as such tools have become powerful tools and of its greater flexibility and lower cost related with engine experiments.
- Second, the effect of fuels has been studied with an objective of achieving homogeneity, improved Combustion phasing and operation range.
- Third, the ability to prolong the functioning range to higher loads; even to full loads, for diesel engines, the potential stratification strategy to prolong the HCCI functioning range to high loads, and lower temperature combustion (LTC) diluted by EGR.
- Fourth, optical diagnostics has been applied widely to reveal in-cylinder combustion processes.

In order to simultaneously decrease PM and NOx emissions, low temperature combustion (LTC) is showing promising results (Amin et al., 2015) without compromising on fuel conversion efficiency. To achieve low-temperature combustion, many researchers have adopted HCCI, premixed charge compression-ignition (PCCI), reactivity-controlled compression-ignition (RCCI) and partially premixed combustion (PPC).

Whatever may be approach, the way fuel-air mixture forms and makes the difference in attaining the above-said techniques. Though the HCCI has potential to achieve LTC but it has limitations with respect to combustion phasing and full load-bearing capabilities, etc. Providing sufficient time for thorough mixture formation through early fuel injection has been employed. Due to challenges posed by HCCI, it is finding difficult to implement in automotive applications (a major source for ICEs). In this direction, the RCCI combustion notion has been evolved. This technique adopts two type of fuels-one a dual-fuel which is a high-reactivity fuel (e.g., diesel, and biodiesel) and a low-reactivity fuel (e.g., alcohol fuels, gasoline and natural gas). These types with one being port injected and other being directly mixed using PPC concept. By spatial stratification among the two fuels and varying ratios of two fuels, combustion phasing is controlled and the combustion duration is controlled. While emphasizing on the use of RCCI concept, clear differences among three combustion concepts were presented.

AFs such as alcohols (especially higher blends containing Methanol and Ethanol), biofuels, and natural gas are attractive candidates for RCCI engines. Better load expansion in RCCI function compared gasoline is observed. RCCI combustion with hydrated ethanol has given higher gross indicated efficiency (GIE) and low PRR (i.e., ringing intensity GIE and has resulted near-zero NOx emissions. Even though, low PM and NOx emissions are achieved through RCCI, higher levels of HC and CO emissions have been noticed, however, a mix of thermal management and oxidation catalyst would reduce these emissions too.

Steven (2001) presented that HCCI concept is cross breed as it combines the good features of SI and CI engine together. He opined that HCCI may make a brake through as it offers dual benefit of fuel consumption and low levels of obnoxious emissions (NOx-PM). At same time he observed that to obtain ultra-low levels emissions, it works on lean mixtures and hence would confine to low-load operation. Till a full-fledged and all-load capability HCCI engine concept was developed, it is the best choice to use in dual-fuel mode or on hybrid-electric configuration since in IC engine mode the engine could operate in HCCI mode under low load and low-speed condition. He finally concluded that before arriving at the HCCI design, thorough investigations on the chemical kinetics of fuel-oxidation and fluid-mechanics phenomena associated with mixing and burning that control are essential to get HCCI – high-risk technology as a viable option in future.

9.2   LEAN-BURN ENGINES

Generally, SI engine operate very near stoichiometric AFs yielding very low load carrying efficiency and heavy CO emissions. In the event of reducing fuel consumption and HC& CO emissions, there is a growing trend in the development of engines that use very lean mixtures. Such engines more or less use stratified charge concept and many number of automobiles to attain high fuel efficiencies by using of lean-burn engines. To decrease spark-ignition engines fuel economy lean burn is found to be effective. These type of engines work with an overall AFs of 20–25 (Ø = 0.7). This is much leaner compared to the combustion in a homogeneous mixture engine.

However, adoption such mixtures would be good for reduction of HC&CO emissions but inferior with NOx emissions as 3-way catalytic converter efficiency will be poor at this equivalence ratios. To make 3-way converter effective even with too lean mixtures, alkaline rare earth metals, palladium, iridium in addition to catalysts platinum and rhodium is also developed for lean burn engines while limiting temperatures for combustion so that production of NOx emission is kept within manageable limits.

Experimental trials were also made to develop SI engines which operate AFs up to 40:1. To make it viable for good start of combustion, a rich mixture close to the spark plug, high swirl, and squish, and spark plug of large normal electrode gap of high voltage have been incorporated.

Many studies of lean burn concept have realized with the usage of natural gases as its stoichiometric AF is near to 18:1. By the use of different throttle positions, fuel injecting timings, spark igniting timings and various AFs. Characteristics of combustion and emissive of a lean burn natural gas fuelled spark-ignition engine were studied by Cho and He (2008). The study was performed on an engine with port fuel injection with a CR of 9.3 and bore to stroke ratio of 1.14.

The performance of SI conventional engine, in general, is a strict function of mixture ratios and spark timings in respective test cases. Cho and He (2008) observed lowest THC and CO emissions fewer than one set of conditions whereas lowest NOx emissions are at different conditions. The lean-burn concept encourages use of alternate fuel especially gaseous fuels with better economic usage of fuel enhances by 10 to 15% approx. and thus emissions are also reduced by 10 to 15%.

9.3 LOW HEAT REJECTION ENGINE (LHR) CONCEPT

The diesel fuelled engine enjoy high load capability, high fuel thermal efficiency and low fuel consumption. In general, the energy balance diagram is drawn to observe the energy distribution and it depicts the various losses also. Out of the 100% chemical energy supplied in the form of fuel to an engine, energy distribution of a typical diesel engine is depicted in Figure 9.6.

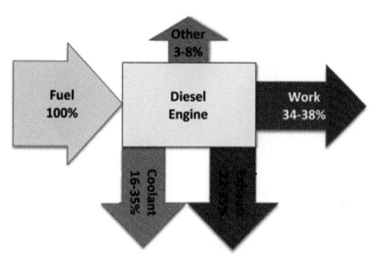

FIGURE 9.6   Energy distribution in typical diesel engine.

It can be seen that keeping useful about 40% of the energy given to engine in terms of fuel, about 60% is leaving the engine as loss in accordance with second law of thermodynamics. Of this, about 30% is going towards coolant losses. Many studies have been explored to tap this loss to extract sizeable amount into useful energy. If this is reduced, the engine chamber can be maintained hot giving a scope for use of low grade fuels and help in utilizing low cetane fuel thereby reducing detrimental effects of ignition delay.

Therefore, since the heat, rejection is reduced and such a concept is popularly called as low heat rejection (LHR) engine which aims for the reduction of heat that transfers to the cooling system and that energy is converted to work which is useful. LHR engines have main advantages like better fuel economy, decrement of emissions from CO, HC, and PM gases, improvement in engine life, operating ability of low octane fuels and increment of exhaust gas exergy and lower noise in combustion because of reduction in increasing rate of pressure. The losses in diesels other than the available energy destroyed by the combustion process are the lost available energy during exhaust blow down, the heat transfer to the coolant, the mechanical friction, and unused exhaust energy.

The concept has been investigated by providing insulation to engine components such as valves and inside of cylinder head, cylinder lines, and piston crown top. A typical use of coated component; piston, for studies is depicted in Figure 9.7.

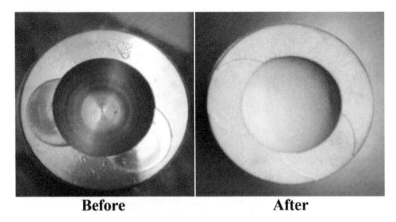

Before   **After**

FIGURE 9.7   Photograph of diesel engine piston with ceramic coating.
Source: Reprinted from: Hasimoglu (2008).

By providing insulation to such intricate engine components wherein the heat is continuously in contact during the engine operation, the chamber is maintained hot enough to allow reduction chemical delay; the diesel fuel will quickly take part in combustion. Thus, the chamber can be regarded as "fire deck." The insulating materials which are used for LHR concept are thermal barrier coatings (TBCs). The heat transferring from the cylinder wall to the cooling system is greatly decreased because of insulation to the cylinder wall which alters the diesel engine combustion characteristics. Materials such as few ceramics, partially stabilized zirconia (PSZ) stabilized with calcia or ytria are normally used for the purpose. The coating is provided on the said components in very small thicknesses (few microns) by using plasma spray process.

Studies have also been done by Prasad et al. (2000) incorporating air-gap piston, as two-part piston with the upper part of piston crown made up of Superni-90 (a material with low thermal conductivity) is fixed to piston crown which leaves or provides with an air gap of 3 mm. This air and low thermal conductivity material provided good insulation for improving the combustion of esterified jatropha oil (biodiesel). They observed lower fuel consumption and lower NOx emissions.

Kamo and Bryzik (1984) extensively studied the use of TBCs and concluded that with TBCs, the heat lost through exhaust gases has increased and the energy in exhaust gases can be recovered with the use of turbocharging or turbo compounding. These effects are also established by Alkidas (1987) and Thring (1986). At the present time, there is a major effort to find ways to insulate the engine parts, particularly the piston, head, and exhaust port, by use of ceramics or high temperature metals. The use of increased part temperatures creates problems in decreased volumetric efficiency (VE), lubrication difficulty, and serious thermal stress problems. However, increased temperature levels may decrease particulates through improved in-cylinder oxidation and can improve the fuel tolerance of the engine. The gains in efficiency from insulation are small unless the exhaust gas energy is recovered by use of turbo-compounding. Turbo-compounding is expensive and thus may practical only for large systems or military applications.

Ethanol-diesel fuel blend has been tried out by Hasimoglu (2008) in a LHR engine without modifying any operating parameters of the engine. An investigation has been made in direct injection (DI) turbocharged diesel engine with an application of LHR by analyzing smoke of NOx emissions reduction with ethanol-diesel fuel mixture effect.

Engine torque increases with the increase in NOx emissions at any conditions in which maximum emissions occur at both high and low speeds of LHR diesel engine. In LHR ethanol engine, there is a 71.3% decrement of NOx emissions at all engine whereas in LHR diesel, there is 22.1% decrement of NOx emissions at all engine torques. Generally, as engine torque increases smoke emissions also increases. However, these emissions do not vary at medium and low engine loads but there is a variation at high loads of engine for both speeds of 2200 and 1200 rpm. For LHR diesel, the smoke emitted is reduced to 24.5% at high loads of engine but increases by 60.5% at medium engine loads whereas for LHR ethanol, smoke emission is reduced from 42.8 to 70% variation during all engine loads at 1200 rpm.

9.4 MODIFIED ATKINSON CYCLE OR MILLER CYCLE

Thermodynamic cycles are used for evaluating the theoretical performance of engines. The engine performance of gasoline and diesel engines nearly approximated using Otto cycle and Diesel cycle respectively. Over a period of time there were development to bring the Otto cycle performance near to actual engine performance and hence new thermodynamic performance cycle have been evolved. One such cycle is Miller cycle developed by Ralph H. Miller (1890–1967). Miller developed the cycle based on the Atkinson cycle, also known as complete expansion or over expanded cycle. However, he realized that Atkinson cycle couldn't be made practical owing complex mechanism involved for providing differential CR and expansion ratio. In earlier days all efforts were focused on improvement of Otto cycle performance and in such a category, the Miller cycle took a shape and was appreciated by many as it was simpler than Atkinson cycle.

Theoretically, the Otto cycle efficiency or air standard efficiency of Otto cycle is considered as a strong function of the CR and this cycle follows isentropic processes for both the expansion and CRs. Moreover, in an Otto cycle, the CR and expansion ratios are same.

In over-expanded cycles of engine, the effective expansion ratio is greater than the effective CR meaning that work spent is lower than work produced and thus efficiency increases for the heat input. In order to implement Miller cycle concept practically, he implemented two strategies viz.; early inlet valve closing (EIVC) and late inlet valve closing (LIVC).

In both the petrol and diesel engines, Miller cycle has been introduced. The benefits of Miller cycle in petrol engine includes reduction in pumping losses at improved and part load efficiency as well as knock mitigation whereas in diesel engines primarily used to control NOx emissions at the higher engine loads.

Mainly the effect of LIVC and EVIC is temperature reduction at end of compression stroke. The use of high geometric CRs which yields an efficiency benefit and longer expansion ratio is enabled by low temperatures.

The idea of valve timing usage for controlling effective CR is not conceived by Ralph Miller. A discussion was evidenced about this in 1927 report saying that there is a limit to knocking in aviation engines by the usage of low octane fuels (Figure 9.8).

Miller had an interest for limiting TDC temperatures by usage of Inlet valve closing time. He demonstrated inlet valve timing mechanism variations which allow IVC varying with engine loads in which at the end of compression stroke for controlling in cylinder temperature. For naturally aspirated and forced induction spark-ignition and diesel, he claims his ideas. Power density increment was his motivation. In his 1954 patent, as load increases has to decrease the end of compression temperature staying within the limits of material properties, at full load condition the engine could burn more fuel. Intercooled or boosted engines are targeted specifically. His 1956 patent specifically targets to Petrol engines was intended for avoiding pre-ignition and maintains at high geometric CR allowing fuel/air ratio rich in quantity.

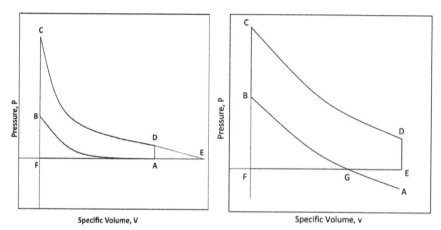

FIGURE 9.8 Comparison of Atkinson cycle and miller cycle variants.

Figure 9.8 compares the Atkinson cycle and Miller cycle variants. While Miller mentioning that both the late and early intake valve closings seemed to be preferring earlier intake valve closing there is still increasing in volume of cylinder because intake charge cools by additional expansion after intake valve closing and this referred to as "internal cooling."

Generally, Atkinson cycle engines are the engines with late intake valve closings. So some prefer to take these engines to as a reference as these have late intake valve closings and are naturally aspirated. However, there is no reference to closing valve timings by the James Atkinson's original patents but refers to an engine in which it's cycle completes one single revolution with crankshaft mechanism and crankshaft that allows high expansion ratio than CR. There is no mentioning of intake valve closings timing management for achievement of this effect.

A credit is deserved by Atkinson for being as first to recognize benefits with different expansion and CRs where as Miller deserves a credit for devising a recipe in which a set of objectives are achieved that is relevant to modern ICEs.

Thus, it would be justifiable if manifestations of over-expanded cycles are referred that rely on variable intake valve closing timing for their implementation as Miller cycle engines—whether forced induction used or not and does not matter whether they are spark-ignition or CI. The application of Miller's ideas successfully made commercially while it is unknown if Atkinson's mechanism was ever done so.

Applications appearing in the 1990s such as the introduction of Mazda's 2.3 L KJ-ZEM in 1993 for passenger car applications was an early gasoline engine version. Also the power produced by Nilgata is a medium speed diesel engine starting in late 1990s, the 32FX. Another application that received huge response around those times is large stationary gas engines. The increased potential for efficiency and power density are responsible for these early applications. The availability of reliable variable valve timing hardware was not yet ready for these various applications and they relied on fixed LIVC or EIVC.

In the early 1990s, for some IMO Tier 1 marine engines, an interest came for NOx emissions reduction from diesel engines by applying Miller cycle. This relatively mild "Miller effect" is used by some of these engines and thus it could do so with fixed valve timing. Further NOx emission reductions would require a more advanced Miller effect and thus variable intake valve closing timing to address the engine starting challenges and low loads. Some of the engines that followed that type were Caterpillar's 2004 on road C11, C13, and C15 variant model engines. Also, a similar approach was adopted by marine engines of medium speed that came into force in 2010 for IMO tier 2 NOx limits.

The efficiency benefits of LIVC strategies in passenger car gasoline engines were attractive for hybrid vehicle engines. In 1997, Toyota's 1st generation Prius adopted this type and subsequent generations of the Prius have continued to use this same technology. In 2007, a naturally aspirated SI engine is introduced by MAZDA, the MZR 1.3 L, for the Japanese market for a non-hybrid vehicle and with fixed LIVC. Starting in about 2012, the pressure for further reduction of fuel consumption saw in non-hybrid light-duty gasoline engines has wider applications of LIVC. For many of these applications in which some already had cam phasers incorporating the Miller cycle was a relative measure of lower cost. Due to additional costs, the engines that have been slower to adopt Miller ideas are heavy-duty diesel engines. Many engines in which cam phasers are not used are low duty diesel engines.

Both experimental and computational works have been done to implement the concept and study the efficacy of the same. It applied to both diesel and gasoline engine versions CFD simulations to investigate the Miller cycle's potential applied to high speed direct injection (HSDI) Diesel engines, facing the emissions reduction challenge enforced by the near term regulations (to comply with Euro VI norms).

In fact with the same value rates of EGR and AFR, the strong reduction of combustion temperature there by abating of NOx emissions is the Miller technique's valuable benefit compared to a traditional cycle. Practical application of the Miller cycle and number of critical issues in the study are addressed by them. They aimed at to assess the limits and the potential of the Miller cycle technique, more than a specific engine configuration development.

A 2.8 L 4-cylinder turbocharged engine (VM Motori make) is chosen and a comprehensive experimental campaign is carried out by them using both the partial and full loads. The authors are allowed calibrating two types of numerical models by the experimental data in which one is for combustion process simulation (CFD-3D) and the other for whole engine analyses (0/1D). In terms of the both fuel consumption and emissions in the European driving cycle, a reliable comparison is provided between the modified engine according to Miller cycle and the basic engine by the integrated usage of these computational tools. A reduction of soot and NOx of 60% and 25% respectively was found and a reduction in 2% fuel efficiency. The abating of NOx can be increased further, since it is demonstrated that the higher rates of EGR (Enrico Mattarelli et al., 2013) can be tolerated by the engine which is operated according to the Miller cycle.

With the adaption of supercharging system which is made up of a mechanical compressor and a geometry variable turbocharger as well as with the usage of full variable intake valve actuation device (both) connecting serially, the design of Miller cycle application has been done.

When the mechanical compressor's support is not necessary, it can be disengaged. A comprehensive experimental campaign on the standard engine has been carried out at both the partial and full load for calibration of two types of CFD tools: one is KIVA-3V code which is a customized version (multidimensional analysis) and other is DI diesel simulation software (GT-power). In both the cases, only when it is needed for specifying Miller cycle's certain features, the calibrated numerical model has been modified and has been found as good agreement with those experiments. For addressing the engine setup management strategies and the choice of the valve actuation laws, modifications in GT-power has been done for more detailed combustion analysis and KIVA was employed.

It seems that the Miller cycle concept is one major alternative of Otto cycle for the determination of its increased efficiency potential and the determination of its network power in the ICE which is spark ignited by shortening the compression process in relative with the expansion process by late or early close of the intake valve. The supercharged Otto engine analysis is adopted for the Miller cycle's operation. The normally aspirated engine pirated version suffers a penalty in the power output and showed an advantage in efficiency by the Miller cycle. Even though the supercharged Otto engine adopting this cycle version has no efficiency advantage, the increased network output is provided with propensity reduction to engine knock problem.

They were of the opinion that the Miller cycle has no inherent potential for increasing efficiency, but by its usage pressure boost increase in turbo-charging or supercharging can actually increase work output with decreased danger of pre-ignition and less air pollution and that the cycle design can be achieved using sensitivity analyses. It is possible for the Miller cycle to be the extreme limits for the industry of ICEs.

9.5   MULTIPLE OR SPLIT INJECTIONS

In order to satisfy the stringent emission norms, many leading manufacturers and researchers are adopting the novel techniques for the development of clean diesel engines. In the process of development of an optimum design with low emissions and high performance, computational design and simula-tions are found to be more promising by the researchers unlike experimental setup which is both time and cost inefficient.

Of late, multiple or split-injection scheme has been traversed by many to accurately control the fuel injected per cycle and also to reduce emissions. The present paper deals single-zone phenomenological model development for the combustion process in both conventional and a split injection diesel engine. In the process of consideration of heat loss effect and mixture aver-aged gas constant, a numerical solution of energy equation is assessed. Pre and main injections were designed using physics-based models for ignition delay, fuel injection, diffusion, and premixed heat release rates. The present study also inferred on influence of EGR, ambient conditions and fuel injec-tion timings on pollution from diesel engine. In this design process, two different multiple-injection levels were implemented to observe the effi-ciency of the techniques and found that EGR levels were changed from 0 to 20%. It is observed that with incorporation of split injection NO_x emissions were reduced to greater extent without affecting the engine performance unlike multiple injection system. It is concluded that the incorporation of split-injection system in diesel engine with electronic controls could be a better option for reducing exhaust emissions (Figures 9.9 and 9.10).

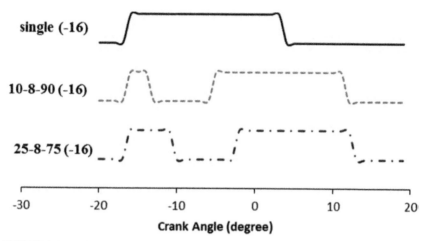

FIGURE 9.9 Schematic model showing injection schemes.

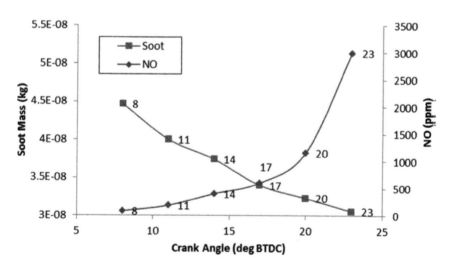

FIGURE 9.10 Comparison of predicted and NO-soot emissions over a range of injection timings. Numbers in the figure indicate the injection timings (BTDC).

A comprehensive nonlinear, zero-dimensional model for DI diesel engine cycles has been studied with prominence laid on the important procedures taking place inside the cylinder and mathematically describing them with the help of semi-empirical, empirical, and physical laws. They felt that the model inferred about performance improvement and engine exhaust emissions reduction. They studied numerically by taking the effect of boost pressure and EGR. They varied the percentage of amount of fuel injected into two phases like 10-8-90, 25-8-75 (Figure 9.9). The dwell and start of injection maintained constant at 8 and 16, respectively. It is concluded that lower the first phase of the amount of fuel, lower are peak temperatures and thus the lower NO_x were observed. However, this has given rise to lower power and is compensated by increase in boost pressure. Between the two split-injection systems, the one with low pilot quantity will have low NO_x emissions They found that Split-injection system is best, severing the purpose of reducing the NO_x emissions without losing the engine's performance. The split system introduced a trade-off between and NO_x emissions (Figure 9.10).

For this purpose, a computer code was developed in C++ and validated with the experimental results. Then a split technique was incorporated into the model. It is noted that for a case of 75(8)25 retarded to injection timings of 12^0BTDC, 8^0BTDC, and 4^0BTDC from 16^0BTDC has resulted in 23.7%, 19.6%, and 16.2% in soot emissions from the baseline soot emissions. Split injection of 75(8)25 retarded to injection timings of 12^0BTDC, 8^0BTDC and 4^0BTDC from 16^0BTDC has resulted in 34.3%, 70.8%, and 76.2% in NO_x emissions from the baseline NO_x emissions. With the comparison of conventional techniques of retarded timings and EGR, split strategy was observed to yield better results and it could be emphasized that effective abatement of harmful emissions can be achieved through split injection technique.

For IDI and DI diesel engines, split injection is a promising strategy to simultaneously mitigate NO emissions and soot provided injection timing is optimized. However, an optimum injection strategy of split injection for IDI and DI diesel engines has been always under investigation. In the present paper CFD code, FIRE was used in order to study the influence of the split injection on pollution and combustion of IDI and DI diesel engines. In the process of optimizing split injection system, three different schemes, in which 10, 20 and 25% of total fuel injected in the second pulse, have been considered. The delay dwell between injections is varied from 5°CA to 30°CA with the interval 5°CA (Figure 9.11).

Advancing or retarding the injection timing cannot decrease the soot and NO_x trade-off by itself. Hence, split injection is needed. For the 75% (25) 25% mixture case, soot and NO_x emissions are lower than the other cases as premixed combustion which is the major factor for NOx formation is comparatively low. Soot oxidation rates increased as the second injection became more lean and hot. The lowest Soot and NO_x emissions are related to the 75%-20–25% and 75%-15–25% cases. Optimum case was found to be 75%-20–25% case considering the highest average of soot and NO$_x$ reduction.

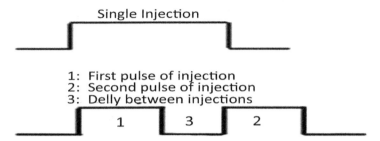

Single Injection

1: First pulse of injection
2: Second pulse of injection
3: Delly between injections

1   3   2

FIGURE 9.11   Schematic diagram of single and split injection systems.

There is a growing trend in the use of AFs especially biodiesel in diesel-fuelled CI engines (Myung Yoon Kim et al., 2006):

- Since the majority of the multi-cylinder engines are provided with electronic fuel injection (EFI) systems with a specific common rail system injection. For such systems, multiple injections or split injection systems can be readily adopted.
- The split system may also be explored for single cylinder engines as it gives more flexibility in fuel consumption.
- The split system is advantageous for engines with DPFs as it permits effectively to implement regeneration technique.

9.6   STRATIFIED CHARGE CONCEPT

If we look into the historical development of ICEs, in earlier days, majority of innovations were done on the gasoline engines. As these engines were mainly homogeneous charge engines, carbureted, and hence prone for poor part-load efficiency and high CO and NO$_x$ emissions. The part-load efficiency can be overcome by employing differential AFs under different load conditions. The idea was given to the development of stratified charge engine concept. Therefore, in the 1970s much effort was devoted to the development of various stratified charge engines. By the 1980s, however, high-compression lean-burn systems had been the main practical outcome in this direction. With increasing pressures for fuel economy as a means of reducing CO_2 output, interest in stratified charge has gained interest by the engine research community in the early 1990s.

The conventional homogeneous charge engine, namely, engine that works on the Otto cycle has undoubtedly worked well with fairly good full load efficiency but plagued with the problems of fuel economy, mechanical loading, AF utilization, and combustion emissions. In a report to the EPA, Ricardo engineers presented that the two major variables, specific heat ratio, and compression greatly influence the thermal efficiency of the IC engine. In many of the practical cycles, the piston speed will be slow at TDC compared to rate of heat releasing, so they are considered as Otto or nearly Otto cycles. The specific heat ratio is directly influenced by mixture strength and equivalence ratio of the working gas. The lower the equivalence ratio of the working gas, higher the thermal efficiency.

Efficiency relationship for the theoretical Otto cycle:

$$\eta_{Otto} = 1 - \frac{1}{r_c^{\gamma-1}}$$

Therefore, from the above relation, it found that thermal efficiency becomes maximum whenever the equivalence ratio is minimum at constant CR.

With the variation in fuel delivery, the engine's power output is controlled. The Diesel engine can be operated successfully at high CRs and low equivalence ratio, at light loads because of nature of the ignition process and inherent stratification of the fuel. In the homogeneous charge gasoline engine the inherent knock phenomena lowers the CR and ignition requirements affects the equivalence ratio.

Definition and Classification: All engines that are not homogeneous are, by default, stratified in some way or other. Due to injection of fresh charge and ejection of gases, even the homogeneous charge engines are stratified in some way.

In order to avoid the confusion between diesel and hybrid stratified charge engines Bamescu (28) modified DISC (direct injection stratified charge) to SADI (spark assisted stratified charge). Haslett et al. (1976) developed a sound definition which says that "a stratified charge engine is an engine where the combusting gases are not, by design, homogeneous throughout the combustion chamber but rather, separated either physically or aerodynamically into igniting, combusting and working regions having differing air to fuel ratios for the purpose of altering the combustion and emissions characteristics of an equivalent homogeneous charge. The igniting and combusting charge is then characteristically different from the working gas which may be air, re-circulated exhaust gas or a combination thereof." The igniting and combusting charge is then characteristically different from the working gas which may be air, recirculated exhaust gas or a combination thereof.

The basic concept of providing varying AFs was achieved either by carburetion alone or combination of carburetor and fuel injector together. The carburetor stratification was achieved by adopting two carburetors: one is for providing lean mixture and another for rich mixtures. The difficulty was to timing of two carburetors and hence many OEMs have preferred to use combination of carburetion and fuel injection (Figure 9.12).

There are many ways to complete such stratification and combust the mixture out of which three of them are illustrated in Figures 9.13 and 9.14 serve as examples.

- The Ford PROCO Engine (Single combustion chamber, early injection).
- The Porsche Engine (Two combustion chambers, fuel injection).
- The Texaco Combustion Process (Single combustion chamber, late injection).
- The Honda CVCC engine (Divided Combustion Chamber with Separate Inlet Valve).
- The GM-ACS engine (Timed Port Injection, Axially Stratified Engines).

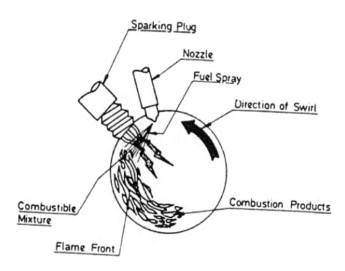

FIGURE 9.12 Texaco combustion process.

The following are the advantages and disadvantages of stratified charge engines.

9.6.1   OVERALL ADVANTAGES

1. The theoretical efficiency of un-throttled stratified charge engine is higher in turn increases the fuel economy than that of the conventional engine. However, this is not true for actual engine incorporated with pre-chamber, due to heat transfer and throttling effects and with single chamber engine, due to misfire and incomplete combustion. When compared to conventional engines only the axially and late injection stratified charge engines and have shown higher fuel economy provides the ignition and combustion are properly done.

2. With increase in quantity of air in cylinder of un-throttled stratified charge engine the CO emissions got reduced.

3. Pre chamber engines releases low NO_x emissions when compared to other type of stratified engines and conventional engines.

4. Neglecting the emissions and efficiency, some stratified charge engines have shown multi-fuel capability and operate the speed at corresponding load.

5. Stratified charge engines start up easily unlike conventional engines.

9.6.2   OVERALL DISADVANTAGES

1. Thermal efficiency of the engine is lower than that of the conventional SI engine.

2. Due to incomplete combustion of lean mixtures or misfire of charge at light load the HC emissions of un-throttled stratified engine is high.

3. Stratified charge engines generate high noise compared to conventional engine.

4. Stratified charge engines are costlier than the conventional gasoline engine as they are more complex and heavier.

5. Stratified charge engines are not the solution for emissions.

9.6.3   *DIRECT IN-CYLINDER STRATIFIED CHARGE INJECTION (DISC)*

In this system, the injection is done at low pressure and it begins at the early crank angles of compression process, increasing the evaporation time and mixing of fuel. Under these circumstances, uncontrolled initiation of combustion could occur, so spark-ignition is required.

The concept, to be successful, requires the mixture at the time of ignition to be distributed (stratified) in an appropriate manner. Control of the fluid mechanics of both air and gas flows is critical. As far as the injection itself is concentrated, problems of droplet and film evaporation, mixing, and wall impact are as important as in current systems. The evaporation rates of mixture increases as the cylinder warms up. Nevertheless, the DISC system may not completely eliminate lag-induced mixture variations during transients particularly under cold conditions if there is any direct impingement of fuel on the cylinder wall or piston crown.

9.7   WATER-IN-DIESEL EMULSIONS

Researchers have developed both in-cylinder and post-combustion approaches for effective reduction of harmful emissions from diesel engines especially NOx and soot emissions that sometimes demanded retrofitting of the existing engine. Since high temperatures during combustion are responsible for NOx formation, addition of water to fuel was thought of a means of reducing NOx emission. This can be a solution based on fuel and has been achieved through various ways such as using an isolated injector to inject water into the cylinder, water is sprayed into the air intake and water with diesel emulsion.

It has been proved that introduction of water into the engine generates the cooling effect which reduces the peak temperatures inside the cylinder that causes the reduction in emissions. Though the approach of water addition yields lower NOx emission but it reduces the power output of an engine as the heat value of water-diesel mixture is less and the tendency of the water and fuel to be separated. If the parts are directly in contact with water, it causes corrosion of the parts. On the other hand, the introduction of water in to the diesel fuel breaks the bigger particles of fuel into smaller particles and helps in complete combustion; this method is known as micro-explosion. The addition of water also seems to play its part by reducing soot particles during the combustion process; this can be accredited to micro-explosion phenomenon. The physical and chemical kinetics of the combustion process were altered by the presence of water vapor content in the fuel. Water vapor lowers the temperature of combustion and alters the reactants chemical composition by which the reduction in NOx and PM formations in the cylinder is observed.

According to the percentage amount of water in the diesel, subsequently, the exhaust emissions from the engine will be reduced. Hence, in order to enhance the efficiency of the engine and reduce the NOx and PM emissions, water-in-diesel emulsion (WiDE) fuel observed as an AF. Even though number of studies is done with addition of water to diesel, no concrete proportion has been finalized. This can be attributed to the incomplete understanding of the behavior of WiDE under combustible condition and the complexity of the operating parameters of engine.

An emulsion is heterogeneous mixture of two different liquids (water and oil) out of which one will be in dispersed phase and the other in continuous phase such emulsions are known to be two-phase emulsions. There are two basic forms in it, when oil droplets are encased within the water it is called an oil-in-water (O/W) emulsion and when water droplets are encased within the oil it is called as water-in-oil (W/O) emulsion.

In order to obtain a stable emulsion there are three conditions to be satisfied they are (i) the two liquids must form a heterogeneous mixture, (ii) sufficient turbulence must be created in order to disperse one liquid into the other, and (iii) an emulsifier or a combination of emulsifiers should be used.

The compound which reduces the surface tensions between two mutually insoluble fluids is called as Emulsifier. They contain both hydrophilic groups (the heads) and hydrophobic (the tails). For generating best emulsions, lipophilic (oil liking)—hydrophilic (water liking) balance (HLB) score is developed. Water-in-oil-emulsion is given by Low HLB score whereas oil-in-water-emulsion is with high HLB values and these values of HLB range between 1 and 20. The desired qualities to be possessed by the emulsifier used in generating water-in-diesel should burn with ease and it should not result in components like soot, sulfur, nitrogen, and they should not have any sort of effect on the physical and chemical properties of the fuel.

The best emulsifiers that can be used are from the aliphatic hydrocarbon family. Generally, the quantity of the emulsifier that has to be added ranges from 0.5–5% by volume. The stability of the diesel emulsion is influenced mainly by the technique used for emulsification, the time taken for emulsification, stirring speed (or ultrasonic frequency), concentration of emulsifiers, fraction of water by volume (dispersed phase) and viscosity of continuous phase (diesel oil). The experimental work by Chen and Tao carried out experiments to study the effect of the dosage of emulsifier, stirring speed, emulsifying temperature, and oil-water ratio on the stability of water-in-diesel emulsion that was generated by mechanical agitator (also called Homogenizer). They concluded that the rise in oil to water ratio, stirring speed, and duration had desirable effects on stability, but a negative effect was noticed with an increase in emulsifying temperature.

Lin and Chen reported that an enhancement in the performance of the engine with reduction in CO emissions was noticed when the emulsion was produced by using ultrasonic vibrator when compared to that produced using mechanical agitator. While on the other hand, the study carried out by Lin and Wang on the effect of speed of mechanical homogenizer machine, it was concluded that the speed has a definite noticeable effect on the size of the liquid droplets. It was noticed from the obtained outcomes that the diameter of the liquid droplets decreased with the increase in speed of the stirring process. Apart from this observation, it was also noticed that the selection of emulsifiers, choosing a suitable frequency of the agitator and the duration of agitation have a considerable importance in the formation of stable emulsions.

As a pre-requisite for a fuel to become good AF to compression-ignition engines (CI engines), use of such a fuel should result in short ignition timing, high cetane number, a suitable fuel viscosity and volatility.

These physical and chemical properties have a great impact on the fuel injection process and the combustion process as well. For instance, the density of fuel has a considerable effect on the fuel mixing process that occurs in the combustion chamber and the resulting viscosity would affect the injection process.

The distribution of a proper droplet size is an important factor to be considered for an emulsion. A stable emulsion consists of properly distributed droplet size which preferably ranges from 0.05 to 1 μm an appropriate HLB value of the emulsifiers is a key factor.

When the emulsions are prepared, one should immediately use it in engines, otherwise due to denser and lighter fraction; the problem of sedimentation would occur and result in separation of water and diesel. Many studies have been done to observe the emulsion stability and compare between (a) unstable emulsion, (b) unstable emulsion with water layer and (c) stable emulsion. It is felt that visual observation could be a better way and standard procedure to determine the stability of the emulsion.

At room temperature, the mixed emulsions were maintained steadily in clear glass bottles and the water-in-diesel emulsion's stability was observed for a period of 2–12 weeks. During this observation process the emulsion changes because of few processes which are time dependent among which Ostwald ripening flocculation, creaming, coalescence, and sedimentation are important. Sedimentation and creaming are also known as de-emulsification processes, in which separation of emulsion into two emulsions occurs. This phenomenon is due to the presence of external force field which is usually gravitation. Whenever the gravitational pull on the droplets exceeds the thermal forces of the droplets, Brownian motion starts which results in the development of a concentration gradient in which the larger droplets settle at either top of the emulsion or at the bottom of the emulsion based on their density measured relative to water.

Many researchers performed experimental studies to observe the effect of water-diesel emulsion and also water injection into the intake manifold on performance, combustion, and emission characteristics of a DI diesel engine under similar operating conditions. By maintaining static injection, timing of 23⁰BTDC, in a typical stationary type engine, at constant speed of 1500 rpm at different outputs and water to diesel ratio for the emulsion was 0.4:1 by mass. The same water-diesel ratio was maintained for water injection method in order to assess both potential benefits.

In the first phase, water-diesel emulsion, prepared by using surfactant HLB: 7 was injected during compression stoke in a conventional manner in order to assess the emission characteristics, combustion, and performance of the engine. In the second phase, during suction stroke an auxiliary injector was used to inject the water into the intake manifold of the IC engine. An ECU (electronic control unit) was incorporated to control the injector operations. From the experimental result, it was inferred that the NO emissions was reduced remarkably with the implementation of these two methods (injection and emulsion). At full load, about 40% reduction in was observed NO emission in both cases. But, NO emission reduction is lesser with injection than emulsion at part loads. Also, 25% reduction in smoke emission however, higher CO and HC levels were observed with emulsion than water injection. As regards NO and smoke reduction, the emulsion was superior to injection at all loads. Peak pressure, ignition delay, and maximum rate of pressure rise were lesser with water injection as compared to the emulsion.

It was concluded that the brake thermal efficiency is reduced at all outputs below diesel values with water injection due to poor combustion. At high outputs, the brake thermal efficiency with the emulsion is significantly above the values with water injection. It is even better than base diesel operation at full load.

The ignition delay is much higher with the emulsion as compared to water injection. The peak pressure and maximum rate of pressure rise are also higher with the emulsion due to the high ignition delay. The diffusion combustion phase is prominent with water injection than the emulsion. On the whole, water-diesel emulsion is more effective in improving full load brake thermal efficiency and lowering NO and smoke levels. Water-diesel emulsion results in higher ignition delays, peak pressures, and rates of pressure rise.

Through this study, it was established that emulsion method is superior in simultaneous reduction of NO and smoke emissions at all loads than injection method (Subramanian). Also, many studies on emulsions have been undertaken to study the effect of surfactants have been tried out.

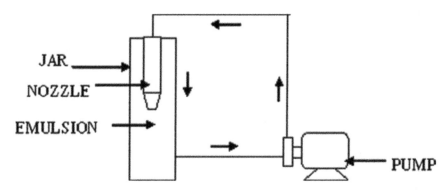

FIGURE 9.13 Emulsion preparation apparatus.

The studies with emulsions revealed that the NO and smoke emissions with emulsified fuel are far less than with neat diesel fuel.

The role of water-in-diesel emulsion and its additives on diesel engine performance and emission levels: retrospection.

In order to reduce the emission gases from the diesel engine, significant effort were made in terms of using water-in-diesel (W/D) emulsion fuels. They have the capability of increasing the performance levels of engine by simultaneously reducing the particulate matter and NO_x emissions. Besides, the principle of W/D emulsion fuel, physio-chemical properties, stability of the emulsion fuel, and their impact of W/D emulsion fuel on, performance, emission, and combustion characteristics. The changes in the emission characteristics and engine performance which the incorporation of nano-additives within W/D emulsion fuel and nano-fuel synthesis were discussed. Finally, through this survey optimization of engine parameters, the confirmation of nanoparticle size in nano-fuels and controlling it from engine exhaust were studied.

Prof. B. Hopkinson introduced the method of water introduction into the Diesel engine combustion zone. The sole motive of this idea was to enhance the thermal efficiency of the diesel engine, to reduce the exhaust emissions and to achieve a better intercooling in the gas engine. Generally, three important ways are implemented to inject water into the diesel engine: (i) DI of water into the combustion chamber using a normal injector, (ii) fumes of water introduced into the engine intake air, and (iii) in emulsion form.

The experimental investigation of all the above methods confirmed that, diesel-water emulsion fuels are the most appropriate which allows the attainment of desirable characteristics even without retrofitting of the engine. In addition to this, the micro-explosion of water particles during the combustion process helps to enhance the atomization and mixing of air-fuel mixture.

A heterogeneous mixture of two liquids in which one of the liquid is in dispersed phase within the continuous phase is known as emulsion. High energy emulsification methods in the field of surface active agents (or) surfactants are used to generate an emulsion. These high-energy methods include mechanical agitation, high-pressure homogenizers, ultrasonic, and supersonic vibrations. The ultrasonic and supersonic vibrations have a negative effect on NO_x and smoke emissions are concerned when weighed against the emulsions generated by mechanical agitation.

As the emulsion fuels slowly get segregated into two immiscible phases by droplet-droplet coalescence, flocculation, and creaming, surfactants are added to stop the destabilization of the emulsion and also to attain kinetic stability to the emulsion fuel. Surfactants lower the surface tension between water and fuel molecules thereby propelling towards attaining stability to the emulsion.

From the above conclusions, the following suggestions are derived for future involvement in the domain of W/D emulsion fuel. The optimization of the process parameters of W/D emulsification such as stirrer speed, water concentration, etc.; can be done to attain better physical and chemical properties with maximum stability.

The size of nanoparticles introduced and their dispersion characteristics over emulsion surface are to be checked by appropriate measurements.

The use of advanced mixing technique, an experimental investigation (SAE 2007-24-0132) and emulsifying agent enabled the production of stable emulsion of up to 30% water in diesel for up to one week. The water/diesel emulsion was stable for long enough time for transportation and use. In some of the water/diesel ratios, the emulsion was stable for about 4 weeks. The physical properties of the stable W/D emulsions in terms of density, surface tension, and viscosity were measured and investigated. And the water droplets distribution within the diesel phase. The addition of water in diesel generally affected the engine combustion noise, brake power output, and specific fuel consumption.

9.8   CYLINDER DEACTIVATION (CDA)

Cylinder deactivation (CDA) is a popular and effective technology to improve spark-ignition engines' efficiency at part load. SI engines capability of reducing the pumping losses by deactivating a fraction of cylinders during part loads and operating the cylinders in active mode at higher loads thereby realizing higher efficiencies is appreciable. This approach of CDA may serve as an alternative solution or may be used in conjunction with the other efficiency enhancing measures such as variable valve actuation (VVA) and engine downsizing. The deactivation of the cylinders is done by seizing the intake and exhaust valves and in order to minimize the pumping losses, the 'gas spring' nature of the entrapped charge comes in handy. This chapter focus the effects and possible benefits of CDA on a four-cylinder turbocharged downsized gasoline engine equipped with multi-air VVA system were experimentally investigated, with a clear aim of reducing the pumping losses which is not in reach with the conventionally employed early intake valve closure (EIVC) strategies. As the MultiAir VVA system is resistant to the deactivation of exhaust valve, an innovative approach was used in which internal exhaust gas recirculation (iEGR) in the inactive cylinders is carried out which reduces the pumping losses in a considerable fraction. This technique was observed to be effective for low speed and lower load operating region of the engine map, below 3 bar BMEP and 3000 Rpm, by generating a 30% reduction of pumping losses and a 8% improvement in fuel conversion efficiency relative to the EIVC unthrottled load control.

By not using, an engine is the best way of saving fuel, is the concept behind CDA, and has been implemented by few automakers to help improve fuel economy and reduce emissions. This method is specific to multi-cylinder engines.

The trade name of this concept differs from one manufacturer to other, but the overall concept is the same: It's just similar to testing the engine using Morse test wherein one of the cylinder are made inoperative, i.e., such a cut-off cylinder in a multi-cylinder doesn't contribute to total power developed. Hence, the power developed by the engine is reduced as cut-off cylinders don't get any fuel.

Deactivation is mostly used on V6 or V8 engines, where, in principle, it reduces the engine's displacement when it functions: Bigger-engine power when all cylinders are activated and smaller-engine fuel economy when some are shut off. Some automakers prefer to use small turbocharged engines instead, which force in extra air and fuel to provide more power when needed. Essentially, deactivation is a larger engine that can act like a smaller one, while turbocharging is a smaller engine that can perform like a larger one. (Some automakers combine deactivation and turbocharging on their engines, as well).

When some cylinders are deactivated, no air is going in or out of them, so there's no pumping loss. Beyond that, as the engine automatically compensates for those "missing" cylinders, it creates less of an intake-exhaust pressure difference. This reduces pumping loss in the active cylinders, making them work more effectively. Although they're helping to move the deactivated pistons, since they're all attached to the crankshaft, the engine is still working more efficiently overall.

This concept is possible and controlled by the engine's engine control unit (ECU) or engine control module (ECM). As soon as more power is needed, such as when you accelerate, the system brings the deactivated cylinders back online. The transition is usually so smooth that it's almost impossible to detect. It is estimated that CDA can reduce fuel consumption and emissions by 4-10%.

This system gives a flexibility to work on low load to high load from minimum to maximum cylinder at a time respectively.

Most engines with CDA turn off half of them at a time, such as an eight-cylinder that switches to four cylinders. Honda's system, which it calls Variable Cylinder Management, can switch a V6 engine to run on three or four cylinders, depending on what's best for the driving conditions. General Motors is also contemplating to put into market by 2019.

The system of CDA was first introduced by Cadillac in 1981. It was called Modular Displacement, and it could switch a V8 engine to run on six or four cylinders. Mitsubishi brought in a four-cylinders-to-two version shortly afterwards. The electronics and fuel systems of the day weren't up to the task, and neither company kept it very long. Today, there aren't many disadvantages to deactivation systems, other than they add some cost and complexity to the engine.

And some vehicles now include a start/stop function on their engines. While this used to be exclusive to hybrids, it's now showing up on conventional gasoline vehicles, and even on some light-duty diesels. Automakers are adding it to further reduce fuel and emissions numbers.

When you come to a full stop with your foot on the brake, such as sitting at a red light, the engine shuts off. The vehicle's lights, stereo, and climate control continue to operate, and certain conditions must be met, including ambient and engine temperature. The engine automatically restarts as soon as you take your foot off the brake.

Many a researcher has studied the physics behind both experimentally and computationally. The concept of engine CDA is also called a variable displacement engine or engine downsizing. The simulation-based investigation is motivated by a desire to enable unthrottled operation at part load, and hence eliminate pumping losses. Other salient features of the proposed engine are turbocharging and CDA. The CDA combined with variable displacement further expands the range of unthrottled operation, whereas turbocharging increases the power density of the engine and allows downsizing without the loss of performance.

A variable displacement turbocharged engine (VDTCE) concept enables operation was proposed by Fernando Tavares et al. They developed an engine model in AMESIM using physics-based models, such as thermodynamic cycle simulation, filling, and emptying of manifolds, and turbulent flame entrainment combustion.

Therefore, the VDTCE is integrated with a hybrid power train to mitigate issues with engine transients and mode transitions.

A predictive model of the power-split hydraulic hybrid driveline is created in SIMULINK, and it was integrated with the engine. The design and control attributes may be taken care by this integrated simulation tool prior to the determination of the potential of fuel economy of the power train comprising a hydraulic hybrid driveline and VDTCE engine.

A system capable of operating and controlling the torque-speed range throughout the engine usage in unthrottled state. Controlling the load by deactivating cylinders, reduced displacement, and/or, simultaneously keeping the throttle wide open produces very significant efficiency gains at low-load, but there is a limit below which the throttling would still be necessary. The nominal engine displacement is 2 L, and the variable displacement mechanism allows reduction by a factor of 2. The incorporation of the engine with the driveline and vehicle dynamics is done in SIMULINK, thus enabling the development and implementation of the supervisory controller.

Given the properties of the hydraulic components are unique unlike electric counterparts, a novel strategy was required. In particular, while the energy conversion and power density of components are comparatively high, the relatively low energy density of the hydraulic accumulator requires careful management of the state-of-charge.

During the normal operation, the modulated control maintains a low level SOC in order to maximize the capacity for regeneration and in order to obtain maximum efficiency and avoid unfeasible regions for un-throttled operation, concentrate son engine operating points around the optimal BSFC trajectory. They compared the conventional baseline containing 3.6 L five-speed automatic transmitted naturally aspirated engine to the PS-HHV power train with a VDTCE in order to assess its behavior and the fuel economy potential. At relatively low speeds, the engine operates smoothly in the PS-HHV, without high-frequency torque fluctuations. The average efficiency of the VDTCE over the transient schedule is more than twice that of the conventional baseline; hence, most of the vehicle fuel economy improvement can be attributed to exceptional part-load efficiency of the VDTCE. However, due to its ability to maintain the engine in the specified speed/load area and avoid the regions of unfeasibility, the power-split hybrid system is the key enabler. For the mid-size passenger vehicle, the investigation on the advanced power train configuration is a high-performance work option. Therefore, although estimated fuel economy improvements are appreciable, they do not represent the ultimate capacity. With the relaxation of performance constraints, there is possibility of further downsizing the engine and adjusting the hybrid system design. The further extension of the VDTCE study to incorporate a realistic assessment of mechanical losses associated with the variable displacement mechanism and possible benefits of using a variable CR is apparent. The torque variation with deactivation of cylinder is illustrated in Figure 9.14.

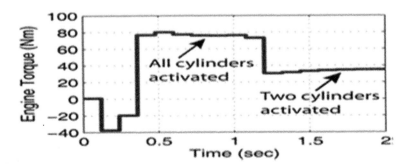

FIGURE 9.14   Variation of average engine torque with time under cylinder deactivation.

Improving the economy of partial-load fuel of 4-cylinder petrol engine with combination of variable valve timing and cylinder-deactivation through double intake manifolds.

Part load fuel economy should be improved greatly for the SI engines in passenger cars which run mostly in urban areas in part-load conditions.

In the middle to low load region the fuel economy could be improved by using CDA and variable valve timing (VVT). By deactivating the intake and exhaust valves of inactive cylinders with the help of current CDA technologies, the system cost and complexity increases. An innovative CDA method is proposed for 4-cylinder SI engines through double intake manifolds (CDAM). By combining the VVT and CDAM along with LIVC (late intake valve closure), a load control idea was investigated. At 4000 rpm and 2000 rpm they observed that the fuel economy was improved by 3.1%–9.2% and 5.5%–17.6%, pumping losses were decreased by 24.5%–35.3% and 58.9%–65.6% in CDA respectively. With the application of double intake manifolds, the fuel economy on the full-cylinder mode was improved by 0.68%–2.95%. Even though the improvement in fuel economy is less compared to the CDAV method, CDAM method is cost effective and less complex in nature.

Baseline engine and the CDAM engine with double intake manifolds have been established. The simulations models are used know the efficacy of the strategy in improving the fuel economy. Some conclusions have been achieved in the following:

1. For low speed range, the effectiveness in the improvement of fuel economy is almost same in CDAV and CDAM using load control strategy, whereas CDAV method has shown the efficient results in high speed range, by deactivating the intake and exhaust valves in the inactive cylinders. For urban area driving applications, CDAM is better.

2. The uniformity in the intake process of 4-cylinders of the CDAM engine is acceptable because of the presence of double intake manifolds and this can be augmented for better results in future. In the FCM load range the intake pressure increases resulting in the reduction of reducing the pumping losses.

3. In the CDA mode, re-optimizing of VVT timings and incorporation of LIVC strategy was performed which resulted in decrease in the pumping losses and improved fuel economy.

4. The efficacy of LIVC operations can be extended with the CDA application which can further reduce the pumping losses in low load range.

5. The CDAM method doesn't require the special mechanisms for the major engine modification and CDA. The double intake manifolds are cost effective and can be controlled easily.

9.9 APPLICATION OF NANO-PARTICLES IN REDUCTION OF HARMFUL EMISSIONS

Particle of dimensions in 1×10^{-9} and 1×10^{-7} m range (or nanopowder or nanocluster or nanocrystal) or particle having at least one dimension less than 100 nm is said to nanoparticle. Ultrafine particles falls between 1 and 100 nm in size, and fine particles are between 100 and 2,500 nm, and coarse particles falls range between 2,500 and 10,000 nm.

Conventional materials formed from nanoparticles have different properties compared to actual. This is typically due to their greater surface area per weight compared to larger particles which causes them in developing unique properties compared to others molecules. Nano-fluids are a liquid suspension of nano-particles which is diluted.

Nano-fluids have enhanced thermo-physical properties according to previous data which is due to increasing of nano-particle's volumetric fraction. Some of the properties are viscosity, thermal conductivity, convective heat transfer coefficients, and thermal diffusivity when compared with oil or water.

Coolants, lubricants, Engine oils, automatic transmission fluids, and other synthetic high-temperature heat transfer fluids—have poor heat transfer properties. These could be enhanced by nano-fluids by addition of nano-particles. The nano-fluids with addition to base fluids have been used to enhance performance of radiators (decreasing frontal area by 10% by Argonne researchers) and also found to increase combustion characteristics when added to fuel. Nanofluids improved the performance of vehicular Brake fluids.

Studies were also undertaken to effectively reduce NOx emissions from engines (Tayfun et al., 2015). They conducted a study on oxygen addition containing nine different nanoparticle additives for NOx emissions. They carried out variable speed test under full load conditions. For this purpose, zinc oxide (ZnO), aluminum oxide (Al_2O_3), iron oxide Fe_2O_3, magnesium oxide (MgO), titanium oxide (TiO_2), silicon oxide (SiO_2), nickel oxide (NiO), nickel-iron oxide (NiFe$_2$O$_4$), nickel-zinc iron oxide ZnO.5NiO.5Fe$_2$O$_4$, and nickel-iron oxide (NiFe$_2$O$_4$) were added to diesel fuel in three different proportions of 25, 50, and 100 ppm. The study was aimed at identifying an optimum additive and dosage. Hence it can observed that NOx emissions can minimized by the addition of nanoparticles (Figure 9.15).

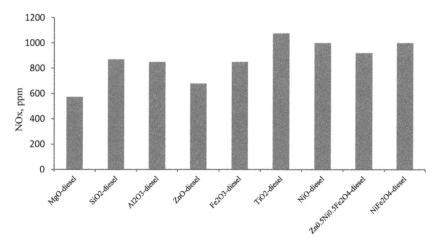

FIGURE 9.15 Comparison of NOx emissions with oxygenated nano-particle-diesel blends, with 100 ppm dosage under full load and 2800 rpm condition.

The study revealed that NOx emissions in diesel-fuelled engine are found to decline with the addition of all nine types of the studied nanoparticle dosages excluding Al$_2$O$_3$ nanoparticle. The peak average reduction was observed by MgO nanoparticle of addition dosage of 100 ppm as 16.7%. The peak average increase was observed by Al$_2$O$_3$ nanoparticle of 25 ppm as 4% addition dosage. NOx emissions by Al$_2$O$_3$ can be eliminated by using EGR due its high EGR tolerance. Hence, results of oxygenated nanoparticles excluding aluminum-based nanoparticle indicates potential of reducing NOx emission in diesel engines. Trends with different dosages shown differently but the trends shown in Figure 9.15 when engine operated under full load and speed of 2800 rpm condition with 100 ppm of dosage maintained uniformly.

Experiments were also done on engines fuelled with biodiesel and studied the effect of addition of nanoparticles. A minimal amount of carbon-coated aluminum (Al@C) nanoparticles of 30 ppm mass fraction in dosage is added. The engine emission characteristics have resulted in the declining of NOx and CO emission with marginal increase in THC emissions. Effect of additive nanoparticles on different properties of fuel and enhancing them optimistically by dosage level are studied. Emission reductions are observed by adding of cerium oxide to the diesel fuel in different proportions (Sajith et al., 2010; Selvan et al., 2009).

KEYWORDS

- **exhaust gas recirculation**
- **homogeneous charge compression-ignition**
- **indicated specific fuel consumption**
- **ringing intensity**
- **thermal barrier coatings**
- **velocity magnitude**

REFERENCES

Ali, M. S., Islam, S. N., Sapuan, S. M., & Megat, H. M. (2001). Effects of compression ratio on turbulence kinetic energy and dissipation in the suction stroke of a four stroke internal combustion engine. 4th International Conference on Mechanical Engineering, December 26–28, 2001, Dhaka, Bangladesh/pp. III 165–172.

Alkidas, Alexandras C. (1987). Experiments with an Uncooled Single-Cylinder Open-Chamber Diesel. *SAE Transactions 96*, 280–290.

Amin Paykani, Amir-Hasan Kakaee, Pourya Rahnama, & Rolf D. Reitz (2015). Progress and recent trends inreactivity-controlled compression ignition engines. *International J. Engine Research*, 1–44.

Assanis Dennis N., & Scott B. Fiveland. (2016). A four-stroke homogeneous charge compression ignition engine simulation for combustion and performance studies. SAE Paper No.2000–01–0332.

Cho, H. M., & He, B. Q. (2008). Combustion and emission characteristics of a lean burn natural gas engine. *International Journal of Automotive Technology, 9*(4), 415–422.

Christensen, M., Johansson, B., & Einewall, P. (1997). Homogeneous Charge Compression Ignition (HCCI) Using Isooctane, Ethanol, and Natural Gas – A Comparison with Spark Ignition Operation, SAE Paper 972872.

Currana, H. J., Gaffuria, P., Pitza, W. J., Westbrook, C. K. (2002). A comprehensive modeling study of iso-octane oxidation. *Combustion and Flame, 129*(3), 253–280.

Enrico Mattarelli, & Valeri I. Golovitchev (2013). Potential of the Miller cycle on a HSDI diesel automotive engine Carlo Alberto Rinaldini, *Applied Energy 112*, 102–119.

Ganesh, D., & Nagarajan, G. (2010). Homogeneous charge compression ignition (HCCI) combustion of diesel fuel with external mixture formation. *Energy, 35*(1), 148–157.

Hasimoglu, C. (2008). Exhaust emission characteristics of a low-heat-rejection diesel engine fuelled with 10 percent ethanol and 90 percent diesel fuel mixture. *Proc. ImechE, 222*, Part D: *J. Automobile Engineering JAUTO633 F IMechE*.

Haslett, R. A., Monaghan, M. L., McFadden, J. J. (1976). Stratified Charge Engines, *SAE Paper No. 760755.*

Kamo, R., & Bryzik, W. (1984). Cummins/TACOM Advanced Adiabatic Engine. *SAE Transactions, 93*, 236–249.

Kitamura, Takaaki, Takayuki Ito, Yasutaka Kitamura, Masato Ueda, Jiro Senda, & Hajime Fujimoto (2003). "Soot Kinetic Modeling and Empirical Validation on Smokeless Diesel Combustion with Oxygenated Fuels." *SAE Transactions, 112*, 945–63.

Krishna, B. M., & Mallikarjuna, J. M. (2011). Effect of engine speed on in-cylinder tumble flows in a motored internal combustion engine - an experimental investigation using particle image velocimetry. J. Appl. Fluid Mech, 4(1), 1–14.

Mack J. Hunter, Daniel Schuler, Ryan H. Butt, & Robert W. Dibble (2016). Experimental investigation of butanol isomer combustion in Homogeneous Charge Compression Ignition (HCCI) engines. *Applied Energy 165*, 612–626.

Maurya Rakesh Kumar & Mohit Saxena (2018). Characterization of ringing intensity in a hydrogen-fueled HCCI engine. *International Journal of Hydrogen Energy 43*(19), DOI: 10.1016/j.ijhydene.2018.03.194.

Ming Jia, & Maozhao Xie (2007). Numerical simulation of homogeneous charge compression ignition combustion using a multi-dimensional model. *221*(4), 465–480.

Mingfa Yao, Zhaolei Zheng, & Haifeng Liu (2009). Progress and recent trends in homogeneous charge compression ignition (HCCI) engines. *Progress in Energy and Combustion Science 35*, 398–437.

Myung Yoon Kim, Jee Won Kim, Chang Sik Lee, Je Hyung Lee (2006). Effect of Compression Ratio and Spray Injection Angle on HCCI Combustion in a Small DI Diesel Engine. *Energy Fuels, 20*(1), 69–76.

Onishi S., Jo S.H., Shoda K., Jo P.D., & Kato S. (1979). Active Thermo-Atmosphere Combustion (ATAC) – A New Combustion Process for Internal Combustion Engines, SAE Paper 790501.

Özgür, T., Tüccar, G., Uludamar, E., Yılmaz, A., Gungor, C., Ozcanli, M., Serin, H., & Aydin, K. (2015). Effect of nanoparticle additives on NOx emissions of diesel fuelled compression ignition engine. *International Journal of Global Warming, 7*, 487–498. 10.1504/IJGW.2015.070051.

Sajith, V., Sobhan, C. B., & Peterson, G. P. (2010). Experimental Investigations on the Effects of Cerium Oxide Nanoparticle Fuel Additives on Biodiesel, *Advances in Mechanical Engineering,* Hindawi Publishing Corporation.

Selvan, V., Arul Mozhi, Anand R. B., & Udayakumar, M. (2009). Effects of cerium oxide nano particle addition in diesel and diesel-biodiesel-ethanol blends on the performance and emission characteristics of a CI engine. *Journal of Engineering and Applied Sciences 4*(7), ISSN 1819–6608 ARPN.

Steven Ashley (2001). A Low-Pollution Engine Solution, *Scientific American,* 90–95.

Thring, R. H. (1986). Low Heat Rejection Engines, *SAE Paper No. 860314.*

Thring, R. H. (1989). Homogeneous Charge Compression Ignition (HCCI) Engines, SAE Paper 892068.

Vara Parasad, C. M., Rama Mohan, K., Murali Krishna, M. V. S. (1999). Performance of a low heat rejection diesel engine with air gap insulated piston. ASME Transactions. *J. Eng. Gas Turbines Power. 121*(3), 530–539.

Yu, R., Bai, X. S., Hildingsson, L., Hultqvist, A., & Miles, P. C. (2006). Numerical and experimental investigation of turbulent flows in a diesel engine (No. 2006–01–3436). SAE Technical Paper.

Zhao Hua (2007). HCCI and CAI engines for the automotive industry Woodhead Publishing Limited, Abington Hall, Abington, Cambridge CB21 6AH, England.

Index

Printed and bound by CPI Group (UK) Ltd, Croydon, CR0 4YY

23/10/2024

01777703-0017